AF379312

Piero Nicolini

Editor

Touring the Planck Scale

Antonio Aurilia Memorial Volume

 Springer

Editor
Piero Nicolini
Dipartimento di Fisica
Università degli Studi di Trieste, and Istituto
Nazionale di Fisica Nucleare (INFN)
Sezione di Trieste, Trieste, Italy

Johann Wolfgang Goethe University
and Frankfurt Institute for Advanced
Studies (FIAS)
Frankfurt am Main, Germany

Centre for Astro, Particle and Planetary
Physics
New York University
Abu Dhabi, United Arab Emirates

ISSN 0168-1222   ISSN 2365-6425 (electronic)
Fundamental Theories of Physics
ISBN 978-3-031-76065-5   ISBN 978-3-031-76066-2 (eBook)
https://doi.org/10.1007/978-3-031-76066-2

This Springer imprint is published by the registered company Springer Nature Switzerland AG
The registered company address is: Gewerbestrasse 11, 6330 Cham, Switzerland

If disposing of this product, please recycle the paper.

My father sits and thinks

*–Darius Aurilia, early 1980's when, as a
child, was asked about his father's profession*

Preface

The initial discussions about a book on quantum gravity date back to February 2012, when I met Aldo Rampioni, at that time Springer publishing editor, at the Spring Meeting of DPG (German Physical Society) in Göttingen. Many things have changed since then, including physics. For instance, there has been terrific progress in particle physics and astrophysics, related to the discovery of the Higgs particle, the imaging of an event horizon, and the detection of gravitational waves. Nevertheless, the main ideas we sketched in Göttingen have not become outdated. The most important open question in physics is the understanding of the very nature of space and time, matter, and energy under extreme conditions. Such conditions take place at a scale, called the Planck scale, at which quantum mechanics and general relativity can no longer be kept separated.

The prominence of Planck scale physics is sometimes downplayed because its effects are expected to show up only at an energy regime far beyond the reach of current experimental facilities. Such an assumption is, however, too simplistic and somehow misleading. The understanding of the physics at the Planck scale is the cornerstone of a description of the universe at fundamental level. Without such a description, it is only possible to use an array of effective theories whose validity is limited to a certain regime. Furthermore, it is also risky to dismiss the effects coming from the Planck scale for the simple reason that experimental observations could suddenly become much more accurate in the near future.

Despite the topic being of primary importance, the idea of a book remained dormant for about 5 years because other projects had priority. For example, from 2013 to 2016, I edited for Springer the proceedings of the Karl Schwarzschild Meetings in Frankfurt. On June 2, 2018, Douglas Singleton broke the news about the unexpected passing of Antonio Aurilia the day before. I, therefore, thought of honoring his memory by dedicating an edited volume along the lines of what I had informally promised to Springer some years earlier.

As a first task, I contacted Aurilia's former coworkers as well as scientists who were aware of his work. Following Aurilia's principles, I decided that freedom would be the watchword of the book. Therefore, I gave authors carte blanche, i.e., unlimited power to choose the topic of their contribution. The only request was to address the yet unknown physics around the Planck scale, namely everything that can happen between the electroweak scale and trans-Planckian energies. Accordingly, I also decided to use the word "touring" in the title.

The book has four main parts. The first part contains an extensive review of Aurilia's contribution to the theory of strings and membranes. It also contains information about the most important events in his life. The second part is devoted to the relationship between stringy matter and microscopic black holes. The third part deals with aspects of Planck scale phenomenology in cosmology, particle physics, and condensed matter physics. The fourth part is the epilogue of the book. Reading the book requires some background in theoretical physics, while some sections of the first part can be appreciated by a non-expert audience.

Editing and writing the current volume has required an unprecedented effort to overcome an array of problems. First, from 2019 to 2022, I experienced a sort of "academic nomadism", similar to what happened to Aurilia during a phase of his career: I actually held positions in Frankfurt, Abu Dhabi, and Trieste with nontrivial stress associated with relocation. Second, the COVID-19 pandemic drastically impeded professional as well as personal daily routines. At the end of 2021, I was not simply overworked—rather I was stuck in the middle of bureaucratic paralyses. For several months, the conclusion of the present book never reached the priority on my daily schedule. I was a victim of procrastination. I, therefore, decided to invest time in learning how to rigorously organize my daily activities. For instance, I started becoming more efficient in managing my email inbox. I began to regularly schedule a slot of time in the evening to plan the subsequent day. I changed my meal time as well as my workout routine.

Despite all the above efforts, the final step to conclude the volume has been the most demanding in terms of working time. I spent a few months gathering historical and biographical details about Aurilia's life. In addition, it took me another 6 months to read all of his scientific papers. However, the main source of delay is connected to my increasing interest in what I was writing about. I loved the man, and I fell in love with his character, while writing about him.

To conclude, I would like to thank Marina Forlizzi, Springer's executive editor for physics and astronomy, for her support, encouragement and patience in many phases of the completion of this book. I am also grateful to Aldo Rampioni for the initial contact with Springer. I would like to thank Roberto Balbinot, Bernard Carr, Gia Dvali, Patricio Gaete, José A. Helayël- Neto, Kai Lam, Harvey Leff, Robert Mann, Anais Smailagic, Paul Townsend, and their collaborators for their submissions. Many thanks go to Jonas Mureika and Douglas Singleton, both for their contributions and assistance in editing the manuscript. To Euro Spallucci, Aurilia's closest collaborator, goes all my gratitude for sharing his thoughts and recollections about his time with Aurilia. Finally, for their help, support, and source of information, I am grateful to the family of Antonio Aurilia. In particular I thank Antonio's wife Elizabeth, his daughter Alex as well as his sisters Anna Maria, Elena, and Liliana.

Frankfurt am Main, Germany/Trieste, Italy   Piero Nicolini
July 2023

Contents

Contributors

Niayesh Afshordi Perimeter Institute, Waterloo, Ontario, Canada;
Department of Physics and Astronomy, University of Waterloo, Waterloo, Ontario,
Canada

Erick Aiken Physics Department, California State University Fresno, Fresno, CA,
USA

Natacha Altamirano Perimeter Institute, Waterloo, Ontario, Canada;
Department of Physics and Astronomy, University of Waterloo, Waterloo, Ontario,
Canada

Roberto Balbinot Dipartimento di Fisica dell'Università di Bologna and INFN
Sezione di Bologna, Bologna, Italy

Michael Bishop Mathematics Department, California State University Fresno,
Fresno, CA, USA

Bernard J. Carr School of Physics and Astronomy, Queen Mary University of
London, London, UK

Gia Dvali Arnold Sommerfeld Center, Ludwig-Maximilians-Universität, Munich,
Germany;
Max-Planck-Institut für Physik, Munich, Germany

Alessandro Fabbri Departament de Física Teórica and IFIC, Universidad de
Valencia-CSIC, Burjassot, Spain

Patricio Gaete Departamento de Física and Centro Científico-Tecnológico de
Valparaíso, Universidad Técnica Federico Santa María, Valparaíso, Chile

Elizabeth Gould Perimeter Institute, Waterloo, Ontario, Canada;
Department of Physics and Astronomy, University of Waterloo, Waterloo, Ontario,
Canada

José A. Helayël-Neto Centro Brasileiro de Pesquisas Físicas, Rio de Janeiro,
Brazil

Kai S. Lam Department of Physics and Astronomy, California State Polytechnic University, Pomona, CA, USA

Harvey S. Leff Department of Physics and Astronomy, California State Polytechnic University, Pomona, CA, USA;
Physics Department, Reed College, Portland, OR, USA

Robert B. Mann Department of Physics and Astronomy, University of Waterloo, Waterloo, Ontario, Canada

Jonas Mureika Department of Physics, Loyola Marymount University, Los Angeles, CA, USA;
Kavli Institute for Theoretical Physics, University of California, Santa Barbara, CA, USA

Piero Nicolini Dipartimento di Fisica, Università degli Studi di Trieste, Trieste, Italy;
Institut für Theoretische Physik, Johann Wolfgang Goethe-Universität, Frankfurt am Main, Germany;
Frankfurt Institute for Advanced Studies (FIAS), Frankfurt am Main, Germany;
Center for Astro, Particle and Planetary Physics, New York University Abu Dhabi, Abu Dhabi, United Arab Emirates

Douglas Singleton Physics Department, California State University Fresno, Fresno, CA, USA

Anais Smailagic INFN, Sezione di Trieste, Italy

Euro Spallucci Dipartimento di Fisica Teorica, Università di Trieste and INFN, Sezione di Trieste, Italy

Paul K. Townsend Department of Applied Mathematics and Theoretical Physics, Centre for Mathematical Sciences, University of Cambridge, Cambridge, UK

Antonio Aurilia and the Extended Objects

Remembering Professor Antonio Aurilia, Quantum Field Theorist Par Excellence and Inspiring Teacher

Kai S. Lam

Abstract In this chapter I briefly recall the time I spent with Antonio Aurilia over the last three decades at Cal Poly Pomona.

I have known Antonio for over thirty years. We first met as colleagues in the Physics Department at Cal Poly Pomona, and remained so until he retired from the department several years ago. In my mind he was not only a most respected colleague but also a dear friend, actually more like an elder brother and mentor, always ready with a warm kind of bemusing sarcasm, an unforgettable brand of Italian humor, and above all hard hitting and heartfelt words of wisdom.

He stood for the conviction that substance was always above appearance, a kind of dedication and good-intentioned forthrightness that sometimes engendered friction, and an honesty without any trappings of superficial pleasantries. But there was also humor, warmth, good nature, and genuine caring in abundance. Somehow, in his presence, I always felt I could speak my mind, even though we disagreed sharply. And if I voiced something brashly, something that might border on confrontation, I felt we could always come to terms afterwards, solely on the strength of our friendship, trust, and mutual respect.

Even though I am not conversant in his field of specialty I have always felt that in his academic pursuits and professional strivings he represented a beacon of light, a clear vision, and a tenacity that would not be swayed by temporary fashion or the latest wrinkles in the development in his field. Nor would he cow easily to mere 'big name' authority. Indeed he would proclaim not infrequently that so-and-so (eminent physicist) may not always be right.

If I may summarize, for Antonio, the search for the deep mysteries of his field (quantum field theory), the integrity of the physics (and astronomy/astrophysics) courses taught in our department, the academic welfare of students, and the long term development of the department, always came first. He firmly believed that there was no contradiction among these goals, and thus spared no efforts, and minced no

K. S. Lam (✉)

Department of Physics and Astronomy, California State Polytechnic University, 3801 W. Temple Avenue, Pomona, CA 91768, USA

e-mail: kslam@cpp.edu

P. Nicolini (ed.), *Touring the Planck Scale*, Fundamental Theories of Physics 219, https://doi.org/10.1007/978-3-031-76066-2_1

words when expressing his views, in his efforts to further them, even though they may have been perceived by him to be at variance with the administration/management sector of our institution. I recall distinctly that he once, in a department meeting, remarked that rather than waiting for the administrative axe to fall on the allotment of department funds which would negatively impact our curricula, the faculty should take some action instead of moping around like 'sitting ducks'. He had just the right phrase of characterization for everything.

But it was not all talk. On several occasions, Antonio actually served with diligence, or sought to serve, in various university level action committees that he thought he might be able to provde positive and constructive input. He was a man of not only strong words but also definitive action.

Antonio's untimely departure came as no less than a shock. It has robbed me of a dear friend, a wise mentor, and a valued colleague. It has also, I dare say, deprived his field of study—quantum field theory—of a dedicated and productive practitioner whose contributions will be long remembered and honored.

Antonio Aurilia—Colleague and Friend

Harvey S. Leff

Abstract My first encounter with Antonio Aurilia was in a long distance phone call. Subsequently, I helped to bring him from the University of Toronto to California State Polytechnic University-Pomona (Cal Poly Pomona), where he had a long and distinguished career as a teacher and researcher. This led to a 30+ year friendship with Antonio and his wife Elizabeth. In this chapter, I share some of my recollections.

1 Phone Call From Toronto

Back in 1984, the Physics Department at Cal Poly Pomona had a resignation by a faculty member who was the sole teacher in a nuclear physics laboratory. As the department Chair at that time, I advertised a tenure-track faculty position through the American Institute of Physics and the American Association of Physics Teachers.

We received many applications, among which one was from Antonio. Examination of his resumé showed that his background, though outstanding, did *not* include any experience teaching a nuclear physics lab, which required a knowledge of multichannel analyzers and other equipment that is unfamiliar to most theorists. Thus, I placed his application at the bottom of the heap, along with those of other theorists.

Soon thereafter, I received a telephone call from Antonio, who was working at the University of Toronto at that time. In his thick Italian accent, which I found charming, he explained that he believed he was qualified to teach *any* and *all* undergraduate physics courses, and was very willing to do so! Of course, that included nuclear physics labs.

His claims reminded me of a "snake oil" salesman, and I tried to terminate the conversation quickly. However, he went on to explain himself to me in a remarkably

H. S. Leff (✉)
Department of Physics and Astronomy, California State Polytechnic University, 3801 w. Temple Avenue, Pomona, CA 91768, USA
e-mail: hsleff@cpp.edu

Physics Department, Reed College, 3203 SE Woodstock Blvd., Portland, OR 97202-8199, USA

P. Nicolini (ed.), *Touring the Planck Scale*, Fundamental Theories of Physics 219, https://doi.org/10.1007/978-3-031-76066-2_2

lucid, disarming, and compelling way. I had sufficient experience to recognize a "real physicist" when I heard one speak,[1] and Antonio seemed to be just that.

2   Visit to Cal Poly Pomona

As one of our most distinguished candidates, despite his lack of nuclear physics lab teaching, Antonio was invited to visit the campus. This gave us the opportunity to get to know him, to see him in a class-room setting, to hear him present a seminar on his research, and to meet and have discussions with our faculty. One of my "techniques" in evaluating candidates was to meet with them in the morning, have lunch and dinner with them (usually with a group of faculty and spouses), to invite them to our home after dinner for more conversation, and to watch carefully how uniform their dispositions and attention spans were over the 12+ hours. Antonio displayed good endurance and an admirably even temperament throughout.

2.1   *In-Class Presentations*

Antonio was asked to teach one introductory lecture and one upper-division lecture. In both, he performed well, and evoked a good response from our students, who sensed that he was a remarkably learned man with excellent communication skills.

The upper-division class he taught was electricity and magnetism. He presented the divergence and curl of vectors and the topic of electric and magnetic fields with style, grace, and his unique insights. I rated his teaching to be well above the quality level of most of our applicants over the years.

2.2   *Seminar and Meetings with Faculty*

In his physics seminar, he explained his rather abstract, highly-mathematical research to a diverse audience of faculty and undergraduate students. He was able to successfully make the *essence* of his research reasonably clear, which was no mean feat.

Antonio also spent hours meeting with faculty one-on-one or in small groups. He impressed us all as a highly accomplished physicist and a dedicated, talented teacher.

---

[1] To me, a "real physicist" lives and breathes physics, a quality that quickly becomes evident in serious conversation.

2.3 Result of the Visit

In each of our encounters with Antonio, it was clear that he spoke, taught and thought like a "real physicist." I was reminded of the adage, "If it talks like a duck, walks like a duck, and paddles in the water like a duck, it must be a duck." In Antonio's case, I concluded that he must be a "real physicist," which was consistent with my initial assessment during our introductory long-distance phone conversation.

As a consequence of his impressive visit, the Physics Department was pleased to offer a faculty position to Antonio. At that time, he was a citizen of both Italy and Canada, but was not a citizen of the United States. He could not begin the faculty position in the United States unless he was granted a Green card.

3 The Green Card

I appealed to Cal Poly Pomona University's higher administration to have one of its lawyers handle the application process to the Immigration and Naturalization Services (INS). I was informed that the university no longer gave such help to departments and that I would have to solve that problem at the department level.

Antonio was an extraordinary candidate, which propelled me to spend nearly a year dealing with the INS to help him obtain a green card. The application process was a bureaucratic maze, and the intent of the items requested was not always clear. When I ran into a stumbling block, I phoned one or more immigration attorneys, whom I found in the phone book (we still had those in 1985). I would ask them their fees and their experience level, and for advice on how to deal with the specific section that had stymied me. Their fees of about $2500 at that time were beyond our limited budget, so I used the bits and pieces of information accumulated in those phone calls to complete the application process.

The university's hiring process required us to obtain letters of recommendation. One of those was a glowing recommendation from the Nobel Prize recipient, Abdus Salam, who was then at what is now the Abdus Salam International Centre for Theoretical Physics in Trieste, Italy.

I made a strong case for Antonio as the most well-qualified candidate for our faculty position, and although it took one year, he received his green card. Antonio Aurilia joined the Cal Poly Pomona Physics Department the next academic year.

4 Career at Cal Poly Pomona

4.1 Teaching

Antonio Aurilia served as a highly distinguished and popular teacher of our undergraduate physics courses (there is no graduate program). Moreover, he maintained a consistently strong research program, even in the face of heavy teaching loads—typically 2 lecture courses and 3 lab courses in each of the three quarters, every academic year. Because we have no graduate TA's, our faculty teach freshman labs and grade lab reports. Antonio did this with integrity and earned the high respect of his students.

He fulfilled his promise to teach a variety of physics courses, though we managed to find an experimentalist to teach the nuclear laboratory. Because Cal Poly Pomona regularly assesses and evaluates its faculty, I had the opportunity to be an evaluator, attending an occasional laboratory and lecture taught by Antonio. His laboratory teaching was excellent. As a theorist, he made sure he did each full experiment before the class began. He knew the various pitfalls, and was able to forewarn students how to avoid them. In lecture courses, despite his theoretical bent, he loved doing classroom demonstrations.

In one class I visited, he demonstrated projectiles with a demo we called "shoot the monkey." It employed a toy monkey hanging from a magnetic catch. Across the room was an air "gun" that shot a wooden pellet at the toy monkey. As the pellet left the gun's barrel, the magnetic catch was released. If the shooter, in this case, Antonio, aimed directly at the monkey, the fired pellet and the dropping monkey would have the same downward acceleration, and the pellet would hit the monkey.

Of course, not all physics demos worked as planned, and on that day, Antonio's aim was a bit off. The pellet missed the falling monkey all together, and hit the metal venetian blinds covering a window. The blinds crashed into the window hard enough to break the glass. Had the demo worked properly, the pellet would have hit the toy monkey and the class would have applauded. However, hearing the glass break brought stone silence. After a characteristic pause, Antonio said in his charming Italian accent, "Did you hear what I heard?" After another brief pause, the class broke out into hysterical laughter. I suspect many of those students will remember the demo over the years and might even remember the physics principles involved!

In his teaching of upper division lecture classes, Antonio did a superlative job. With his deep understanding of both physics and mathematics, he provided unique insights in addition to illuminating the fundamentals. He particularly loved to teach our History of Physics class in which was given *carte blanche* to cover those aspects that interested him most. I regret that I did not have the opportunity to attend any of those lectures, which I know would have been a unique learning experience.

4.2 Research

Although I am not an expert in the area of quark confinement, it is clear that Antonio tackled some of the most difficult problems in theoretical physics. He published many significant articles in Physical Review D over the years. My own examination of these as a novice shows them to be thoughtfully conceived and skillfully executed. I leave discussion of his research to others more qualified than I.

4.3 Faculty Governance

Antonio attended Physics Department meetings and reserved his comments for moments when he could add something truly helpful. He was far from the most talkative participant, but was always one of the wisest. He also served for years as a representative in the Cal Poly Pomona faculty senate, and on various *ad hoc* search committees. His even temperament, good judgment, and articulate explanations made him an ideal person to have present at any meeting.

5 Retirement

After his retirement from his Cal Poly Pomona professorship, Antonio and his wife Elizabeth (Liz) moved to Portland, Oregon (Fig. 1). About eight years earlier I had moved to Portland with my wife Ellen. We four had enjoyed one another's company when we all lived in California. Our then teenaged twin sons even helped them move furniture to a new home. Ellen and I rekindled that relationship with Liz and Antonio in Portland. We visited each other's homes and went to restaurants and concerts together. Socially, although Antonio was generally a very pensive, deep-thinking person, he had an excellent sense of humor, and we four had many good times together. Antonio was politically savvy, and we enjoyed good discussions about U.S. and European politics.

6 Death

On June 2, 2018, I received an email from Liz, informing me that Antonio had died unexpectedly that morning. He had undergone surgery for an injury suffered in a fall a couple of weeks earlier, but had been expected to recover. With his untimely death we lost a valued friend of more than three decades. His memory is still strong within us and will remain so as long as we live. Regarding Antonio's death, as Liz wrote

Fig. 1 Elizabeth and Antonio in our California backyard. Courtesy of Alexandra Aurilia Stephenson. All rights reserved

in her sad email message, "There are no words are there?" Our friendship with Liz lives on, and through her, we still feel a connection with Antonio.

From Hadronic Bubbles to Quantum Gravity: Chasing the Infinite with Antonio Aurilia

Piero Nicolini

Abstract The present work discusses the development of the physics of extended objects, along the path of Antonio Aurilia and his collaborators. In parallel to a description of the main events of Aurilia's life, the central part of the work gives an overview of the applications of the theory of strings and branes in a variety of contexts, that span hadronic physics, solid state physics, black holes, cosmology and physics at the Planck scale. The final section addresses the issue of the role of theoretical physics in modern society, along with Aurilia's views on this topic which he expressed on various occasions.

1 Introduction

Imagine a world without electricity, where the only source of light are the flames of candles. Before the second industrial revolution, which took place starting from the 1870s, the use of candles was simply the norm. From a modern perspective this sounds awful. People's daily lives would be very different or even impossible. In reality, the lack of electricity is more than an exercise of the imagination. This could have been a pretty realistic scenario, in case Maxwell had not lived. Now, the natural question arises of who was this Maxwell.[1] Indeed, it is easy to forget those who worked with abstract ideas or concepts long before practical applications of these

---

[1] The reference is to James Clerk Maxwell, the Scottish scientist who has formulated the theory of classical electromagnetism by means of a system of differential equations.

P. Nicolini (✉)
Dipartimento di Fisica, Università degli Studi di Trieste, Strada Costiera 11, 34151 Trieste, Italy
e-mail: piero.nicolini@units.it

Institut für Theoretische Physik, Johann Wolfgang Goethe-Universität, Max-von-Laue-Str. 1, 60438 Frankfurt am Main, Germany

Frankfurt Institute for Advanced Studies (FIAS), Ruth-Moufang-Str. 1, 60438 Frankfurt am Main, Germany

Center for Astro, Particle and Planetary Physics, New York University Abu Dhabi, P.O. Box 129188, Abu Dhabi, United Arab Emirates

P. Nicolini (ed.), *Touring the Planck Scale*, Fundamental Theories of Physics 219,
https://doi.org/10.1007/978-3-031-76066-2_3

ideas reach the average person. The present contribution deals with a case similar to Maxwell's, namely the work of a scientist who dedicated his life to finding solutions to fundamental questions. His name was Antonio Aurilia.

Aurilia's scientific work concerned the transition of string theory from a phenomenological model for hadrons to a theory of everything. The present work provides detailed explanations of Aurilia's main papers and analyses their impact on fundamental physics. For example, Hawking suggested that Aurilia's idea about the existence of a topological field in the Universe could lead to the solution of the cosmological constant problem [1].

The chapter also tells about the life of a scientist and his era. In the 1970's Aurilia benefited from belonging to a generation of researchers during what has probably been the last golden era of theoretical physics. For instance, his position with Salam at the ICTP, Trieste in 1974 is an emblematic example of an experience in an enormous laboratory of ideas. The same can be said for his stay at CERN in 1980. The historical analysis also shows that Aurilia's life was marked by two main features. First of all, Aurilia had an inner drive for the infinite, for the unknown, for what lies on the edge of the humanly intelligible. He also had an innate ability to perceive physical phenomena, well before understanding them mathematically. This emerges markedly in his papers, whose texts sometimes seem to describe a dream. Secondly, Aurilia was, for most of his career, in a state of permanent escape, interspersed by short periods where he seemed to be "in the right place at the right moment". Such a restlessness was probably due to a strong yearning for scientific independence, but also to a series of life changing events that occurred either on one side or on the other of the Ocean. Flashbacks on the work of great scientists from the past, such as Kirchhoff, Planck and Boltzmann, forge the main thesis of the final part of the chapter: The life of theoretical physicists has not improved at all over time mostly because the society underestimates the importance of their work.

Aurilia was not only a great scientist, but also an educator, science historian and thinker. For instance, in the paper [2], he briefly conveyed his vision about goals of theoretical physics:

> At its most fundamental level, research in theoretical high energy physics means research about the nature of mass and energy, and ultimately about the structure of space and time. It may even be argued that the whole history of physics, to a large extent, represents the history of the ever changing notion of space and time in response to our ability to probe infinitesimally small distance scales as well as larger and larger cosmological distances.

Aurilia's legacy is, therefore, manifold. A section dedicated to the "Trieste-Toronto-Pomona triangle" describes the network of scientists that stemmed from Aurilia's collaborations at institutions where he held faculty positions. Another section offers a summary of Aurilia's thoughts about the post–Cold War society and current trends in science and education. The concluding section addresses the issue of the future of mankind and the role of theoretical physics in the forthcoming technological society.

Fig. 1 View of the apartment building in Via Caracciolo, 24, Naples. Courtesy of Salvatore Samuele Sirletti. All rights reserved

2 Early Life

On the evening of May 14, 1942, a shooting star of unprecedented brilliance flew over the Gulf of Naples: Antonio Aurilia was born in Materdei, a quarter of Naples—see Fig. 1. He was the fourth and youngest child of Clemente Aurilia (1907–1949), an employee in the footwear sector, and Assunta Ligesto (1909–1961), a housewife. Both parents had belonged to Neapolitan families for generations. They were humble, but they stayed resolute in offering their children the support for a university education. Antonio's birth was celebrated in the family with unprecedented joy. The rare astronomical event marked the arrival of the first boy in the family after three girls.[2] The Aurilia family faced economic hardships in particular after the death of the father. At that time, Antonio was just seven years old. The three elder sisters were soon forced to find employment and had to interrupt their studies. Already as a child, Antonio showed his talent for studying and a never ending wish to know and learn new things. He was raised Catholic and believed in God. From 1956 to 1961, he attended the Liceo Classico Gian Battista Vico in Naples—see Fig. 2. In parallel to his high school studies, he enjoyed playing soccer, swimming, rowing and boxing. During the final years at the Liceo Vico, he developed his passion for science and decided to study physics at the university. When Aurilia's mother passed away in April 1961, the sisters committed themselves to support him during his academic studies, in order to fulfill their parents' wish.

---

[2] Aurilia's parents recalled that they actually saw the meteor that day.

Fig. 2 View of the Liceo Vico, Naples. Courtesy of Salvatore Samuele Sirletti. All rights reserved

2.1 Undergraduate Years

Aurilia began his undergraduate education at the University of Naples in November 1961. It was a tough time for him. His mother has just passed away and he could not focus on lectures. With the encouragement of the faculty of the local theoretical physics group headed by Eduardo R. Caianiello, he returned to his duties as a student of the Laurea degree in Physics.

At that time, the Laurea was the highest degree awarded in Italy conferring an internationally unrecognized title of "Dottore", namely Doctor. Such a long standing anomaly of the Italian educational system ceased only in the 1980s when actual PhD programs were introduced.[3] In the CV in Fig. 3, 4, 5, 6, Aurilia still refers to his Laurea as the "Italian doctorate", even if, in his PhD transcript of the University of Wisconsin, such a degree is recognized as a BS only—see Fig. 7. Conventionally the Laurea gave access to postgraduate schools, called "Scuole di Perfezionamento", prototypes of modern PhD schools, offering the only professional training at that time.

For his graduation, Aurilia had to submit a Laurea thesis [3]. During this challenging time with the loss of his mother, Aurilia's acquaintance with Hiroomi Umezawa was particularly important, as it turned out to be the event that changed his life and career. Indeed the famous Japanese scientist, who has been a professor at the University of Naples since 1964, ultimately became Aurilia's Laurea supervisor.

---

[3] This explains why eminent scientists from Italy, like the Nobel Prize laureates Fermi, Segrè, Natta and Rubbia, did not hold any PhD.

CURRICULUM VITAE

Name: Antonio Aurilia

Address: Department of Physics, Syracuse University,
                        Syracuse, New York

Date of Birth: May 14, 1942

Place of Birth: Naples, Italy

Marital Status: Married

Education: Ph.D. in Physics - University of Wisconsin, U.S.A.,
                        August, 1970.

                        "Laurea in Fisica," - University of Naples, Italy,
                        November, 1966.

Employment: 1972-1974 - Post Doctoral Fellow-Research Associate,
                        Department of Physics, Syracuse University, Syracuse,
                        New York, U.S.A.

                        1970-1972 - Post Doctoral Fellow, Instructor, Theoretical
                        Physics Institute, University of Alberta, Canada.

                        1967-1970 - Research Assistant, Department of Physics,
                        University of Wisconsin, U.S.A.

                        1966-1967 - Italian National Research Council Fellow,
                        Theoretical Physics Institute, University of Naples,
                        Naples, Italy.

Other Research
Experience: Summer 1973 - Argonne National Laboratories, Illinois,
                        U.S.A.

                        Summer 1972 - Canadian National Research Council
                        Fellowship, Banff, Alberta, Canada.

                        Summer 1971 - N.S.F. Fellowship, University of
                        Colorado, Boulder, Colorado, U.S.A.

                        Summer 1970 - N.S.F. Fellowship, Brandeis University,
                        Mass., U.S.A.

                        Summer 1969 - N.S.F. Fellowship, University of
                        Colorado, Boulder, Colorado, U.S.A.

Fig. 3 Aurilia's CV dated June 18, 1974. Courtesy of Elena, Annamaria and Liliana Aurilia. All rights reserved

The content of the Laurea thesis, that resulted in Aurilia's first two publications [4, 5], was about the formulation of a consistent higher spin field theory, a topic that today finds applications in Chern-Simons matter theories and gauge-gravity duality (see Fig. 6). For these results, Aurilia received a first-class Laurea degree (110 *e lode*

PAPERS BY DR. AURILIA

PUBLISHED

1. Comment on the Breakdown of Dilatation Invariance (with Y. Takahashi), Phys. Rev. D. 6, 2294 (1972).

2. Spontaneous Breakdown of Symmetry and Gauge Invariance in a Relativistic Field Theory (with Y. Takahashi and H. Umezawa), Progr. Theoret. Phys., 48, 290 (1972).

3. The Spontaneous Violation of Dilatation Invariance in Relativistic Field Theories (with N. Papastamatiou, Y. Takahashi and H. Umezawa), Phys. Rev. D 6, 3066 (1972).

4. Dilatation and Dimensional Transformations (with Y. Takahashi and H. Umezawa), Phys. Rev. D 5, 851 (1972).

5. A Role of Gauge Fields in the Spontaneous Breakdown of Symmetry (with Y. Takahashi and H. Umezawa), Lettere al Nuovo Cim., Vol. 1, 603 (1971).

6. Theory of High Spin Fields (with H. Umezawa), Phys. Rev. 182, 1682 (1969).

7. Projection Operators in Quantum Theory of Relativistic Free Fields (with H. Umezawa), II Nuovo Cim. X, 51A (1967).

UNPUBLISHED

1. Dynamical Symmetry Breaking and Manifest Covariance in Gauge Theories (submitted for publication).

2. Dynamical Symmetry Breaking in Gauge Theories: Superconductivity and Particle Physics (submitted for publication).

3. Invariant Relative Velocity, (with F. Rohrlich, submitted for publication).

4. Equivalent Field Theories and the Bound State Problem (submitted for publication).

5. Action-At-A-Distance Quantum Field Theory (with F. Rohrlich, submitted for publication).

Fig. 4 Aurilia's list of publication attached to the CV dated June 18, 1974. In the list of unpublished papers, the manuscript n. 5 with Rohrlich has never been published. Aurilia published just the manuscript n. 3 in collaboration with his postdoctoral sponsor during the two year appointment in Syracuse. Courtesy of Elena, Annamaria and Liliana Aurilia.

6. Model for Unified Strong and Electromagnetic Interactions (with
    F. Rohrlich, Internal Report).

7. Ph. D. Thesis: Self Consistency in the Theory of Quantized Fields
    (Advisor H. Umezawa).

8. Thesis for the Italian Doctorate: Theory of High Spin Fields (Ad-
    visor H. Umezawa, in Italian).

PROFESSIONAL REFERENCES:

G. Dell' Antonio: Istituto De Fisica Teorica, Universita' De Napoli,
Napoli, Italy. Present Address: Department of Physics, New York
University, New York, N. Y., U. S. A.

F. Rohrlich, Department of Physics, Syracuse University, Syracuse, N. Y.,
U. S. A.

Y. Takahashi, Department of Physics, University of Alberta, Edmonton,
Alberta, Canada.

H. Umezawa, Department of Physics, University of Wisconsin, Milwaukee,
Wisconsin, U. S. A.

M. Wellner, Department of Physics, Syracuse University, Syracuse, N. Y.,
U. S. A.

Ph. D. Thesis Abstract:

     By combining the ideas of Nambu in his approach to superconductivity
with the techniques of Q. E. D., we presented a gauge theory of spontaneous
symmetry breaking which, contrary to current practice, does not involve
spinless boson fields in the fundamental Lagrangian. The theory consists
only of massless Fermi and vector meson fields and leads, via the Higgs'
mechanism, to a physical spectrum of solely massive particles. This is
the relativistic generalization of the well-known "plasmon phenomenon" in
superconductivity first noticed by Anderson. The results were published
in refs. (2,5).

Fig. 5 Aurilia's Professional references in 1974, including Umezawa from the time in Milwaukee, Takahashi from the time in Edmonton and Rohrlich and Wellner from the time in Syracuse. The list contains Dell'Antonio, because he must have been Aurilia's formal supervisor during the brief stint at the Scuola di Perfezionamento in Naples. The two actually never collaborated together. Courtesy of Elena, Annamaria and Liliana Aurilia. All rights reserved

su 110) in physics in November 1966 after completing a thesis titled "Theory of high spin fields".

Thesis for the Italian Doctorate: Abstract

   We introduced a systematic method for constructing quantized field
theories for free particles of arbitrary spin and finite mass. The method
yields a single wave equation, which implies all the usual subsidiary conditions, including symmetry or antisymmetry properties, so that the wave
function in question has exactly the number of components appropriate to
its spin value. The wave equation automatically reduces to the Klein-
Gordon equation for particles of integral spins and to the Dirac equation
for particles of half-integral spins. The results were published in
refs. (6,7).

Honors and Awards

   Italian degree was granted "Summa cum Laude".

   Member of The Sigma Xi Society.

Professional Society Memberships

   American Physical Society

   Canadian Association of Physicists

   American Association for the Advancement of Science

Fig. 6 Last page attached to Aurilia's CV dated June 18, 1974, including a summary of the Laurea thesis. Courtesy of Elena, Annamaria and Liliana Aurilia. All rights reserved

2.2 *Doctoral Studies with Umezawa in Milwaukee*

Soon after the Laurea, Aurilia enrolled himself to the local Scuola di Perfezionamento and was awarded a fellowship by CNR, Italy's National Research Council.

   In the fall 1966 things had, however, changed to some extent. Umezawa had left Naples to take up a position as a professor at the University of Wisconsin-Milwaukee (UWM). He proposed that Aurilia follow him to the US and start a PhD there. Although attracted by the offer, Aurilia did not want to leave his sisters alone,

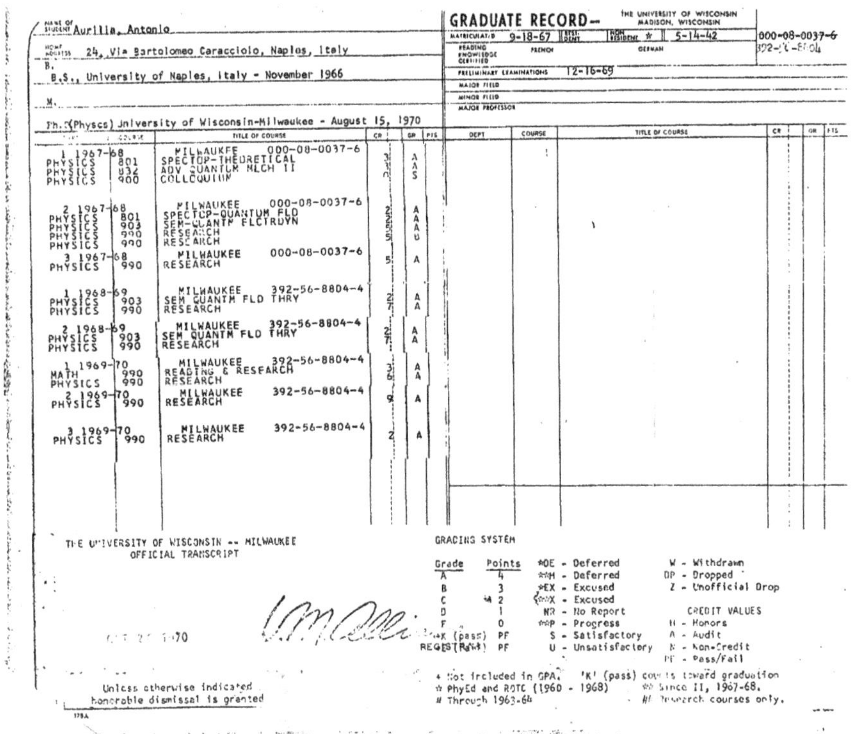

Fig. 7 Aurilia's graduate school record at UWM. The home address is still Via Bartolomeo Caracciolo, 24, Naples. The Italian Laurea degree is recognized as a BS. The matriculation date is September 18th, 1967, the doctoral preliminary examination took place on December 16th, 1969 and the PhD defense on August 15th, 1970. Aurilia earned 63 graduate credits. He attended 3 lectures, 3 seminars and 1 colloquium corresponding to 14 credits for the so called core courses. The rest is labeled as independent study PHYSICS 990 and MATH 990. Aurilia maintained a GPA of 4.0 for the core courses (there is only a B for a PHYSICS 990). The structure of the PhD course at UWM has not changed much since then, including the computation of credits and the number of terms. Interestingly the codes of the courses are still the same [6]. Courtesy of Elena, Annamaria and Liliana Aurilia.

since they had supported him so much, giving him the opportunity to freely make his own choices by sacrificing theirs.

During the first year at the Scuola di Perfezionamento, without the guidance of his Laurea advisor, Aurilia began to feel isolated and lost. He could not learn much from the school he was attending, or at least not as much as he expected. In general, he did not like university life in Italy, nor the leading political views in academia at that time. His mind and heart felt with pain that there were no good prospects for a career as a theoretical physicist in Naples in particular and in Italy in general. Aurilia did not want to be anyone's lackey. He refused to pursue small, temporary, and often illusory goals. He soon realized that the only way to save his life and career was to accept Umezawa's offer. Otherwise, he would probably have succumbed and disappeared.

In considering pros and cons of a PhD abroad, Aurilia may have been influenced by his official supervisor at the Scuola di Perfezionamento. From Fig. 6, this was Gianfausto Dell'Antonio, who, at that time, had just come back to Naples from a Fulbright fellowship at the Institute for Advanced Study, Princeton [6]. For this reason, albeit reluctantly, Aurilia made up his mind to leave Italy.

In September 1967, Aurilia, with $300 in his pocket (equivalent in purchasing power to about $2,500 today), boarded the plane to Milwaukee. There, he had to face new difficulties: He had to look for a home, get used to a new country, a new culture and do research to get his PhD. He also recalled that Umezewa had very little English proficiency. At that time Aurilia was also just learning what later became nearly perfect English. This meant that the discussions about physics between the two took place only through calculations on the board! The requirements for the PhD at UWM—see Fig. 7—have not changed since that time. Even the codes of the graduate courses are the same [7]. Aurilia had to earn academic credits. This must have been a very new system for him, taking into account that credits were introduced in Europe more than 30 years later, i.e., in 1999. Despite such troubles, Aurilia, in letters written to his sisters, described the life in Milwaukee with subtle humor and self-irony.

Some of the aspects of the personal and scientific interaction with Umezawa are witnessed by Aurilia's own words, he wrote as a preface of a paper dedicated to the memory of his supervisor [8]:

> Umezawa was especially fond of classical music (and the whole array of musical equipment that goes with it). Schumann was one of his favorite composers, "because of his craziness," in Hiroomi's words. Like Schumann's musical scores, Umezawa's "physical scores" are sometimes difficult to comprehend, but one cannot fail to appreciate their originality, as I did, when as an undergraduate student of Hiroomi at the University of Naples I was confronted with his first book on quantum field theory.
>
> In 1966 Umezawa left Italy to join the Physics Department at the University of Wisconsin-Milwaukee. In 1967 I followed him, and, of course, my life changed forever. I still remember vividly the many nights spent at Umezawa's house discussing physics, politics, history, philosophy, etc., until the early hours of the morning, with the sounds of classical music in the background. In retrospect those long, ponderous conversations in broken English (interrupted by a stream of delicious Japanese snacks exquisitely prepared by Mrs. Umezawa) were formative as well as informative and amounted to a worldly experience destined to have a profound and lasting influence on a young graduate student suddenly transplanted from Naples to a Japanese home in Milwaukee, Wisconsin.

The focus of the PhD research was on the Nambu–Jona-Lasinio (NJL) model for superconductivity (see Fig. 5). Such a model describes the dynamics of nucleons and mesons similarly to Cooper pairs for superconductivity. Nowadays the NJL model is considered a possible substitute for QCD in the low energy limit. Contrary to QCD, however, the NJL model predicts a spontaneous gauge symmetry breaking leading to the appearance of massless pseudoscalar bosons in the Lagragian. The goal of Aurilia's work was to overcome the problem of massless degrees of freedom and obtain a physical spectrum of solely massive particles.

THE REGENTS OF

THE UNIVERSITY OF WISCONSIN

ON THE NOMINATION OF THE FACULTY
HAVE CONFERRED UPON

ANTONIO AURILIA

THE DEGREE OF

DOCTOR OF PHILOSOPHY

TOGETHER WITH ALL HONORS, RIGHTS, AND PRIVILEGES BELONGING
TO THAT DEGREE. IN WITNESS WHEREOF, THIS DIPLOMA BEARING
THE SEAL OF THE UNIVERSITY IS GRANTED.

GIVEN THIS FIFTEENTH DAY OF AUGUST IN THE
YEAR NINETEEN HUNDRED SEVENTY AND OF THE
UNIVERSITY THE ONE HUNDRED TWENTIETH.

Fig. 8 Aurilia's PhD certificate dated August 15, 1970. Courtesy of Elena, Annamaria and Liliana Aurilia.

During his graduate studies, Aurilia also benefited from the collaboration and advice of Yasushi Takahashi,[4] another eminent Japanese scientist and Umezawa's friend, who moved to the University of Alberta, Edmonton in 1968. Such a collaboration is reflected in the author list of two publications of the Milwaukee period including both Japanese scientists [9, 10].

Aurilia was also awarded two visiting fellowships from the National Science Foundation (NSF) to spend summer stays at the University of Colorado (1969) and at Brandeis University (1970). In August 1970, Aurilia received his PhD after completing a doctoral dissertation titled "Self consistency in the theory of quantized fields" [11]—see Fig. 8. By that time, he had already secured a position of postdoctoral researcher in the group of Takahashi at the University of Alberta.

3 Academic Career

Aurilia's career can be summarized in two phases. The first one, from 1974 to 1986, is characterized by a sort of academic nomadism. The international reputation, an acquired global mindset and cultural agility, the freedom from teaching duties led Aurilia to embark on a long journey through well known institutions in North America

---

[4] A well-known quantum field theory relation between correlation functions is named after him, namely the Ward-Takahashi identity.

and Europe. Such institutions include the International Center for Theoretical Physics (ICTP), Trieste, Italy, Imperial College, London, UK, the Theory Division at CERN, Geneva, Switzerland, Chalmers Institute of Science and Technology Gothenburg, Sweden, the University of Toronto, Mississauga, Canada and the Dublin Institute for Advanced Studies (DIAS), Dublin, Ireland. Aurilia's scientific wandering was in part motivated by the search for better career opportunities to face new responsibilities as a husband and father.

The second phase, from 1986 to his retirement in 2011, covers the time at the California State Polytechnic University, Pomona. Here, Aurilia showed he had talent not only for research but also for teaching. The contributions from two of Aurilia's colleagues, Harvey Leff and Kai Lam, offer an interesting description of Aurilia's commitment as a teacher—see [12, 13].

3.1   1970–1974: Postdoctoral Appointments in Edmonton and Syracuse

In fall 1970, Aurilia moved to Edmonton to start his first postdoc position in Takahashi's group. Such a move resulted in another life changing event. Aurilia met his future wife Elizabeth Adams at the University of Alberta. At that time, she had just started a teaching position there in the Art Department, coming from England. They stayed in Edmonton together for two years. The job market for theoretical physicists was already difficult in the early 1970s. Aurilia was offered another postdoctoral position at Syracuse University with Fritz Rohrlich. So the two headed to Syracuse and by this time they were living together. Elizabeth was offered a visiting artist position in the Art Department of the same University. The two had very good time there and decided to get married on December 1st 1972, in the Court House in Syracuse (Figs. 3 and 4).

The work in Edmonton with Takahashi was still influenced by Umezawa. It focused on the consistency of field theories against dilatation transformations [14]. As a development of his PhD work, Aurilia analyzed the breakdown of dilatation invariance as a possible dynamical mechanism to give mass to particles in the presence of a "new" Goldstone field, the dilaton [15, 16].

The time in Syracuse was, on the other hand, quite different. Aurilia enjoyed the life at the Physics Department, which was more vibrant than that at the University of Alberta. From the reference list attached to his CV (see Fig. 5), it is clear that Aurilia interacted not only with Rohrlich but also with other scientists such as Marcel Wellner. On the research side, this postdoctoral position was a sort of change of research line and a preparation for future investigations. Rohrlich himself was shifting his attention from his past activities on electrodynamics with Julian Schwinger and Hans Bethe to topics in general relativity and string theory with Louis Witten. As a result, Aurilia's work in Syracuse resulted in three manuscripts. The first one, covering pedagogical topics on special relativity, was published in 1975, after the end of the postdoctoral contract [17]. Another manuscript with Rohrlich was about the analysis of the Einstein-Podolsky-Rosen paradox within (non-local) quantum field theory. This work, however, was never accepted for publication [18]. The third and final manuscript with Rohrlich was never submitted to a journal. It was an internal report of the University of Syracuse about a unified description of electromagnetic and strong interactions [19]. This work proved to be the precursor of Aurilia's subsequent activity on the problem of particle confinement (Fig. 6).

During the fall of 1973, Aurilia applied to several assistant professor positions, but the job market has not improved in the US with respect to the previous year. The whole world was hit by a recession following what was later called the "first" oil crisis. It was a hard time for science and for young theoretical physicists in particular. Aurilia was finally offered a further fixed term position, starting in fall 1974. This time, however, he had to go back to Europe, a fact that turned out to be another life changing event. In retrospect, a decisive role in such an operation may have been played by Rohrlich himself, who had an extensive network of contacts. In addition, Rorhlich was at that time a Fulbright lecturer [6] at the University of Graz, Austria, an institution in the vicinity of the town where Aurilia would have settled down in that fall.

3.2   1974–1979: Return to Europe and Tenure in Trieste

The postwar Trieste was marred by social and economical instabilities. Unemployment, large scale emigration, military occupation, permanent threat of invasion and a general uncertainty about the future of a region torn by the Iron Curtain seemed to come to end in 1954, after the Allied Military Government over the Free Territory of Trieste was replaced by the Italian Government civil administration. During the 1960s, Trieste profited from improved living standards resulting from Italy's strong economic growth, known as Italian economic miracle. The town, however, kept its own characteristics of a crossroads between the Italian, German and Yugoslavian worlds, or more generally between Eastern and Western Europe. The improved economy and the multicultural background were decisive factors for the creation, in 1964 (i.e. in the middle of the Cold War) of the International Center for Theoretical Physics (ICTP) in Trieste.[5] The Pakistani theoretical physicist, future Nobel laureate Abdus Salam became the first director of the ICTP, while Paolo Budinich, a professor of theoretical physics at the University of Trieste, was appointed as deputy. Salam, while serving as ICTP director, kept working as a professor at the Imperial College London. This fact created an enormous advantage for the ICTP and the University of Trieste, which had a direct communication channel to one of the departments,

---

[5] The ICTP was founded under the aegis of the International Atomic Energy Agency (IAEA) and the United Nations, following the generous financial support of the Italian Government [20]. The mission of ICTP is to foster the growth of physical and mathematical sciences in developing countries.

where the hottest topic at that time (i.e. the formulation of gauge theories) had its inception [21].

In September 1974, Aurilia landed in Trieste to start his postdoctoral appointment in the group of Salam and his close collaborator John Strathdee.[6] Despite the international reputation of the ICTP and the great opportunities that could materialize by working in Trieste, Aurilia was, however, a little bit reluctant at the beginning. He felt that he had made a big step to come to America and this was a step backward, which of course it was not. Indeed the decision to move to Trieste turned to be the right one since Aurilia obtained a permanent position there less than a year later.

3.2.1 From ICTP Research Associate to INFN Tenured Researcher

In hindsight, it appears that Aurilia's permanent hiring is due to an intricate combination of events, factors and facts of scientific and non-scientific nature.

Undoubtedly Aurilia, during the first year at the ICTP, must have made a good impression on Salam, who would continue to write recommendation letters for him during the following years. In addition, Elizabeth was pregnant at that time. This could have led Salam to consider Aurilia's case with extra care. Secondly, the size of a campus often has an impact on how workplace relationships develop. At that time, the Miramare campus had just one building (corresponding to one half of what is now called Leonardo building), that was the home of three scientific institutions, namely the aforementioned ICTP, the Institute for Theoretical Physics of the University of Trieste, and the Trieste Division of INFN, Italy's National Institute for Nuclear Physics. Such cohabitation of a relatively small number of scientists—affiliated with a relatively large number of institutions—in a single building implied that communication between decision-makers was facilitated and could occur by word-of-mouth. In such a context, Aurilia rapidly made a name for himself. A third factor, was the "academic dynasty", sometimes known as "faculty inbreeding", when a negative connotation of this custom is emphasized [24]. In practice, this means that all the faculty members of a given institution are former students of a professor, recognized as the founding father. This phenomenon was the norm in Europe during those years and in part still exists.[7] Trieste in 1974 was no exception. Budinich was the academic father of all theorists in Trieste and the *éminence grise* (i.e. the wire-puller) at the University and INFN. Such an oligarchic structure resulted in a formidable

---

[6] Takahashi and Strathdee were collegues at the Dublin Institute for Advanced Studies. In 1958, they published two papers together [22, 23]. It is likely that Takahashi's reference for Aurilia landed on the Salam's desk owing to the "filter" of Strathdee.

[7] For example, all the faculty at the time of my arrival in Frankfurt, i.e. 2009, belonged to the academic family of Walter Greiner. He had been so influential that even today in Frankfurt we all study and teach from his textbooks. When I was asked to teach Thermodynamic and Statistical Mechanics at the New York University in Abu Dhabi, I discovered with some surprise that the official textbook was Greiner-Neise-Stöcker. I was proud to inform my students that I knew the first author of the book personally. It was fun to tell them that I had spoken to the third author on the phone before class!.

advantage for Aurilia. To obtain a permanent position, he had to convince just two or three people at most, or maybe not even those. Having Salam on his side was simply enough. And indeed, this is what happened.

According to what Aurilia personally told me during one of our conversations in a cafe in Claremont, California, it was Budinich who actually moved the pieces on the chessboard. And it could not have been otherwise, since in the power structure he was, as ICTP deputy director, in an intermediate position between Salam and the rest of the faculty at the University and INFN. This means Budinich had the interest in pleasing Salam by accommodating his requests in relation to Aurilia's hiring, and at the same time he had an easy job in implementing the plan with the Italian faculty. Indeed, the other two institutions on campus were both headed by one of his trusted colleagues, Giuseppe Furlan. The latter is the third man of this story and he could have contributed to the decision to hire Aurilia at INFN rather than at the University.

During the year 1974, there was a certain scarcity of job openings in theoretical physics at the University of Trieste.[8] For example, two of Furlan's former students, Franco Legovini and Massimo Tessarotto, [9] had to take a detour and applied for assistant professor positions in a related discipline, namely mathematical physics, at the Institute for Theoretical Mechanics. In addition, for Furlan, a true born Triestiner, the University was the "natural domain" of the locals, while Aurilia, with his Neapolitan background and American experience, was still considered a stranger or almost a foreigner. The INFN, on the other hand, operated nationwide and Furlan was the chairman of its national board for theoretical physics.[10] Maybe Furlan thought that the hiring of Aurilia could have been a way to ease the duty of his office by getting support from southern Italian research groups. Another motivation for the hiring at the INFN could simply have been the lack of young faculty at a division that was founded just two years earlier.[11]

The thing is that Aurilia interacted with Furlan since his arrival at the ICTP. As evidence, Aurilia wrote the following text in the acknowledgement section of his only paper with the ICTP affiliation [27]:

> I should like to thank Professor G. Furlan for valuable suggestions and comments on the manuscript. I should also like to thank Professor A. Salam, the International Atomic Energy Agency and UNESCO for hospitality at the International Centre for Theoretical Physics, Trieste, and Professor F. Rohrlich for hospitality at Syracuse University, where part of this work was done.

---

[8] This was the case despite the loss of two faculty in 1972, namely Claudio Rebbi, who moved to CERN and later to Boston, and Giordano Bisiacchi, who died in a car accident.

[9] I promised Legovini a free copy of the present book. Sadly and unexpectedly, he passed away in March 2020. Tessarotto has been my postdoctoral sponsor, during my time at the Department of Mathematics in Trieste. Legovini and Tessarotto have been classmates both at the high school and at the university.

[10] We recall that Furlan was actually the deputy of another Triestiner, Claudio Villi, who was the national President of INFN from 1970 to 1975. [25].

[11] On April 21, 1972, the Trieste Division of INFN was founded. Luciano Bertocchi, the future deputy of Salam, was appointed director [26].

Such a work, published in Nuclear Physics B, can be called a "three institute paper", since the content of the research was designed in Syracuse, the drafting of the manuscript and its submission (dated December 9, 1974) occurred during the ICTP spell, and the publication took place on June 16, 1975 when Aurilia was already at INFN. It is a particularly important paper being Aurilia's first single author paper and ninth peer reviewed publication overall. The content is still influenced by Umezawa's thinking. It addresses a development of Aurilia's PhD thesis about the formulation of spontaneously broken gauge theories without involving a boson gauge field. In particular, Aurilia showed that the Schwinger mechanism could describe either massive spin-one particles or, in the non-relativistic case, plasma oscillations of a superconductor.

Gradually Aurilia became involved in social activities with the local theoretical physicists, in particular with Furlan and his associates. For instance, he became a key player of the so called Monday Group (Italian: *Il Gruppo del Lunedì*), a team of soccer players of the Department of Theoretical Physics, that included Furlan and Legovini.[12] Two further events marked this time. The first one was the birth of Aurilia's son Darius (April 1975), with the consequent need of a larger apartment. Aurilia, however, had difficulties in finding an adequate location at that time in Trieste. While recalling his struggles, Aurilia once told me: "In Trieste, I had a (permanent) job, but I did not have a large enough home. For a while, I and Elizabeth had to move to (the nearby town) Monfalcone".[13] The trouble with the location ultimately became one of the factors that led Aurilia to consider the possibility of finding a job elsewhere.

The second event was a change of research line. This occurred due to the collaboration with two new colleagues. One of them is the aforementioned Legovini, who was himself changing research lines after having worked with Furlan on the decay of the neutral pion. The other one was John Wheeler's former student and then ICTP scientist Demetrios Christodoulou. Aurilia's new scientific interests were aligned to a rapid change in theoretical physics worldwide. Starting from 1974, the hot topics were not confined to the nuclear/particle physics realm, but now began to include classical and quantum gravity, cosmology and black holes.

3.2.2 1974 Revolutions and Initial Work on the Extended Objects

Aurilia's arrival in Trieste occurred in coincidence with a series of events that determined a revolution in physics. Due to the extent of change, it is probably more appropriate to speak of "revolutions", since at least two of them marked the year

---

[12] On a wall of the Leonardo building there were some pictures of a legendary soccer game between the Monday Group and a team of ICTP associates.

[13] Aurilia recalled in one of his e-mails that they rented apartments or houses also in Duino, Sistiana and Santa Croce di Trieste. Each place was, however, unpractical for commuting to Miramare, where Aurilia had his office.

1974. The premises for what happened were connected to the status of high energy physics in the preceding decade.

During the 1960s, the increase of collision energy up to 30 GeV and the improvement of detection technology with bubble chambers allowed the exploration of new kinematic regions in particle and nuclear physics [28], [29, p. 24]. The main goal of all the experiments of this epoch was the search for fundamental constituents of nucleons. At that time, the quark model was already proposed by Gell-Mann [30] and Zweig [31, 32]. Quarks, however, could not be experimentally isolated and, therefore, were not considered as actual particles nor reliable concepts for a consistent description of strong interactions. At the same time, initial results emerging from nucleon scattering supported the idea of the existence of families of particles having the same "internal" quantum numbers, e.g. baryon number, isospin, strangeness, etc. [33]. More specifically, particles of a given family were observed to be aligned along the Regge trajectories, i.e. straight lines in the spin vs. mass squared plot (also known as Chew-Frautschi plot [34]),

$$J = \alpha' M^2 + \alpha_0, \tag{1}$$

with a slope $\alpha' \approx 1 \text{ GeV}^{-2}$. As a second feature, scattering reactions were characterized by a strong suppression of transverse momenta leading to a "spaghettification" of the event. Another peculiarity was the duality of the s- and t-channels in certain scattering events (e.g. pion-nucleon), corresponding to a symmetry of amplitudes under interchange of Mandelstam variables, s (invariant energy squared) and t (momentum transfer squared). In practice, force mediators in one channel and short lived resonances in the other channel (e.g. the ρ-meson) were virtually the same physical phenomenon [35, 36]. Such a background resulted in a radical change of the interpretation of high energy physics, most notably in a hard criticism of the concept of fields and the associated model of point-like particles. At length scales of the nucleon, the very basic idea of elementary particles was rejected or dismissed as meaningless. Within the so called "nuclear democracy" (also known as the "bootstrap model" or S-matrix theory), no individual particle had a special role and only the Regge trajectory was considered as a fundamental object.

Remarkably, Gabriele Veneziano in 1968 accommodated all the above features in a compact formula for scattering amplitudes able to capture the $s - t$ symmetry [37]. The generalization of this work to multi-particle amplitudes ultimately led to the introduction of a one–dimensional object [38–40], later called string[14], whose vibration pattern could represent a particle and its excited states. In the same years, however, deep inelastic lepton–hadron scattering experiments with high momentum transfer at Stanford Linear Accelerator Center (SLAC) provided indications about the existence of point-like objects in nucleons [41, 42]. This fact reinforced Feynman's parton proposal [43] and resulted in the first stab in the heart of the S-Matrix/hadronic string paradigm. A decisive step forward in favor of the traditional quantum field theory description of nuclear forces was, however, made only five years later. In 1973,

---

[14] The aforementioned Claudio Rebbi actually introduced the word string in such a context [29].

Gross and Wilczek discovered the asymptotic freedom for non-abelian field theories [44]; a gauge theory for strong interactions was formulated by Fritzsch, Gell-Mann and Leutwyler in the form today known as QCD [45]. Such theoretical developments received experimental corroboration in 1974, when the J/psi meson was for the first time detected, during the so called "November Revolution" [46, 47].

Despite the huge progress in particle physics, the year 1974 should rather be remembered for another event, that, in hindsight, has become the most important revolution of that period: Hawking, for the first time, employed quantum mechanics to successfully describe a gravitational system. In doing so, he achieved several "firsts", that marked the beginning of a new era in science. He showed the possibility for a black hole to emit particles [48]. He showed that such a quantum emission is thermal, namely a black hole behaves like a black body [49]. By calculating the temperature of a black hole, he also corroborated Jacob Bekenstein's preliminary results on black hole thermodynamics [50]. More importantly, Hawking's work ignited the fire that eventually burned many of the, at the time existing, barriers between sub-fields in physics. Indeed, the Hawking radiation is the only topic able to simultaneously intersects the interest of physicists working in particle physics, astrophysics, statistical mechanics and information theory. Many of the ideas developed by Hawking during this epoch still offer compelling explanations for debated issues in physics. For instance, one of the contributors to the present book, Bernard Carr, was Hawking's PhD student at that time. His work with Hawking on primordial black holes remains, after almost 50 years, a viable explanation for dark matter [51], in part due to the lack of signals beyond the Standard Model.

In a nutshell, the Hawking radiation started the "Hawking revolution". It was not just the beginning of a new field that nowadays we call quantum gravity, but also the start of a new way of thinking, the start of an "interplay" between gravity and gauge theories.[15] In the 1960s, traditional QFT seemed to be giving way to the new S-matrix theory. By 1974, however, the trend was reversed: Hadronic string theory was dying and QFT, in the form of QCD, was on the rise. In the same year, Hawking's work on evaporating black holes shifted the focus of the discussion to the Planck scale, 20 orders of magnitude above the strong interaction energy. In the end, this was a kind of rescue maneuver for string theory, which had to find a new *raison d'etre*. Indeed, in the years that followed, string theory was resurrected as a candidate theory of everything, including quantum gravity. It is therefore not very surprising that the Hawking revolution influenced research groups worldwide,[16] including those operating at the ICTP. Even if the major efforts of theoretical physicists were still directed to the understanding of strong interaction, black holes and cosmology gradually became the new focus of the research activities. Salam himself got interested

---

[15] Today such an interplay has a working example in the AdS/CFT correspondence or more generally in the Gauge-Gravity duality paradigm. At that time, the interplay was reflected by Aurilia's group composition, namely Aurilia himself (gauge theorist), Legovini (particle physics) and Christodoulou (gravity/differential geometry).

[16] For the academic year 1974–1975, Hawking was a Sherman Fairchild Distinguished Scholar at the California Institute of Technology (Caltech). He announced the discovery of black hole radiation during a major seminar in the presence of Richard Feynman [52].

in gravity. In 1975, he founded in Trieste, with Wheeler's former associate Remo Ruffini, a series of conferences on gravitation and general relativity named after the Swiss mathematician Marcel Grossmann [53].

Despite the experimental indications from SLAC and some theoretical difficulties (e.g. the requirement of a higher dimensional spacetime and the emergence of a spin-2 field), the death of the hadronic string was not sudden. In 1974, QCD had taken the lead but it still had problems.[17] To explain some decays, one had to invoke additional quarks that escaped detection. For this reason, there has been a time of transition, during which physicists tried to combine stringy ideas within the QCD paradigm. The main goal was to explain the major unsolved problem of strong interactions, namely confinement. The by-product of such a mix of approaches was the possibility of keeping the good properties of both them, namely the duality and the Regge trajectory for processes with small t and the asymptotic freedom for hard scattering [29, p. 123]. Along such a line of reasoning, Yoichiro Nambu proposed in 1974 a confinement scheme [55], in which quarks were interpreted as magnetic monopoles (along the lines of the Nielsen-Olesen vortex-line model for type II superconductors [56][18]), while their confinement was explained by a permanent stringy bond, by means of a formalism similar to that proposed by Kalb and Ramond [58]. Nambu's work is particularly relevant for the current discussion because it represented the first of the two premises for Aurilia's new research line. The second premise was the MIT bag model. Such a formulation was simply based on the assumption that color fields have to live within a finite region [59].

In summary, taking into account all the events of the year 1974, it is no wonder that Aurilia himself recalls this period with particularly intense words [8]:

> The year 1974 represents another turning point in my life and in my relationship with Umezawa. In that year I returned to Italy, at a time when the world was feeling the convulsions of the end of the Vietnam War, the Nixon era, etc.. By then I was married, with our first son on the way. From the narrow and strangely detached vantage point of high-energy physics it was the time when the discovery of special "stringlike" solutions in quantum field theory was attracting the attention of most theoretical physicists because of their potential use as simulators of meson resonances. At that time gravity played no role in particle physics, and general relativity was the playground of a distinctly narrow academic community of mathematical physicists. At any rate, with different motivations and methodologies both Umezawa and I became interested in the general properties of strings and other relativistic extended objects.

3.2.3 Formulation of the Theory of Hadronic Bubbles

The collaboration with Legovini and Christodoulou produced seven papers, five consecutive Physics Letters [60–64] followed by two papers on Journal of Mathematical Physics [65, 66].

---

[17] In 1978, Salam was still considering the "strong gravity" paradigm as a viable alternative to QCD [54].

[18] Incidentally an extension of the Nielsen-Olesen model has been the subject of my Laurea thesis at the University of Trieste in 1997, under the supervision of Roberto Iengo [57].

For the first time, Aurilia was the most experienced researcher in a collaboration, being senior to Legovini by four years and to Christodoulou by nine. But it was not just a matter of age. Aurilia was definitively the leader of the group: He was the one who forged new ideas, the one who published single author papers, the one who worked on separate research projects with each of the other two. Christodoulou was the mathematician of the group. He provided the formalism for the physical ideas Aurilia had in mind. Legovini was the "translator".[19] He guaranteed a smooth transition between Physics and Mathematics and an efficient communication within the group overall. Indeed, Legovini had experience both in Mathematical Physics and in phenomenological aspects of meson decays, two essential ingredients for the kind of investigation the trio was going to do.

The main goal of their research activity was the search of an efficient mechanism for confinement within the formulation of a bagged QCD [64]. The trio aimed to overcome some of the shortcomings of the MIT bag model by offering a consistent geometrical and gravitational perspective for the finite region where color fields were supposed to live. Their formalism was based on an extension of string dynamics [55, 58, 67] to a higher dimensional extended object.

At the crux of the matter there is Aurilia's idea about the unified description of strong and electromagnetic interaction, whose initial traces are already visible in his unpublished paper with Rohrlich [19]. The proposal is as follows. In two dimensions electrodynamics is confining, because the Coulomb potential is linear with the interaction distance. This corresponds to the Schwinger model for point-like objects [68]. To preserve the confining properties, the four dimensional extension of the Schwinger model should involve a higher rank gauge field, $A_{\mu\nu\rho}$, namely:

$$j^\mu A_\mu \text{ on } \mathcal{M}^2 \longrightarrow j^{\mu\nu\rho} A_{\mu\nu\rho} \text{ on } \mathcal{M}^4. \tag{2}$$

For tensor consistency, the interaction term should involve a higher rank current, namely $j^{\mu\nu\rho}$. In other words, the increase of the rank of the gauge field directly implies the presence of an "extended object" in the interaction Lagrangian. Indeed $j^{\mu\nu\rho}$ describes a membrane, namely a relativistic hypersurface or bubble. Accordingly, the "Maxwell field" carries four indexes, i.e., $F_{\mu\nu\rho\sigma}$. In [62], the formalism has been generalized to a gauge field $A_{\mu_1\ldots\mu_{n-r}}$ in n dimensions, with explicit derivation of the analogue of the Lorentz force and the Maxwell field equations. The particle corresponds to the case $r = n - 1$, the string to the case $r = n - 2$ and the bubble to the case $r = 1$. Therefore, the bubble is an extended object of dimension $n - 1$ and the Maxwell field is a differential n-form. For dimensional reasons, the Maxwell field can have just one degree of freedom, corresponding to an electrostatic like field. The bubble cannot generate any magnetic or radiation field. Surprisingly, the potential between two distant surface elements of the bubble linearly depends on their distance, a fact that confirmed Aurilia's prediction. In simple terms, the relation between two

---

[19] I have been the person that informed Legovini about Aurilia's death. During that phone call, Legovini recalled some details of the collaboration that are in part summarized in the present section.

dimensional electrodynamics and bubble surface elements in n dimensions can be visualized by observing that the two dimensional solution of Laplace's equation

$$V(x) = Ax + B, \tag{3}$$

corresponds to the potential between two parallel charged planes. Notice that planes in a three dimensional space have just one dynamical degree of freedom, along their normal line, similarly to the bubble.

The strong point of the proposal is that the resulting Maxwell field is confining in the bubble interior. In contrast to the MIT bag model, Aurilia's bubble can "breath", namely its surface is a true geometric entity with dynamical evolution governed by $T^{\mu\nu}_{\text{matter}}$. The latter is a component of the total energy momentum tensor of the field-bubble system,

$$T^{\mu\nu} = T^{\mu\nu}_{\text{matter}} + T^{\mu\nu}_{\text{field}}, \tag{4}$$

with $\nabla_\mu T^{\mu\nu} = 0$. In the limit $F_{\mu_1 \dots \mu_n} \to 0$, the energy momentum tensor $T^{\mu\nu}$ describes a free relativistic membrane. On the other hand, the surface of the bag in the MIT model is dynamically determined by the confined quark field and by a pressure term that is put in by hand in the energy momentum tensor.

At this stage, Aurilia could still be considered a gauge theorist with interests in string theory and nuclear physics. His interest in classical and quantum gravity had not yet developed, despite the collaboration with Christodoulou and the geometrical formalism that the bubble mechanism required. His research portfolio actually evolved only from the year 1983, when he initiated a collaboration with "a genuine son" of the Hawking revolution [69]. Such a collaboration ultimately became so prolific and important that Aurilia himself dubbed it "the Firm" [70].

3.3   1980–1986: International Stardom and Return to Canada

During the time of the collaboration with Legovini and Christodoulou, Aurilia's daughter Alex was born in London (November 1977), while her father was spending a sabbatical at Imperial College. This was the beginning of a time in which Aurilia was still officially working at INFN but was on leave from it. From 1978 on, he did not even have an official residence in Trieste, nor did he have an apartment there. This fact was the source of great uncertainty and exposed an unsolved work-family conflict.

In 1979 Aurilia was a guest scientist in Norman Dombey's group at the University of Sussex, Brighton. In 1980 he obtained a visiting position at the CERN Theory division in Geneva. Due to his publications of this period, Aurilia achieved fame and international recognition. His name attracted the attention of many scientists, including Stephen Hawking [1].

3.3.1   Work with Kobayashi, Nicolai and Townsend

After an eight-year break, Aurilia returned to work with Takahashi. This time there was a third collaborator from Japan, the future Nobel laureate Makoto Kobayashi [71]. The focus of their work was on higher spin field theory, a topic Aurilia had known since 1966 when he was writing his Laurea thesis with Umezawa. In 1980, the motivations, however, changed, since the theory of supergravity was recently formulated [72]. The gravitino, namely the graviton superpartner, is a spin-$\frac{3}{2}$ field and actually coincides with the Rarita-Schwinger field. The latter was known to exhibit causality violation and superluminal propagation when coupled to the photon [73]. In their paper published in Physical Review D, Aurilia, Kobayashi and Takahashi were able to identify the origin of all the above unwanted features: In practice the Rarita-Schwinger Lagrangian acquires the wrong sign in its kinetic term due to the action of one of the two components of the photon field. Although their work did not solve the problem, it helped to clarify that the customary procedure of quantization of the gravitino in the presence of a background field was not correct.

The next paper, designed and written at CERN with Herman Nicolai and Paul Townsend, is probably the best known among Aurilia's publications [74]. The work is again based on the non-dynamical 3-form field $A_{\mu\nu\rho}$.[20] Its topological nature can, however, be relevant in the definition of some integration constants that are not explicitly specified in the field Lagrangian. This is the case of the θ parameter multiplying the CP violating term in the QCD Lagrangian [75], that can be written as

$$F^a_{\mu\nu} F^a_{\rho\sigma} \epsilon^{\mu\nu\rho\sigma} = \lambda^2 F_{\mu\nu\rho\sigma} \epsilon^{\mu\nu\rho\sigma}, \tag{5}$$

namely in terms of the bubble Maxwell field, where λ is a constant with dimensions of mass. More importantly, the paper aims to show that $A_{\mu\nu\rho}$ is connected to a cosmological term. In certain supergravity theories, the 3-form field can be integrated out to leave a surface term that equals a cosmological constant. This is the case of $N = 8$ supergravity in four dimensions. More details about the derivation of this result can be found in the work of Paul Townsend, as part of the present book [76]. Interestingly, Townsend also presents new results in terms of membrane/domain wall vacuum solutions emerging from the aforementioned supergravity Lagrangian containing the 3-form field [76].

Remarkably, Hawking quoted the Aurilia-Nicolai-Townsend work in his 1984 Physics Letters B paper, when he attempted to provide a natural explanation for the small value of the cosmological constant [1]. In Hawking's view, the 3-form field would enter Universe's gravity-matter path integral like any other field. Due to its non-dynamical nature, the Maxwell field $F_{\mu\nu\rho\sigma}$ would provide a background cosmological term, that is not directly connected to any ultraviolet regularization procedure. This would circumvent the known difficulty in describing the cosmological constant as the vacuum energy in quantum field theory. At this point, one can argue that the

---

[20] This means that $F_{\mu\nu\rho\sigma}$ has to be constant, a fact that is consistent with the proposed analogy with two dimensional electrodynamics (3) and the field between two charged parallel planes.

Aurilia-Nicolai-Townsend work was one of the reasons that led Hawking to believe that $N = 8$ supergravity in four dimensions could have been the so called "Theory of Everything" [77].[21]

At CERN Aurilia published a second paper with Townsend about a further application of the 3-form field formalism. This time Takahashi joined the collaboration in place of Nicolai [81]. The topological nature of the $A_{\mu\nu\rho}$ field was used to explain the presence of Goldstone bosons for both the Higgs mechanisms and the U(1) problem of QCD. Today the U(1) problem is explained in a alternative way by the presence of instantons [82].

3.3.2   Faculty in Toronto and New Research Line with Spallucci

The two CERN papers resulted in the latest of Aurilia's publications before leaving Europe. A return to Trieste was not possible. Elizabeth and the children were in London, while living in Trieste without support from family members would have been hard. In addition, there was the long standing problem of finding an adequate apartment. This issue was exacerbated by taking into account that Italian salaries in the research sector were below the international average.[22] The worst part of the position in Trieste was the absence of any perspective of career progression at INFN. It seemed also quite unlikely to switch back to the ICTP or to become professor at the University of Trieste. It was really a stagnating situation similar to what Aurilia already experienced 14 years earlier in Naples before going to Milwaukee. On the other hand, the positive aspect was that Aurilia was free from any teaching duty, a fact that allowed for long term visits. In practice the position at INFN seemed to be tailored to build a future elsewhere.

During the time at CERN, Aurilia was offered the possibility to go back to Edmonton and join Takahashi's group. After much deliberating, Aurilia and his wife decided to accept the offer to return to Canada. In Alberta, Aurilia published three papers from 1981 to 1983. The first two turned out to be the last ones in collaboration with Takahashi [84, 85].[23] Again the subject of the work concerned the generalized Maxwell equations for the antisymmetric tensor gauge fields $A_{\mu_1,\ldots,\mu_r}$, with $r = 1 \ldots r - 1$.

---

[21] $N = 8$ supergravity in four dimensions has additional good properties. It is the dimensional reduced theory of $N = 1$ supergravity in 11 dimensions, also known as CJS theory [78], that is the only known classical 11 dimensional theory with supersymmetry and no field with spin higher than 2. In addition, Witten showed that 11 is the minimal dimension to unify the gauge groups of the Standard Model [79]. The reasoning is the following [33]: U(1) corresponds to a circle ($S^1 \to D = 1$), SU(2) to a sphere ($S^2 \to D = 2$), SU(3) to the complex projective space ($\mathbf{CP}^2 \to D = 4$). Thus, globally one needs 7 extra dimensions. Considering $3 + 1$ non-compact dimensions, one ends up with 11 dimensions. It is not a coincidence that this is also the number of dimensions required by M-theory, since 11 dimensional supergravity is believed to be its low energy limit [80].

[22] Italian academic salaries are still today below the European average both in absolute terms and in purchasing power (see [83]). On the top of that, European salaries are in average below US/Canada salaries.

[23] A third one with Takahashi has never been published [86].

The cases $r = 2$, 3 in four dimensions were expanded with respect to previous publications[24] and new details about the Higgs mechanism for extended objects were introduced. Mass generation was a key element to solve the U(1) anomaly in QCD, along the line of a formulation originally proposed with Townsend one year earlier [81]. As a byproduct, the analysis of the massive case allows for a dynamical interpretation of the "spin jump", a discontinuity of the spin when going from massless to massive theories, previously noticed by Deser, Townsend and Siegel [87].

The third paper from the time in Edmonton [88], was written with Maurizio Martellini, who was a fresh PhD visiting the University of Alberta from the University of Milan [89]. With this collaboration, Aurilia, for the first time, entered the quantum gravity business. Interestingly, the main idea of the paper resembles two other quantum gravity proposals, namely 't Hooft's spontaneuos dimensional reduction [90][25] and Weinberg's asymptotic safe gravity [92, 93]. In practice, the non-renormalizable character of gravity is just a spurious low energy effect, that disappears at the Planck scale. In Aurilia and Martellini's view, it is the gauge group that undergoes a phase transition, called MacDowell-Mansouri mechanism [94], between the two regimes. For energies around the Planck scale, M_P, gravity is assumed to be described by an SO(3,2) invariant Yang-Mills Lagrangian, that is renormalizable, unitary and asymptotically free. For energies below M_P, SO(3,2) is contracted to the group ISO(3,1) via the MacDowell-Mansouri mechanism [94]. In this picture, general relativity emerges as a boundary effect, since the group contraction reduces the kinetic term in the SO(3,2) Lagrangian to a total derivative.

The year 1983 saw another important turning point for Aurilia. Following the job application round of Fall 1982, he was offered an assistant professor position (tenure track) from the University of Toronto Mississauga (UTM). Before settling outside Toronto with his family, Aurilia went to Trieste for a summer visit.[26] It was in this occasion that his scientific interest and production changed, decisively and for the last time.

There are many reasons that led Aurilia to pursue new research frontiers, but certainly the key factor was his meeting with Euro Spallucci. From this point on, they founded the aforementioned "Firm", that produced 30 peer reviewed papers (including 12 Physical Review D and 6 Physics Letters B) in the following two decades. It is therefore not surprising that Aurilia ranked Spallucci as the best one among

---

[24] The case $r = 2$ is equivalent to the Kalb-Ramond B-field [58]. The case $r = 3$ corresponds to the Aurilia-Christodoulou-Legovini's bubble [62].

[25] For further details about the dimensional reduction in quantum gravity see Mureika's chapter in the present book [91].

[26] Due to the brevity of his return, Aurilia had to share the office with a colleague, i.e. Maria Teresa Vallon, who at the time was a PhD student in Iengo's group at the International School for Advanced Studies (SISSA). SISSA was founded in 1978 following the model of the Scuola Normale Superiore of Pisa and the Institute of Advanced Studies in Princeton. At the time, Aurilia was already at Imperial College on leave of absence from INFN. Notice that SISSA was the fourth institution based in or in the vicinity of the Leonardo Building in Miramare, in addition to the aforementioned, ICTP, INFN and Institute of Theoretical Physics of the University of Trieste. It was soon followed by a fifth one, the Third World Academy of Sciences (TWAS), founded by Salam in 1983.

his collaborators.[27] To better understand why Spallucci turned out to be Aurilia's scientific perfect match, we have to recall that Aurilia up to this point had little experience with classical and quantum gravity. Spallucci, on the other hand, belonged to a new generation[28] of young Italian theoretical physicists,[29] whose research background was in general relativity, quantum field theory in curved space, black hole evaporation and quantum cosmology. In other words, Spallucci was a genuine son of the "Hawking revolution". His doctoral advisor was Gallieno Denardo, who had worked with Ruffini on Penrose-like processes for the extraction of energy from a black hole [95, 96]. Another important aspect to consider was Spallucci's capacity of dominating the mathematical formalism to quickly reach the physically meaningful goal.[30] In other words, in the person of Spallucci, Aurilia could find the synthesis of the characteristics of two opposite scientific profiles, the "mathematician" and the "translator", as we previously noticed in the case of Christodoulou and Legovini.

The actual meeting between Aurilia and Spallucci occurred almost by chance. Aurilia held a seminar at the ICTP about the theory of extended objects. At the conclusion of his talk, Aurilia stated an open question in relation to the possibility of using the bubble formalism to describe the physics of neutron stars. Spallucci, who attended the seminar, was intrigued by the topic and went to Aurilia's office to discuss the issue. At this point, Aurilia gave Spallucci some unpublished lecture notes about hypersurfaces coupled to gravity, written by Werner Israel. On this basis, they started their collaboration and they soon realized that the 3-form field could describe not only a hadronic bubble but also the dynamics of Israel's topological defects. The text of the acknowledgement section of the paper [69] reflects the above chain of events:

> We are indebted to W. Israel for making available to us some useful notes on his formulation of shell-dynamics and for pointing out to us Hawking's work of Ref. [5]. Finally, one of us (A.A.) wishes to thank M. Kalinowski and G. Kunstatter for some useful discussion.

This work, initiated in the summer of 1983, resulted in two peer reviewed publications—the already quoted 1984 Physics Letters B [69], and a Nuclear Physics

---

[27] This is literally what Aurilia told me, in Italian, during a conversation at his house in Upland, CA in Fall 2008. He said it without being critical or diminishing the merit of other eminent scientists he worked with, such as Christodoulou, Kobayashi, Mann, Nicolai, Takahashi, Townsend and Umezawa. The collaboration with Spallucci simply outran any other collaboration in terms of scientific impact, namely duration, number of papers and citations.

[28] In Bologna, there was another notable member of this generation, namely Roberto Balbinot, who would become Aurilia and Spallucci's collaborator and my doctoral advisor.

[29] Aurilia was senior to Spallucci by ten years.

[30] Spallucci's essential and sober use of mathematics and capacity of explaining complicated topics in crystalline clear terms were legendary among students in Trieste during my undergraduate years. He was considered by many as the best teacher in the department. Well before I started publishing a number of groundbreaking publications works with Spallucci, it happened to me to hear the opinion of my doctoral supervisor about the collaboration with Spallucci. With admiration and conviction, Balbinot said something like this: "Spallucci is like a machine! When he shows you a calculation on the black board, he starts and... and never stops until he gets to the result... and he always gets to the result! He often concludes his explanations with a "done!" or "that's it!". In doing this, he shows his collaborators how trivial the mathematical calculation has been, even if it is not!".

B [97] paper. Additional members of the collaboration were Legovini and Denardo. The starting point of both papers was Hawking's aforementioned claim about the possibility of using the 3-form field as a cosmological term. From this premise, the content was focused on the derivation of the bubble dynamics in the presence of gravity. This was done without invoking any scalar field coupled to gravity, as in the case of Coleman and De Luccia [98], but by solving gravity-matter field equations descending from the action Aurilia introduced five years earlier with Christodoulou and Legovini [62]. Interestingly, the bubble boundary is unstable and it can either collapse into a black hole or expand, depending on the relation true/false vacuum in the bubble interior/exterior. For the case of zero total energy, the expansion corresponds to Vilenkin's tunneling mechanism, namely the birth of the Universe from "nothing" into a de Sitter space [99]. The borderline solution between the aforementioned two cases is the static configuration, for which the radius of the bubble, its Schwarzschild radius, the true vacuum domain radius and the false vacuum domain radius are coincident. Such a static gravitational object resembles the geometrical model for hadron confinement within the strong gravity paradigm [100]. After all, this is not the only admissible interpretation. For instance, one can assume the inner cosmological constant to be of the order of nuclear matter density. Accordingly, the radius of the static bubble turns out to be of the order of 10 km. This fact corroborated Aurilia's original intuition about the application of his formalism to neutron stars.

The above two papers marked the end of an era and the start of another one for at least three good reasons. To begin with, Aurilia, during the aforementioned summer visit in Trieste, started the preparation of some lecture notes for the courses of the fall semester in Toronto. This was actually the beginning of his teaching career[31] that lasted until his retirement in 2015.

A second fact related to such papers is that they are the last ones with INFN, Trieste as an affiliation. The move to Toronto was supposed to be definitive. Aurilia and his wife were very happy with life in Canada. They bought a house and they applied for citizenship, which was soon granted.[32] After obtaining tenure, it was no longer necessary for Aurilia to stay on leave from INFN. Furthermore, Aurilia had not received any salary from INFN since 1983. Therefore, the affiliation to INFN became purely formal and superfluous.

Next, Aurilia, while serving as an assistant professor in Mississauga, successfully applied to the Natural Sciences and Engineering Research Council of Canada (NSERC) and received a research grant for the project titled "Properties and Applications of Generalized Maxwell Fields". This marked the beginning of a series of successful applications for grants, sabbaticals that continued until his retirement.

However in the end there is an unexpected conclusion to Aurilia's time in Toronto. Despite this good start with the awarding of the grant, things went quite wrong for Aurilia in Toronto. Like a bolt from the blue, he discovered that by 1984 his chances

---

[31] At the beginning of his term in Toronto, Aurilia was assigned the course of numerical analysis, a topic he did not like and for which he had no previous experience.

[32] Aurilia held four passports, namely from Italy (by birth), UK (by marriage), Canada and USA (by naturalization).

for a tenured position had already evaporated. Aurilia was promised a permanent position following the retirement of a faculty in the department. Unfortunately, the professor Aurilia was supposed to replace, suddenly decided he did not want to retire. As a result, the above two papers are not only the last papers with the INFN affiliation, but also the first and the last papers with the affiliation of University of Toronto, Mississauga.

There could have been additional reasons behind the decision for denying Aurilia tenure.[33] Typically in such cases, there is always a list of "too ..." or "not enough...". None of the items of the list is decisive by themselves, but globally they concur to create a negative image. For example, there could have been suspicions about his interest in teaching numerical analysis. In Toronto, during the 1980s, there were already important groups working on solid state physics, atmospheric physics and meteorology whose research activities demanded an extensive use of computer simulations [102]. Aurilia's research line was certainly not aligned with such topics and might have appeared fishy or esoteric to some decision makers. Apart from the grant, to his credit Aurilia had established contacts as well as an actual collaboration within John Moffat's group. Moffat, however, was a faculty at the main campus of the University of Toronto, while Aurilia was at a satellite campus. This separation could have determined additional political/academic reasons that might have worked against Aurilia.

The bottom line result was that Aurilia, at 43, had not yet secured an academic position aligned with his competence and experience. Rather, he was compelled to apply to professorships in the next round of job opening calls. By comparison, Legovini, who already had a permanent position in Trieste, was promoted to the role of associate professor one year later. Christodoulou was already full professor at the University of Syracuse.

Aurilia's personal story can be seen as an emblematic example of the epochal problem related to the career opportunities for aspiring faculty in theoretical physics. Sadly, after more than 35 years since such events, the situation has not improved. Data show that more than 70% of the faculty in the US work as adjunct, i.e. on a contract basis off the tenure track [103, 104]. The global pandemic has certainly exacerbated the problem, but even before the COVID-19 outbreak, there was a clear strategy of replacing retiring faculty with a cheap workforce. The issue of Aurilia's tenure in Toronto discloses another long standing problem of academia since its inception, namely the hiring process. The latter is too often governed by a sort of "principle of minimization",[34] namely the tendency to hire weak or average candidates. This is in part due to fact that talented scientists are often perceived as too independent, unreliable or even inadequate. They raise suspicion because they can potentially challenge the established authority and order. In addition, they often propose original, non-mainstream research lines that are harder to understand and evaluate.

---

[33] As an example see the notable case of Sean M. Carroll [101].

[34] I am deliberately borrowing the term from variational calculus to describe social dynamics and decision making in some academic circles.

3.4 1986–2015: Professor at the California State Polytechnic University, Pomona

The job application round of fall 1984 resulted in three job offers for Aurilia. The most attractive was that from the California State Polytechnic University, Pomona (Cal Poly Pomona). Aurilia opted for it and formally agreed to sign a contract in 1985 with the aim of settling down soon after.

At this point, it is necessary to make a general remark about the importance of the events described here. All the bad luck Aurilia has suffered throughout the years, e.g. the emigration from his native Italy, the troubles with obtaining an apartment in Trieste and the bitter conclusion of the experience in Canada, finally seemed to have come to an end. This did not occur, however, without pain. The move to California was difficult, because of massive bureaucratic hurdles. The faculty at the Physics Department in Pomona had to convince the US Immigration Services that they really wanted (and needed) Aurilia, a non-US citizen at that time, over American applicants. The situation was uncertain and sounded hopeless in some phases, but Aurilia found a person, that was so determined to conduct such a tough bureaucratic battle that in the end the Department succeeded in hiring him. For these reasons, one has to give credit to Harvey S. Leff, the former Physics Department Chair in Pomona, who headed the hiring committee. He can be ranked among the scientists that positively changed Aurilia's career and life, together with Umezawa and Spallucci. Due to his open-mindedness, unbiasedness and perseverance he never deviated from his intended goal of hiring the best candidate. This was done against all odds, including the fact that originally the post was advertised for an experimentalist. For more details about the job interview, the hiring process and the green card, see the contribution that Harvey S. Leff has submitted to the present book [12].[35]

3.4.1 Visiting Scholar at SLAC

The first three years in Pomona were intense for Aurilia. He published an initial paper resulting from the project supported by the aforementioned Canadian grant. Apart from Spallucci, the collaboration involved two young Canadian scientists. One was Randy Kissack, who later turned his interests towards space physics.[36] The other one was a rising star in the field of gravity (and author of a chapter of the present book [107]), namely Robert Mann, who at the time was in Toronto, after having spent a postdoctoral appointment at Harvard University and before moving to Waterloo.[37]

---

[35] For additional information about Leff see [105].

[36] In 1986, Kissack was already affiliated with the Institute for Aerospace Studies of the University of Toronto. In 1993 he obtained a PhD in space and atmospheric physics from the University of Western Ontario, where he later became a faculty member [106].

[37] Actually this was Mann's second stint in Toronto, having obtained his PhD there in 1982 under the supervision of John Moffat [108].

The subject of the work was the development of a full theory for the relativistic bubble dynamics under the action of the 3-form field. The paper is 15 page long, and it offers the derivation of field equations from a variational principle, some points about differential geometry and phenomenological considerations for both cosmology and nuclear physics. The paper would be published in May 1987 in Physical Review D [109], after an initial submission in January 1986. A 14 month review process can be surprising from a modern perspective, but one has to take into account that at that time communications between scientists occurred over the phone or regular mail. This explains why there might have been some difficulties in coordinating the various phases of the project, being that Aurilia was based in California, Mann and Kissack in Canada and Spallucci in Trieste.

The teaching load in Pomona was heavier than in Toronto. To ease the situation, Aurilia submitted a research proposal, titled "Evolution of cosmic network. An analytical and numerical approach" to the U.S. Department of Energy. The project was funded and he obtained a scholarship and release time to visit the Stanford Linear Accelerator Center (SLAC). As a consequence, the first paper Aurilia did based on work done entirely in California—from the development of the main idea to the publication—appeared on Physical Review D in October 1989 [110], three years after the move to Pomona.[38] Spallucci coauthored the paper as well as all the following papers Aurilia published. Interestingly, the third author of the paper was Marc Palmer, an undergraduate student in Pomona at the time. This was a great achievement because it showed Aurilia's capacity to involve the young generation in his research activities. Furthermore, the funding of two additional research proposals[39] paved the way to early tenure and promotion to the rank of professor in 1989.

The second paper written at SLAC is a beauty [111]. It is no longer in the framework of bubble dynamics. It is about the Schwinger model in curved spacetime and its relation to the two dimensional Riemann Cartan theory. By exploiting such an analogy, the Riemann Cartan formalism is interpreted as a quantum theory of the Abelian axial-vector field. In particular, the paper offers a derivation of the effective action for the torsion field by integrating the anomalous Ward identity in four dimensions. The paper is completed by the introduction of the gauge-fixing Wess-Zumino term as well as by a discussion of a variety of admissible forms of the pure gravity Lagrangian. The content of the work is technically involved. It is wide and deep. It provides evidence that Aurilia and Spallucci are a perfect match. Aurilia used analogies between equations to predict things that nobody has thought about or understood before. Spallucci manipulated the mathematical formalism with the ease of a child playing with an abacus.[40]

---

[38] Incidentally, this also was Aurilia's first paper, that was fully prepared with a typesetting software like TeX, the precursor of the current LaTeX.

[39] Both projects were approved in 1988. The first one, titled "Field Theory of Geometric Strings and Membranes" was a Research, Scholarship, and Creative Activity (RSCA) award of Cal Poly Pomona; the second one, titled "Quantum Bubble-Dynamics", was Aurilia's second grant from the U.S. Department of Energy to visit SLAC. An additional RSCA award titled "Quantum theory of dynamical surfaces" was granted soon after Aurilia received tenure.

[40] "Precognition" and "enhanced mental abilities" are among the powers one can find in the literature of superheroes. This time, they perfectly fit for Aurilia and Spallucci in their work [111].

3.4.2 Quantum Bubbles, Higgs Mechanism and New Members in the Firm

The work at SLAC was instrumental for the following two papers, which sought to extend the classical bubble dynamics to the quantum realm.

The initial paper, published in Physics Letters B in November 1990, introduced the idea of a minisuperspace approximation to describe the quantum bubble dynamics in terms of a $(1 + 1)$-relativistic equation [112]. One of the interesting results is that the bubble never collapses to a singular point due to the uncertainty principle. For this reason, the quantum bubble can be seen as the precursor of a family of regular black hole geometries, known as "noncommutative black holes", which I derived 15 years later in collaboration with Spallucci and Smailagic [113–115]. The other important result of quantum bubble dynamics is the existence of a non-vanishing probability of for tunneling to a classical configuration of arbitrarily large radius. This fact can be interpreted as the gravitational analogue of the Klein paradox for electrons scattered beyond a potential barrier [116]. More importantly, it is in agreement with the Farhi-Guth-Guven mechanism for producing the Universe out a small bubble of false vacuum [117].

There is also a historical note regarding this paper due to a technological development. For the first time, the authors' affiliations included the e-mail address, namely:

CBKG004@CALSTATE.BITNET
SPALLUCCI@TRIESTE.INFN.IT.BITNET.

At that time, internet was not yet developed. Communications between users affiliated at universities occurred by a network prototype called BITNET. Such a system became quickly obsolete due to the rise of the Internet protocol suite (TCP/IP). In October 1991, namely one year later, Aurilia and Spallucci already switched to their e-mail addresses which are still in use today. This fact eased the communication issues, reduced the dead time and encouraged the inclusion of additional collaborators from remote locations. This was the case of Roberto Balbinot, who joined the Firm during this time to become a regular collaborator for a period of about 7 years.[41] The initial goal of the new trio was the extension of the quantum bubble dynamics to the case of curved background spacetimes. The related paper was the second publication emerging from Aurilia's stay at SLAC and also the second of a series of four works in a row that Aurilia published in Physics Letters B at the beginning of the 1990s [119].

The third Physics Letters B paper [120] of this series was the extended version of an unpublished essay titled "The dual Higgs mechanism and the origin of mass in the Universe" [121]. For this work, Aurilia and Spallucci were awarded an "honorable mention" in the 1991 Gravity Research Foundation essay contest on gravitation. The peer reviewed version of the paper included Legovini in the list of authors. This

---

[41] Balbinot has contributed to the present book—see [118].

would be the last appearance of Legovini in a collaboration with Aurilia. Legovini ended his research career for good and focused on academic teaching.

The idea behind the calculations of this paper is powerful. The Higgs mechanism is reformulated for the case of an extended object. The bosonic membrane, coupled to the 3-form field $A_{\mu\nu\rho}$, is shown to be invariant under a generalized U(1) transformation. The symmetry breaking occurs in a similar way to the conventional mechanism for particles. This time, however, it is the tension that becomes negative and generates a non-trivial vacuum state. Another difference is that the gauge field $A_{\mu\nu\rho}$ becomes massive but, being non-dynamical, it represents a constant background field. Interestingly, such a constancy determines a constraint that modifies its spin content. Some components of the gauge field vanish while others are still dynamical. The massive gauge field Lagrangian reads

$$\mathscr{L}[A] = -FF - |c|AA \qquad (6)$$

where A is a 3-form field, and $F = dA$ is the field strength. Equation (6) describes the dynamics of pseudoscalars, corresponding to the time-like components of dual gauge field $\star A$. The pseudovector components of $\star A$ are eliminated, corresponding to a spin jump of the gauge field. The balance of degrees of freedom implies that such pseudoscalars correspond to the dynamical content of the original Goldstone bosons after the symmetry breaking. Interestingly the Goldstone bosons are represented in terms of an integral of the Kalb-Ramond field over the membrane.

The main idea is that, in this way, it is possible to explain the lack of detection of the axion field. Within this scenario, axions are the Kalb-Ramond bosons. They cannot be observed because they are hidden in the cosmological constant, resulting from the tachyon condensation. The 3-form field has actually eaten the axions up!

The year 1991 ended with the submission of the fourth manuscript to Physics Letters B [122]. The subject was again the axion field but this time from an alternative perspective. The focus was on an analogy between a soap film over a ring and a Kalb-Ramond gauge potential bounded by a closed string. The excitation of the Kalb Ramond field generates a relativistic membrane. Because of the presence of a global Peccei-Quinn symmetry, the membrane is termed "axionic membrane". Aurilia and Spallucci derived the Hamilton-Jacobi equation for such a system. This publication resulted in Aurilia's first paper appearing in an open access digital repository—see Fig. 9. At that time, this repository was the LANL preprint archive, whose successor is the current arXiv.

Starting from 1992, Anais Smailagic joined the Firm from the ICTP.[42] The research topics slowly became more formal and quantum gravity oriented. Their initial paper [124] was in the framework of Zel'dovich-Sakharov induced gravity [125, 126]. The basic idea was to replace, in the generating functional, the gauge field with a relativistic membrane coupled to a 3-form potential. By considering the low energy expansion of the membrane effective action, one finds the Einstein-Hilbert action with a cosmological term coupled to the non-dynamic F-field. This

---

[42] Smailagic has contributed to the present book with a chapter jointly written with Spallucci [123].

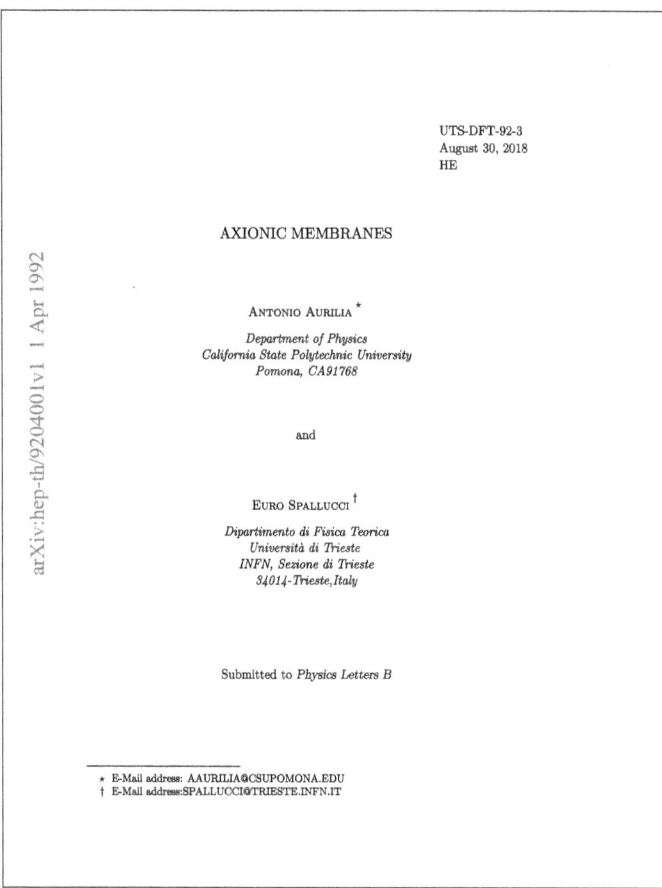

Fig. 9 Aurilia's first preprint posted on a open-access repository. The manuscript was submitted to Physics Letters B on October 1, 1991. It was posted on LANL preprint archive on April 1, 1992. It was finally published on May 21, 1992 [122]. E-mail addresses appear for the first time on the front page. Note the internal preprint classification UTS-DFT-92-3, namely Department of Theoretical Physics, University of Trieste, March 1992. The date August 30, 2018 was not present in the original submission. It refers to an update performed by arXiv. Reprinted from https://arxiv.org/pdf/hep-th/9204001 with permission, ©The Authors. All rights reserved

fact is of great advantage because the constancy of F leads to an almost vanishing cosmological constant. In practice, this paper offers a solution to the cosmological constant problem, namely the discrepancy between the observed value of the cosmological constant and its predictions based on a quantum field theory calculation. To some extent, this result is the concrete realization of Hawking's 1984 intuition about the use of a constant background field F to solve the above long standing issue [1].

A second paper with Smailagic was in the framework of a reformulation of the Nambu-Goto action in terms of non-canonical variables that preserve reparameterization invariance on the wold sheet [127]. The goal was to overcome the drawbacks of the geodesic field in the Nielsen-Olesen model by extending its validity to the case of open strings. The main result of the paper is the formulation of string dynamics in terms of an anti-symmetric tensor field and this allows for the formulation of a Higgs-like mass generation mechanism for the Kalb-Ramond field.

In the meantime, Aurilia published with Spallucci an additional paper in 1992 [128]. The subject was the mathematical equivalence between classical solutions of the Dirac-Nambu-Goto action for the relativistic membrane and the Kalb-Ramond field. The content is mathematically oriented with many references to differential geometry concepts. Legovini is acknowledged for many discussions about the topic of the paper.

3.4.3 Eleven Summers in Upland, California

Starting from 1990s the INFN underwent a general redesign. The institute was already organized in divisions (Italian: *sezioni*) as in most major Italian universities and national laboratories. More or less like today, it had its own personnel (i.e. scientific positions like that which Aurilia had) while university staff could be affiliated to it. Nationwide the INFN covered five research lines, led by "scientific commissions". Scientific commission IV covered theoretical physics.[43] Each division was led by a director, who managed the budget. The amount for theoretical physics was often modest. INFN staff and affiliates could barely afford to attend one national conference a year. The mobility funds each faculty could obtain from his/her university were also very limited. The novelty of the 1990s was the introduction of scientific research networks (Italian: *iniziative specifiche*[44]). The goal of such networks was to foster collaborative work between affiliates from different divisions. For example, affiliates were encouraged to publish joint papers, co-supervise students, co-organize conferences, events, establish collaborations with foreign scientists by inviting them in to Italy or visiting them abroad. Each research network was named after the division, where the network national coordinator was affiliated. Networks were periodically evaluated with an external peer review system and yearly funded.

In such a new panorama, Giovanni Venturi launched BO11, Italy's first research network dedicated to quantum gravity, black holes and cosmology. Apart from Bologna, the network included the division of Trento. The acronym BO11 stood for the name of the front runner division, BO for Bologna, the research area internal classification, 1 for fields and strings, and the serial number of the networks stemming from Bologna, 1 being BO11 the first Bolognese network for fields and strings.

---

[43] As previously stated, Giuseppe Furlan was the national chairman of scientific commission IV from 1974 to 1978.

[44] The literal translation is "special or specific initiatives". The stress is on the free initiative to do research on a specific topic.

After a year in Paris and three years in Palermo, Balbinot came back to Bologna and, in the mid 1990s he succeeded Venturi as head of BO11.

This historical background is necessary to understand that the Firm had in those years the first ever chance to obtain regular financial support from INFN. This was initially possible due to an increase of the overall budget the Trieste division made available for theoretical research. A radical change, however, occurred in 1998, when BO11 was expanded to Spallucci, his associates and the INFN Trieste division.

The means of the collaboration between the Aurilia and Spallucci evolved very rapidly through the years. In 1983, they discussed research on the phone and they exchanged drafts of research papers via regular mail. Ten years later, they both used e-mail and they started meeting on a yearly basis. From 1998, such visits also involved students and collaborators. Spallucci ended up visiting Aurilia eleven times, typically one month each summer, from 1994 to 2004. Spallucci was hosted by Aurilia at his house in 2460 Beck Street, Upland, California. During the time together, they designed projects, discussed physics and performed calculations. The work was done at home (see Fig. 10). They did not need much else to be scientifically productive and original.

Fig. 10 Aurilia's house in 2460 Beck St., Upland, CA 91784. During the hot Californian summers, the discussions between Aurilia and Spallucci took place below this pergola. The working sessions had short breaks during which Aurilia used to smoke Toscano cigars. On other occasions, the work was interrupted by River, a 60-kg-Rottweiler, who challenged Spallucci at rope pulling. It is estimated that 15 papers were published as a result of the work done here from 1994 to 2004. Courtesy of Alexandra Aurilia Stephenson. All rights reserved

3.4.4 Stochastic Formulation of Quantum String Dynamics

In 1994, Ivo Vanzetta became the first undergraduate student from Italy to publish a paper with the members of the Firm. His Laurea thesis under Spallucci's supervision was expanded and became the basis for a manuscript that was published in Physical Review D [129].[45] The topic covered by this work is a "postmodern" formulation of string theory. The word postmodern was borrowed from a paper of Heller and Tomsovic, that discuss old ideas of the stochastic interpretation of quantum mechanics from a new perspective [130]. Accordingly, postmodern string theory means a quantum theory of strings in terms of a stochastic formulation. To reach such a goal, Aurilia and Spallucci took advantage of their previous results on the Hamilton-Jacobi formulation for string dynamics [127, 128].

A peculiar aspect of Aurilia and Spallucci's approach is the use of two field variables, namely a vector gauge potential $A_\mu(x)$ and the string current $W^{\mu\nu}(x)$ to express the string dynamics. From this view point, the Nambu-Goto action becomes only an effective description at low energy, i.e. in the classical regime. The string quantization is obtained by considering a Gibbs ensemble of relativistic strings, rather than a single string, at the classical level. The ensemble represents the starting point for the Hamilton-Jacobi theory, that is the natural formalism to describe the transition between the classical and quantum regime. Following this reasoning, the quantum counterpart of the string ensemble equation reads

$$\left[\, \text{Wheeler} - \text{DeWitt operator} \,\right] \Psi[C] = 0. \tag{7}$$

with

$$\Psi[C] = A[C] \exp\left(\frac{i}{\hbar} \oint_C dy^\mu A_\mu(y) \right), \tag{8}$$

where A_μ is the aforementioned vector potential associated with a family of extremal world sheets. In conclusion, the quantum string is governed by a Wheeler-DeWitt like equation, that controls the probability $|\Psi[C]|^2$ for the (closed) string to have a given spatial shape C.

Such a postmodern quantization actually leads to a quantum geometry picture of string dynamics. This is an alternative to the "premodern" interpretation according to which the string excited states are identified as a family of particles. Interestingly, in the concluding section of the paper, the quantization scheme is connected, via a low energy limit in curved space, to the existence of a cellular structures at constant string density in the Universe (Fig. 11). In other words, the paper provides the quantum mechanical premises for the existence of domain walls and chaotic inflation.

---

[45] This was the first and only scientific contribution of Vanzetta as a member of the Firm. It has been, in any case, a decisive factor for his career. Having a publication record as an undergraduate student opened the doors to graduate studies. Vanzetta decided to accept the offer of a PhD position in brain research at the Weizmann Institute of Science. He obtained the PhD there in 2001 under the supervision of Amiram Grinvald. Vanzetta currently holds a position of permanent scientist at the Institut de Neurosciences de la Timone, Marseille, France.

Fig. 11 The quantum precursor of the chaotic inflationary scenario corresponds to celluar domain structure. Each cell has constant stringy density

Aurilia submitted a second manuscript to Physical Review D in 1994 [131]. This time the third author was Smailagic. The topic was in the framework of p-branes coupled to gravity via a conformal symmetry breaking term. The main result was the emergent structure of the Universe in terms of constant curvature "p-cells". This provided further evidence for the aforementioned scenario based on chaotic inflation and domain wall models.

In 1995 there was an important addition to the Firm. Stefano Ansoldi, joined the collaboration with Aurilia and Spallucci. He was the second undergraduate student from Italy and the fourth overall, after Kissack, Palmer and Vanzetta. Ansoldi ended up by publishing 12 papers with the Firm over a span of 7 years. This work was the basis for his Laurea thesis, his Dissertation and his first postdoctoral appointment.[46]

Starting from the paper with Vanzetta it was clear that the Firm was interested in focusing on two basic elements: stochasticity and quantum geometry. With Ansoldi on board, the barycenter of the Firm shifted towards more mathematical oriented topics. The above two elements were essentially merged in a unique quantum string feature that could be expressed with the theory of fractals.

In the initial paper with Ansoldi, published in Physical Review D [132], the Firm took advantage of the results of the paper with Vanzetta on the Hamilton-Jacobi/Wheeler-DeWitt equation for the classical/quantum string. It was natural to

---

[46] Ansoldi is currently a Ricercatore Universitario Confermato (tenured faculty position at assistant professor level) in theoretical physics at the University of Udine.

aim for a derivation of the closed string propagation kernel. This is what was obtained by requiring consistency with a proper definition of the string path integral:

$$K\left[C, C_0; A\right] = \left(\frac{m^2}{2i\pi A}\right)^{3/2} e^{\frac{im^2}{4A}\left[\sigma^{\mu\nu}(C)-\sigma^{\mu\nu}(C_0)\right]\left[\sigma_{\mu\nu}(C)-\sigma_{\mu\nu}(C_0)\right]} \tag{9}$$

Here $m^2 = 1/2\pi\alpha'$, $\sigma^{\mu\nu}(C)$ is the area element of the loop $C : x^\mu = x^\mu(s)$, and A is the "proper" area of the string manifold, that is invariant with respect to world-sheet coordinate reparametrization. It is worth noticing that the above heat kernel extends the conventional particle kernel by making use of A as an evolution parameter.[47] The net result gives the transition probability for the string to change its shape from C_0 to C. Interestingly, this formalism opened the door to Green's functions and the second quantization of the string.

The following two papers were dedicated to Umezawa, who passed away on March 24, 1995. The first one covers a topic that is new and old at the same time, namely the boson condensation of membranes coupled to the 3-form field [134]. The resulting scenario is that of the membrane condensate confining the gauge field on a layer surrounding a region of ordinary vacuum. This is the analogue of the flux tube lines in a type-II superconductor Cooper pair condensate.

The second paper dedicated to Umezawa was a four author paper, with Ansoldi and Balbinot both being part of the collaboration [8]. The paper has some forewords written by Aurilia about his interaction with Umezawa during his doctoral studies. The subject covers again the classical and semiclassical spherically symmetric objects. This time, however, the scenario goes beyond the case of inflationary bubbles in order to include additional gravitational objects. Indeed bubbles can expand into a Friedman-Robertson-Walker Universe or collapse into black holes. Some bubbles may be connected by wormholes. The paper presents such a plethora of objects as a decay product of the primordial spacetime foam. The related decay rates are calculated by making use of an effective Lagrangian.

3.4.5 Fractal Strings, p-Branes and Quantum Spacetime

In October 1996, Aurilia submitted with Ansoldi and Spallucci a manuscript titled "Hausdorff dimension of a quantum string" to Physical Review D [135]. The work was a generalization of a paper of Abbott and Wise about the dimension of a particle path [136]. Classically a world line is a one dimensional object. Quantum mechanically, however, it is harder to define the trajectory of a particle. Fluctuations become dominant. The particle path looks like a fractal. It is still possible to calculate the path dimension, even if one has to invoke a special definition proposed by Hausdorff. As a result, one finds that the fractalized path has a dimension equal to 2. Therefore, the excited particle state has an increased path dimension.

---

[47] The role of A as the string evolution parameter was originally proposed by Eguchi in 1980 [133].

Based on this premise, Aurilia conjectured a similar scenario for string dynamics. He realized, however, that important differences with respect to the particle case can occur. There has to be a new sort of spacetime fuzziness related to the string tension, namely $\sim (\alpha')^{1/2}$, even in the case of no quantum mechanical effects, i.e. in the limit $\hbar = 0$. Furthermore, Aurilia interpreted the emergence of fractality as a new spacetime feature similar to curvature in general relativity. Curvature is the response of the presence of mass-energy. Fractality is the response of the (increased) resolution, namely higher energy/closer distance, over the spacetime manifold (Fig. 12). By using the string kernel (9), Aurilia and the members of the Firm were able to show that the string world-sheet switches from a classical two dimensional regime to a quantum mechanical three-dimensional regime.

According to Aurilia, the classical spacetime would emerge as a low energy condensate of strings in analogy to the Ginzburg-Landau theory for superconductivity [2]. In practice, the local string density in loop space, $|\Psi[C]|^2$, would turn out to be an order parameter. For loop areas below a certain critical loop area, A_c, strings undergo increasingly shape shifting transitions that destroy the long range correlation of the condensate, namely

$$|\Psi[C]|^2 = \begin{cases} 0 & \text{if} \quad A \leq A_c \\ \propto (1 - A_c/A) & \text{if} \quad A > A_c. \end{cases} \tag{10}$$

This picture is extended to any higher dimensional object. For p-branes the Hausdorff dimension reads

$$D_H = p + 2. \tag{11}$$

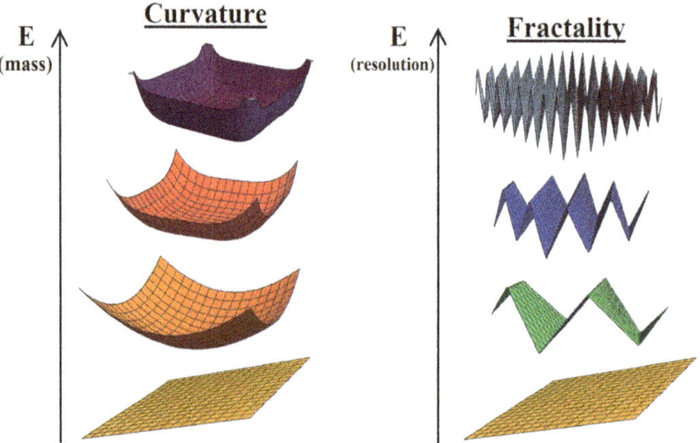

Fig. 12 Curvature and fractality are two indicators of the presence of mass-energy on a background spacetime. According to general relativity, the spacetime reacts to the mass-energy with the presence of curvature. Higher energies also imply higher resolution. Therefore the spacetime looks smooth at large distances but rough and fractalized at short distance [2, 135]

Eventually, A_c is identified with the Newtonian constant G, signaling the regime at which all such geometrical structures boil over and the spacetime condensate dissolves.

3.4.6 Geometric Democracy, "the Planckion" and Self-Complete Quantum Gravity

The year 1997 was still in part dedicated to the old theme of bubble dynamics and the minisuperspace approach. Aurilia published a paper about this topic in Classical and Quantum Gravity [137]. This work turned out to be Aurilia's last collaboration with Balbinot. The content of the paper is centered around an interpretation of Hamiltonian constraints. At the classical level the constraint corresponds to a gauge choice for bubble matching conditions. At the quantum level, the Hamiltonian constraint becomes the Wheeler-DeWitt equation for the bubble physical states. In the meantime, the fractal properties of string dynamics became Aurilia's major interest. Aurilia successfully applied for a further Cal Poly Research, Scholarship, and Creative Activity (RSCA) award.[48] In such a way, he obtained the support for a sabbatical leave for the winter quarter 1998.

The work on fractals generated two additional papers published during a time span of four years. The first manuscript had a pedagogical character and aimed to reformulate the path integral in quantum mechanics following the formalism developed for the string kernel in loop space (9) [138]. In contrast to the conventional formulation, the path integral was written over the entire trajectory history in Hamiltonian phase space. Such an approach allowed for a more transparent interpretation of the fractality of a quantum path due to the uncertainty nested in the canonical pair (q, p).

The research on fractals proceeded with a parallel activity on p-branes. Aurilia published a related manuscript in Physics Letters B with Ansoldi, Smailagic and Spallucci in 1999 [139]. This work was the first attempt to formulate a bubble quantization by determining the vacuum partition function $Z[J]$. The formalism was based on the special case of p-branes having $D = p + 1$, where D is the bulk spacetime dimension. Such a result was later generalized to the case of arbitrary $p \leq D - 1$ in a paper published in Progress of Theoretical Physics [140]. The list of authors included an additional undergraduate student, namely Luca Marinatto,[49] who turned out to be the fifth and final undergraduate student in the Firm. During this time, Aurilia aimed toward a better understanding of the mass generation mechanism at fundamental level. Such a mechanism is conventionally known for particles, i.e. point-like objects that are just the low energy limit of strings or p-branes. It was therefore natural

---

[48] The project Aurilia submitted was titled "Fractal structure of quantum spacetime".

[49] This has been the only paper authored by Marinatto, as a member of the Firm. For his doctoral studies, he was offered a position at the University of Syracuse, but he opted to remain in Trieste. He joined the group of Tullio Weber to work on the foundations of quantum mechanics. After receiving his Ph.D., he had postdoctoral contracts in Trieste with Weber and Giancarlo Ghirardi until 2008. Since 2009, he has been teaching high school physics in Udine.

for Aurilia to address the issue of a Higgs mechanism for extended objects, within a paradigm he termed "principle of geometric democracy".[50] Similarly to what happens in electromagnetism, there exists an electric-magnetic duality for extended objects. This is a transformation between p-branes and $\tilde{p}$-branes with $\tilde{p} = D - p - 4$. The net effect of such a duality is the exchange of the gauge field defined in the "bulk" and "boundary" of the p-brane history. The paper [140] actually presented a path integral formulation and a mass generation mechanism for extended objects in terms of an extension of the Stückelberg mechanism. As a follow up work [141], Aurilia considered the consequences of the *topological symmetry breaking*[51] following the bubble/p-brane nucleation. In the process, the massless non-dynamical gauge field became massive with a spin jump to a dual vector representation, as observed in [121].

The program of geometric democracy continued with another paper published in Physical Review D in the year 2001 [142]. The main result is the derivation of the following propagator for p-branes

$$G_p \left(x - x_0, \sigma\right) = \frac{i}{2M_0} \int_0^\infty ds \, \exp\left\{is\frac{M_0}{2}(p+1)\right\} K_{\text{cm}}\left(x - x_0; \sigma\right) K_p\left(\sigma; s\right)$$

(12)

by means of a minisuperspace-quenched approximation. Such an approximation was borrowed from analogous methods in use in quantum cosmology and QCD. At the classical level, the approximation requires a splitting of the dynamics into a center of mass motion and collective mode deformations of the brane. This splitting is reflected at the quantum level in the factorization of the propagator, which depends on the center of mass kernel $K_{\text{cm}}\left(x - x_0; \sigma\right)$ and the volume kernel $K_p\left(\sigma; s\right)$. Interestingly, the low energy limit corresponds to the low resolution limit in which the object appears point-like. By low energy one means low with respect to the brane tension T_p, i.e.

$$E \ll \left(T_p\right)^{\frac{1}{1+p}}.$$

(13)

This is in agreement with the intuition according to which in the infinite tension limit collective modes do not have dynamics and the system collapses into its center of mass. Carlos Castro coauthored the paper. This was his only contribution as a

---

[50] According to Aurilia, particles no longer have a privileged role. Objects of different dimensions have to be treated on the same footing. This also means that, for a given p, none of the p-branes is fundamental. More radically, this implies that words such as "elementary" and "building block" are devoid of physical meaning. Aurilia's principle (also known as "dimensional democracy") is reminiscent of Chew's nuclear democracy at the dawn of string theory. It is interesting to note that the switch from nuclear to geometric democracy is aligned with the historical evolution of string theory, whose energy domain has been lifted from about 1 GeV to the Planck scale.

[51] The simplest example of topological symmetry breaking is the emission of an alpha particle from a nucleus. For an observer outside the nucleus, the particle world line is semi-infinite. The end-point of such a world line is actually "singular" and leads to a non-conserved current $\partial_\mu J^\mu \neq 0$. Such a current is equivalent to the introduction of a mass term in the action [141].

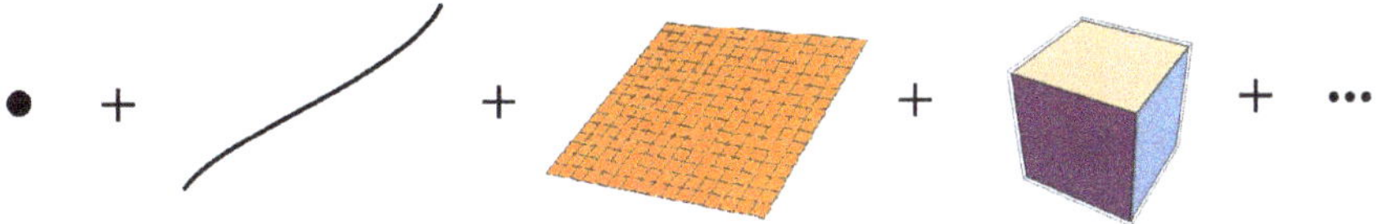

Fig. 13 Aurilia's principle of geometric democracy. None of the above objects is fundamental. Each of them is treated on the same footing irrespective of its dimension p

member of the Firm. It is evident that, in this phase, Aurilia was seeking a mathematical formalism capable of accommodating the dynamics of objects of different dimension. Castro was Aurilia's first contact with the Clifford algebra community. In the following years, Aurilia established an intensive correspondence with other experts in Clifford. algebra, such as William Pezzaglia and Matej Pavšič, though no joint papers emerged from it[52]

One of the remarkable aspects of a Clifford algebra is the possibility of summing heterogeneous quantities, e.g. scalars and vectors and higher rank tensors. Indeed, a generic element A of a Clifford algebra $\mathscr{C}_n$ defined on a n-dimensional vector space V_n reads:

$$A = \alpha + \alpha^i e_i + \frac{1}{2!}\alpha^{i_1 i_2} e_{i_1} \wedge e_{i_2} + \ldots \frac{1}{n!}\alpha^{i_1 \ldots i_n} e_{i_1} \wedge \cdots \wedge e_{i_n}. \tag{14}$$

Any element of such an algebra is called *polyvector* or *Clifford aggregate* [143–145]. A polyvector is a superposition of tensor quantities of different *grade*, namely a scalar, a vector, bi-vector, including an n-vector (see Fig. 13). A generic vector of grade r, namely an r-vector, is called a *multivector*. Geometrically, each multivector is associated with an object of a certain dimension. A scalar is associated with a point, namely an object with no geometric extent. A vector is associated with a line segment, which has length and orientation. A bi-vector is associated with a surface that has area and orientation. A three-vector is associated with an oriented three-dimensional volume. The highest grade multivector is an r-vector, with $r = n$, where n is the aforementioned dimension of the vector space V_n on which the Clifford algebra has been defined. Such unique properties of Clifford algebras fit with Aurilia's principle of geometric democracy, according to which a polyvector A is actually the most faithful representation of the fundamental object of physical interactions.

In 2001 Aurilia was the recipient of a Cal Poly Research Scholarship and Creative Activity (RSCA) award, for the project titled "The search for Quantum Gravity". At this point, the process of scientific transformation that started in 1983 following the meeting with Spallucci was complete. Aurilia was definitively recognized as a

---

[52] During my visiting position in Singleton's group at CSU Fresno, I met William Pezzaglia on several occasions, mostly in San Francisco. He actually invited me to give a seminar at California State University East Bay in 2008. Matej Pavšič was my host at the Stefan Institute in Ljubljana, during a 6 month CNR-NATO fellowship I spent there in 2005. Such visits were my first steps within the network of the Firm.

leading figure in quantum gravity. In hindsight, we can say that Aurilia's ideas about the very nature of a quantum spacetime turned out to be the forerunners of research themes in quantum gravity of the following two decades. This is particular evident by considering Aurilia's scientific impact in research areas such as noncommutative geometry, doubly special relativity, gravitational self-completeness, nonlocal gravity and gravity emergence.

The work of the years 2001–2002 generated three papers [146–148]. The first one titled "Fuzzy dimensions and Planck's uncertainty principle for p-branes" was published in Classical and Quantum Gravity in collaboration with Ansoldi and Spallucci [146]. In the introductory section of the paper, Aurilia conveys his vision about the phenomena at Planck scale with illuminating words, which can be summarized as follows. The spacetime in the extreme energy regime is not even a proper geometry. It is rather a fluctuating quantum background called "pregeometry" from which the physical spacetime *emerges*. Such an emergence is governed by a mapping X between pregeometric quanta, represented by p-dimensional simplices[53] or p-cells, and extended objects in target spacetime, namely

$$X : p - \text{cell} \in \{\text{quantum pregeometry}\} \longmapsto p - \text{brane} \in \{\text{target spacetime}\}. \tag{15}$$

The classical spacetime can be regarded as a p-brane condensate, in analogy to a superconducting medium. When the condensate cools down, one recovers the standard physics of point-like objects. Near the critical temperature, i.e. the Planck energy, such a condensate evaporates in a "gas of branes". In practice, the dimension p is no longer a constant parameter. It becomes a dynamical (quantum) variable, that fluctuates along with the geometry. In support of such a scenario, Aurilia had taken advantage of the previous work on "polydimensional physics" [142]. The quenched minisuperspace p-brane propagator (12) is actually equivalent to a propagator of a point particle, that sees an extended spacetime

$$d\Sigma^2 = dx^\mu dx_\mu + \frac{1}{(p+1)!V_p^2} d\sigma^{\mu_1\ldots\mu_{p+1}} d\sigma_{\mu_1\ldots\mu_{p+1}}. \tag{16}$$

Here dx represents the coordinates of the barycenter of the the brane. The brane extension is governed the $p+1$ additional variables. The multivector $d\sigma$ describes the target space volume of the brane and V_p is the proper volume of the p-dimensional boundary of the braneworld manifold. As long as the volume modes are not excited, the propagator reduced to the conventional point-like object propagator in Minkowski space. Conversely, at larger momenta new dimensions open up due to the extended nature of the object.

---

[53] A p-simplex is a p-dimensional triangle. The case $p = 0$ corresponds to a point, while the case $p = 1$ corresponds to a line segment. A triangle is simply a 2-simplex and a tetrahedron is a 3-simplex.

The line element (16) is, however, just the subspace of the Clifford line element

$$d\Sigma^2 = dx^\mu dx_\mu + \frac{1}{2l^2}d\sigma^{\mu_1\mu_2}d\sigma_{\mu_1\mu_2} + \frac{1}{3!l^4}d\sigma^{\mu_1\mu_2\mu_3}d\sigma_{\mu_1\mu_2\mu_3} + \cdots$$

$$+ \frac{1}{(p+1)!l^{2p}}d\sigma^{\mu_1\ldots\mu_{p+1}}d\sigma_{\mu_1\ldots\mu_{p+1}} \cdots \frac{1}{D!l^{2(D-1)}}d\sigma^{\mu_1\ldots\mu_D}d\sigma_{\mu_1\ldots\mu_D}. \tag{17}$$

This led Aurilia to conjecture the existence of the *Planckion*,[54] a quantum mechanical object that results from the superposition of diverse p-branes. In the paper, one finds both the Planckion action and its polydimensional propagator. More importantly, Aurilia presents the dispersion relation for the Planckion and the related uncertainty relations. For a given p-brane one has:

$$\Delta x \geq \frac{1}{\Delta k_x}\left[1 + \frac{1}{2}V_p^2\beta(k_x^2)^p\right] \tag{18}$$

where k_μ is the center of mass momentum, k_x is its component along the x-axis and β is a constant of order $O(1)$. Remarkably, (18) is an extension of the Generalized Uncertainty Principle (GUP) [150–152]. From the known case $p = 1$ of the string, in whose context the GUP was originally derived [153–155], one can actually set the typical scale of the volume V_p and find a minimal resolution length over the spacetime

$$(\Delta x)_{\min} \sim V_p^{1/p} \sim G^{1/2} \sim L_{\mathrm{P}}. \tag{19}$$

In other words, the Planckion is the only fundamental object according to the principle of geometric democracy. For lengths smaller than the Planck length L_{P}, the spacetime dissolves in a pre-geometric soup of quantum p-cells.

The second paper of the triad of the years 2001–2002 was never published and never submitted for publication. It is Aurilia and Spallucci's essay for the 2002 Gravity Research Foundation competition. It remained almost unnoticed until 2013, when, upon my personal insistence, it was finally posted on the arXiv by the authors. Together with the Planckion, the content of this paper was one of the topics of the discussions I had with Aurilia during my visiting position at CSU Fresno in 2008. My interest in it, however, remained dormant at least until 2010, the year in which Dvali and his associates proposed an equivalent paradigm termed "gravitational self-completeness" [156, 157].[55] At the first Karl Schwarzschild Meeting held in Frankfurt in 2013, Carr independently proposed a new principle termed "black hole uncertainty principle correspondence" [159, 160]. His proposal is similar to Aurilia's original ideas and has several points of contact with the aforementioned

---

[54] The term "Planckion" was customarily used in the Russian literature. Its introduction is, however, due to Treder [149].

[55] Dvali has contributed to the present volume with a chapter about black holes and the quantum N portrait paradigm [158].

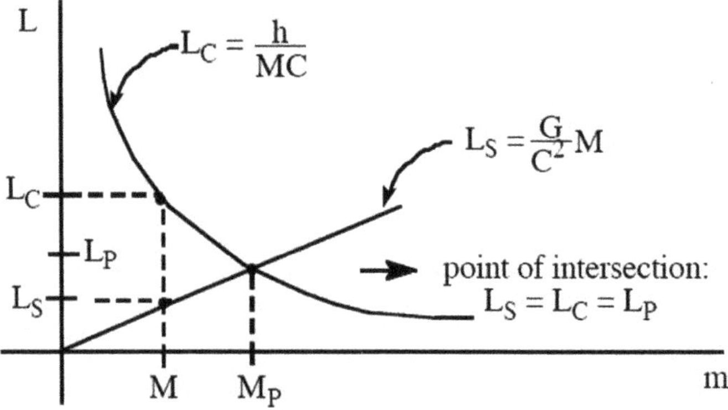

Fig. 14 Original version of Aurilia and Spallucci's length-mass diagram [147]. It is the forerunner of Dvali's self-completeness diagram and Carr's black hole uncertainty principle correspondence diagram. Aurilia drew the diagram and typed the words with the primitive graphic software he had at his disposal in the year 2001. Reprinted from https://arxiv.org/pdf/1309.7186 with permission. ©The Authors. All rights reserved

self-completeness paradigm (to this purpose see Carr's contribution to the present book [161]) (Fig. 14).

The key question of Aurilia and Spallucci's paper is the possibility of Lorentz contracting the Planck length. This is a topic that at the time was studied with great effort along two main directions, i.e. formulations of a relativity theory with two invariant scales (e.g. saturating Lorentz boosts [162], doubly special relativity & κ-Poincaré algebras [163], nonlinear boost operator [164, 165]) and extensions of the Heisenberg uncertainty principle (i.e. the aforementioned GUP [150–152]). Against this background, the line of reasoning in the paper develops in an original way as follows. First, one starts by observing that the Planck units emerge from the combination of $\hbar$, G and c:

$$L_P = L_P(\hbar, \ G, \ c) \equiv \sqrt{\frac{\hbar G}{c^3}} \tag{20}$$

$$M_P = L_P(\hbar, \ G, \ c) \equiv \sqrt{\frac{\hbar c}{G}}. \tag{21}$$

This is an initial hint that the Planck scale is at the confluence of quantum mechanics, gravitation and special relativity. The introduction of a mass M breaks such a confluence but allows for the definition of two additional length scales, namely the Compton wavelength and the Schwarzschild radius:

$$L_C = L_C(\hbar, \ M, \ c) \sim \frac{\hbar}{Mc} \tag{22}$$

$$L_S = L_S(G, \ M, \ c) \sim \frac{GM}{c^2}. \tag{23}$$

We stress here that L_C depends on $\hbar$, but not on G. Conversely, L_S depends on G, but not on $\hbar$. This corresponds to saying that gravity is negligible in particle physics, while black holes are classical objects. Alternatively, one can say that the microscopic world is quantum mechanics dominated and the macroscopic world gravity dominated. The only exception to such a general scheme is represented by the Planckion, namely the particle-black hole system whose size and mass coincide with L_P and M_P. Indeed, black hole formation is the crucial point of Aurilia and Spallucci's paper, a point that was missing or neglected in the literature at that time.

A hypothetical elementary particle with mass larger than the Planckian mass necessarily collapses into a black hole with $L_C < L_P < L_S$. Similarly, for sub-Planckian masses ($M < M_P$), the Schwarzschild radius is always smaller than the Planck length with $L_S < L_P < L_C$, confirming the fact that the size of a particle is set by the Compton wavelength.

The symmetry is, however, not perfect. There is a substantial difference between L_C and L_S. The Compton wavelength is consistent with the Lorentz contraction, because there is an inverse relationship between mass and length—the larger the relativistic mass the shorter the contracted length. This is clearly not the case for the Schwarzschild radius, which is proportional to the mass. The whole problem is connected with the difficulties of applying Lorentz boosts to L_S, since boosts ignore the gravitational effects generated by M. The solution to such a puzzle, comes from a few things about (quantum) gravity we know for sure. The area of the event horizon of a black hole can never decrease. The event horizon is a null surface, whose shape and size cannot be altered by a change of coordinates. Accordingly, a Lorentz boost has to leave the Schwarzschild radius unchanged.

From this perspective the Planckion, being a (quantum) black hole, is no exception. Therefore the conclusion is that, due to the onset of gravity and black hole formation, the Planck scale has to be the end-point, rather than the turning point of the relativistic contraction. The extension of the Heisenberg uncertainty principle naturally follows for the same argument.

The third paper of the aforementioned triad offered the mathematical formalism in support of boost saturation presented in the second paper. The preparation of the manuscript started in 2002, but there have been several versions and updates. Eventually the paper was submitted in 2013 for publication to a special issue of Advances of High Energy Physics titled "Experimental Tests of Quantum Gravity and Exotic Quantum Field Theory Effects". The academic editor who solicited the submission and handled the manuscript for review was myself.[56] As a result, this paper is the only publication having the names Aurilia, Nicolini and Spallucci on it (despite each having different roles). Indeed, Aurilia and I had discussions in 2016 about the possibility of writing a joint paper, but, unfortunately, this never materialized.

The key point of the paper is an extension of Lorentz boosts to the trans-Planckian domain. The first observation is that the Planck mass M_P represents a sort of

---

[56] The other academic editors of this special issue were Emil Akhmedov, Steven Minter and Douglas Singleton. Singleton is also one of contributors to the present book [166].

"extremal" value for the energy, since it separates two phases of matter, particles and black holes. Accordingly, the phase transition has to have repercussions on Lorentz boosts. In particular, it should not be possible to boost a length below the Planck length L_P. To account for such a scenario, Aurilia and Spallucci proposed the following form of length transformation:

$$L(\beta) = L_0\sqrt{1 - \beta^2} + \frac{L_P^2\,\Theta_H(\beta)}{4L_0\sqrt{1 - \beta^2}}. \tag{24}$$

Here Θ_H is the step function.[57] The first term at the r.h.s. is the conventional Lorentz contribution, with L_0 the length in the rest frame. The second term is the new one. It dominates the r.h.s. only when $\beta \approx 1$, being $L_P \ll L_0$ for any experimentally accessible length. For the new term, $L(\beta)$ admits a minimum, that can be found from

$$\frac{dL}{d\beta} = 0 \quad \Longrightarrow \quad \gamma = \gamma_* \equiv \frac{2L_0}{L_P}, \quad L(\beta_*) = L_P. \tag{25}$$

In other words, the new form of Lorentz boost (24) prevents the contraction below the minimal length. The connection with the GUP is now evident. The study of Planckian string collisions revealed a gravitational correction to the Heisenberg uncertainty principle proportional to the Regge slope α'. In quantum gravity, such a slope is nothing but G. Therefore $\sqrt{\alpha'} = L_P$, in agreement with (24).

One can assume that L_0 is either a particle Compton wavelength L_C or a black hole Schwarzschild radius L_S. In the former case, one obtains that the Planck mass is the actual barrier for the mass spectrum of elementary particles, namely

$$L_C > L_P \quad \Longrightarrow \quad M < M_P. \tag{26}$$

In the latter case, the position of the mass is reversed with respect to the case of the Compton wavelength, i.e.[58]

$$L(L_S) = 2GM\sqrt{1 - \beta^2} + \frac{L_P^2\,\Theta_H(\beta)}{8GM\sqrt{1 - \beta^2}}. \tag{27}$$

The above inversion has important consequences for the hoop conjecture, namely the mechanism of matter collapse into a black hole. Indeed, from $L(L_S) > L_P$ one obtains $M > M_P$. In practice there is no inertial reference frame where a particle, whose Compton wavelength at rest is larger than its own Schwarzschild radius, can fall behind its event horizon.

---

[57] For more details see [148].

[58] For ease of notation we are temporarily setting $c = 1$.

The second part of the paper addresses the issue of the nature of the Gibbons' superforce

$$F_{\rm s} = \frac{c^4}{4G}.$$ (28)

This is conventionally interpreted as the Planckian limit of Newtonian forces and electrostatic forces. Aurilia noticed a curious fact in relation to the above equation. Black holes have an unconventional dependence on the volume density that scales with as one over their size squared. This is in marked contrast to the result for an ordinary object, namely a constant corresponding to the local matter density of the object itself. This is so because both the mass and the volume scale with the third power of the size of the object. Conversely, for black holes the only density that turns to be constant is the linear density, namely $M/2L_{\rm S}$. Surprisingly, such a linear density has the same value of the Gibbons' superforce. According to Aurilia this is not an accident. The phase transition at the Planck scale between particles and black holes is a sort of "classicalization". This is a phenomenon introduced by Dvali and coworkers to support of idea of the self-complete nature of quantum gravity. Whatever is the phenomenon under consideration, the region below $L_{\rm P}$ is de facto inaccessible. For instance, Planckian scattering of particles always produce a classical state, i.e. a black hole state, whose size cannot be smaller than the Planck length. Much in the same way, Aurilia interpreted the Planck scale as a *super-unification* scale, where fundamental interactions have the same strength and "classicalize". Indeed, Aurilia noticed the absence of $\hbar$ in the Gibbon's original expression. To reinforce his argument, Aurilia postulated that particles in the phases preceding the transition (i.e. sub-Planckian energies) are described by strings whose tension reads

$$T_{\rm s} = \frac{\hbar c}{2\pi \alpha'}.$$ (29)

Such a tension actually has a maximum in terms of the Gibbons' force. In other words, black holes, by connecting the above two expressions (29) and (28), reveal their stringy nature as well as a classicalizazion phase in the macroscopic, trans-Planckian regime.

3.4.7 Publications After 2002, History of Physics and Unfinished Work

After the papers in quantum gravity, Aurilia briefly came back to the old theme of extended objects and the 3-form field. He wrote two papers in 2003 with only Spallucci as a coauthor. After submission and peer review, they were published in Physical Review D on the same day (May 7, 2004) [167, 168].

The content of these publications represent a continuation of the 2001 paper about the topological symmetry breaking following the bubble/p-brane nucleation [141]. The goal of the investigation was the calculation of the zero point energy due to the quantum fluctuation of $A_{\mu\nu\rho}$. In practice, Aurilia and Spallucci showed that the 3-form field is responsible for the difference of vacuum pressure between the interior and the exterior of a 2-brane. This is the equivalent of the Casimir effect for the hadronic/cosmological vacuum. The paper also offered the calculation of the Wilson loop for the 3-form field coupled to the three dimensional world history of a spherical bubble. The resulting static potential between two antipodal points of the bag is proportional to the volume of the bag itself, in a sort of generalization of the conventional "area law" for quarks. Ultimately, Aurilia and Spallucci aimed to study the repercussions of the fluctuations of $A_{\mu\nu\rho}$ on spinor fields. This was the subject of the second paper [168]. In particular, Aurilia and Spallucci looked for a confinement mechanism that could be derived by coupling the 3-form field to a fermionic current density. For consistency of the field equations, the formalism was extended to the massive case. The net result has been an effective action containing a confining term for the dynamics of the fermion field.

The time after the publication of the above two papers was characterized by personal and professional changes for Aurilia. Between 2004 and 2013 there was a hiatus in his scientific production. This was in part due to the absence of direct contacts with Spallucci, whose last visit in California took place in the summer of 2004. In parallel to this, summers were for Aurilia the ideal period of the year for traveling and visiting relatives. This became increasingly important when his daughter Alex moved to Kansas. By 2004, Aurilia and his wife were already planning to leave California for good after their retirement. They aimed to move to Vancouver Island, in the Canadian province of British Columbia. Both British Columbia and Kansas are about 2,000 km (1,200 mi) from Pomona. This distance can be covered by a 3 h flight or a 20 h drive. Both locations are, however, somehow remote since they are far from the main traffic routes. In one of our e-mail correspondences, Aurilia recalled that going back and forth from Kansas and British Columbia was time demanding and had to be planned very far in advance.

Another factor slowing down the publication rate was probably connected to the research topic itself. The return to the physics of quarks was meant to be temporary, just to close old and unfinished projects.[59] In his heart, however, Aurilia had only one topic, namely quantum gravity. This was evident not only from the contact he had with Spallucci and myself during those years, but also from the interaction with Nathan Tung,[60] a research student at Cal Poly. From fall 2008 to spring 2009, Aurilia and Tung made some progress on the nature of Gibbons' superforce in the framework of a project titled "The Planck Scale and the Unification of Physics".

Quantum gravity, however, needs deep thinking and reflection. It takes years to make actual progress. Aurilia started writing down his ideas in a sort of notebook,

---

[59] Aurilia actually obtained another Research, Scholarship, and Creative Activity (RSCA) award in 2005 from Cal Poly Pomona for the project "Quarks are the building blocks of the Universe: why can't we see them?".

[60] Nathan Tung is very likely the last student (either undergraduate or graduate) Aurilia supervised in some form. Indeed, Tung had completed his Bachelor's degree at Cal Poly Pomona with another professor, some months before his collaboration with Aurilia. As a result, Aurilia was never his official supervisor. Nevertheless, the activity with Aurilia was instrumental in securing Tung a Ph.D. position. The following fall, Tung started his graduate studies at UCLA [169]. After completing his Ph.D. in 2014, he started a career as a lecturer, both at Cal Poly Pomona [170] and UCLA, where he is currently an assistant adjunct professor [171].

and then in a master file. He, however, felt his conjectures were just preliminary and required further discussions with his associates. As a result, he planned in 2010 an eight-week visit to the University of Trieste to meet Spallucci. The visit was part of a tour within the Cal Poly Study Abroad Program. It included visits of the University of Padua, Bologna, Florence and Pisa, following in Galileo's footsteps. Indeed, Aurilia was selected as a College of Science Faculty Leader for the course "History of Physics". In 2007, he visited London and Cambridge within the same program, following in Newton's footsteps.

Unfortunately, due to some organizational issues, the journey to Trieste in fall 2010 was reduced to only two weeks and many of the potential discussions were postponed to another meeting in the future. This occurrence actually did not happen. Aurilia and Spallucci never met again.

3.4.8   The Universe in Ten Weeks

The title of this section is taken from one of Aurilia's courses at Cal Poly. Up to now, the sequence of events has led us to the exploration of an array of research themes, covering nuclear physics, string theory, quantum gravity and cosmology. Aurilia's style is clearly recognizable for the clarity and originality of ideas, intuitions and concepts. To a lesser extent, we know about his teaching activity. At Cal Poly, the teaching load is in general heavy. Aurilia routinely offered a variety of courses, including large enrollment general education courses, lab courses as well as upper division courses for majors.

From the fragmented information one can find on online teaching evaluations of his courses, however, it is possible to have some clues as to the kind of teacher he was [172, 173]. For instance, he is described as a professor with excellent preparation, whose courses have often a high degree of difficulty. As a person, he is described as a "cool guy", with a unique sense of humor and a tendency to criticize society and habits of people. My overall impression is that he never fully immersed himself in the role of teacher. His deep thinking, his analytic skills and curiosity prevented him from stopping being a researcher. Teaching was his daily duty. Research was his mission. To this purpose, it is very interesting to read the following teaching evaluation [174]

> I was hesitant to take Physics 102, Fall Quarter of '98, but, it fit my schedule perfectly. I was not sorry. Dr. Aurilia is such a great teacher. He explains the concepts by the use of interesting experiments, with a great sense of humor so that I don't feel intimidated taking a physics course anymore. For Spring of 1999, I read a flier advertising a new course "The Universe in ten weeks" catered to students with an inquisitive mind, taught by Dr. Aurilia. This course fills GE area 2D. I learned a lot for ten weeks and truly looked forward to going to class. It was fun, with great experiments, a fantastic web site, a disc that comes with the text, Cal Tech seminars, movies and speakers in class. The tests are a little challenging, but he grades fairly. He cares that the students learn, is always available for questions after class. I definitely recommend this course. Dr. Aurilia, as in the past ratings, is the best teacher I've ever had, but everyone knows this. You should make sure to sign up in advance because the seats for his classes fill up quick, and is hard to add later.

For further information about Aurilia and his relation with the students, I recommend the reading of the contributions written by Harvey Leff [12] and Kai Lam [13].

4 Final Years

In 2011 Aurilia became Professor Emeritus and continued to teach at Cal Poly even if at a reduced pace. In 2015, Aurilia and his wife moved to Milwaukie, Oregon (a suburb of Portland) to retire. It seems there was some change of plan throughout the years. As mentioned above, he was initially interested to settle down in British Columbia.

Whatever the reason, Aurilia did not retire from research. From his study, he could contemplate the beauty of Mount St. Helens and other peaks of the Cascade Volcanic Arc, that is located at about 80 km (50 mi) north from Portland. He spent most part of the day thinking about some of the key aspects of quantum gravity and theoretical physics in general, and eventually resumed writing and publishing papers. Most of his scientific production of this period, however, remained unknown to the public. Rather than submitting preprints to the arXiv, Aurilia used to print them out like in the old days and share them with colleagues to get some feed back. He also seldom submitted his manuscripts to a journal for publication. There exist, in any case, two notable exceptions. One preprint was posted on the arXiv in April 2015 [175], just before leaving Pomona for good. It was never submitted to a journal. A second preprint was posted in February 2017 and published in Europhysics Letters in May of the same year [176]. This happened to be Aurilia's last publication, two days before he turned 75.

4.1 Last Publications

The above two papers [175, 176] included new collaborators, namely Patricio Gaete and José A. Helayël-Neto,[61] two regular visitors and associates of the ICTP. The main issue addressed in these papers is the long standing problem of the nature of the strong interaction and its non-perturbative regime.

More specifically, the first paper introduces a possible interplay between Chern-Simons theory and the Schwinger model. Recall that without additional ingredients, Chern-Simons formulations are not suitable for actual non-perturbative calculations. Aurilia was therefore interested in exploring Gabadadze's alternative scenario based on the existence of a long-range force between different instantons [178]. Such a force was expected to provide an explanation for the physics of glueballs, i.e. conjectured non-perturbative gluon bound states. In the paper, Aurilia introduced the calculation of the static potential associated with such a force and showed that it satisfies the

---

[61] Gaete and Helayël-Neto have contributed to the present book. See [177].

Wilson criterion for confinement. In addition, Aurilia aimed to evaluate the quantum effects of the axial anomaly on the potential, also known as "color screening phenomenon". As in the previous publications, the trademark of Aurilia's calculations was the use of the 3-form field $A_{\mu\nu\rho}$. This time, however, the interpretation of $A_{\mu\nu\rho}$ was in terms of the potential for Gabadadze's force:

$$A_{\mu\nu\rho} \equiv \frac{1}{\Lambda_{\mathrm{QCD}}^2} \mathrm{Tr}\left(\mathbf{A}_{[\mu} \mathbf{F}_{\nu\rho]}\right)$$
$$= \frac{1}{\Lambda_{\mathrm{QCD}}^2}\left(\delta_{ab} A_{[\mu}^a \partial_\nu A_{\rho]}^b + \frac{2}{3} f_{abc} A_{[\mu}^a A_\nu^b A_{\rho]}^c\right). \tag{30}$$

Interestingly Aurilia noticed that the above expression coincides with the dual of the Chern-Simons current $K^\mu = \frac{1}{3!}\varepsilon^{\mu\nu\rho\sigma} A_{\nu\rho\sigma}$. Therefore he termed (30) Chern-Simons-Schwinger (CSS) potential. The effective quark potential associated with $A_{\mu\nu\rho}$ is calculated by considering a scalar field in the Lagrangian, i.e. a bosonized form of the Schwinger model in four dimensions. By integrating out the scalar field degrees of freedom, the net result due to the interaction term reads:

$$V \sim \frac{g^2}{g^2 - m^2} \frac{1}{L}\left(1 - e^{-L\sqrt{g^2 - m^2}}\right) + \frac{m^2/2}{g^2 - m^2} L . \tag{31}$$

Here m is the scalar field mass, g the coupling constant of the scalar field–3-form interaction term, and L the distance between two test quarks. As expected, the above potential contains a linear correction that confines quarks at large L.

The paper published in Europhysics Letters in 2017 [176] is actually an extended version of the 2015 preprint [175]. The new part concerns the explicit presentation of Lagrangian terms for the BF theory, namely a topological field theory F in the presence of a background B. For a generic BF theory, the action has the following form:

$$S \sim \int \mathcal{L}\left[\mathbf{B} \wedge \mathbf{F}\right]. \tag{32}$$

After duality transformations and field redefinition, the Langrangian terms are brought to an equivalent Proca-like representation. This is an intermediate step that is instrumental to derive confining potential of the kind (31).

The key point of Aurilia's reasoning is connected to the emergence of a massive mode in the Schwinger model as well as the yet unexplored relation between quantum electrodynamics in two dimension (QED_2) and its four dimensional counterpart(s). While the dimensional reduction from four to two dimension leads to the unambiguous identification of the Schwinger model, the converse is not true. QED_4 is not the unique extension of QED_2.[62]

---

[62] The process of "dimensional oxidation" from two to four has been introduced also in the context of cosmology. The basic result in this case is similar to Aurilia's conclusions. There exist a variety of four dimensional theories that correspond to a formulation of gravity in two dimension [179, 180].

5 Legacy

From 1967 to 2017, Aurilia authored 53 peer reviewed publications and 6 conference papers [181–186]. The peer reviewed papers included 19 manuscripts published in Physical Review/Physical Review D and 13 Physics Letters B. He also published a book review for the the journal "Physics in Perspective" [187]. Aurilia wrote additional papers that were never published. He actually never submitted them to any journal. For instance, at the University of Syracuse he wrote a booklet of lecture notes on field theory [188] and two internal reports [19, 189]. The same fate befell three of his essays for the Gravity Research Foundation competition [121, 147, 190]. He also left unpublished four additional manuscripts [18, 86, 175, 191].

In conclusion, apart from his Laurea thesis [3], his Ph.D. thesis [11] and an unknown number of unfinished manuscripts, Aurilia's scientific legacy consists of 70 papers in total. Such a scientific production can be broken up into six research lines:[63]

1. Theory of high-spin fields
2. The role of relativistic extended objects in Particle Physics and Cosmology
3. Classical and quantum properties of relativistic extended objects
4. Quantum properties of generalized Maxwell fields
5. Fundamentals of Planckian Physics
6. Chern-Simons-Schwinger model of confinement in QCD

5.1 My Take on Aurilia's Work

The legacy of a scientist is not only related to the number of publications, but also to the impact they had in the literature. In practice, it is useful to estimate the "sequel", namely the follow up work an author has inspired with his/her ideas. To this purpose, a popular method is to count the number of citations, namely the papers that quote the work of a certain author. There exist also more sophisticated indicators stemming from the number of citations. For instance, the average citations per paper and the h-index[64] are often used because they are easy to calculate. Such figures as well as other scholarly evaluations are part of "bibliometrics", a branch of knowledge that has taken hold in recent years.

One of the merits of bibliometrics is that it has contributed to a greater transparency in academic assessments, but nevertheless has to be taken with a grain of salt. For instance, data collection has become reliable only after the advent of the World Wide Web. This means that for authors like Aurilia, who published ground

---

[63] The above topics are listed, almost chronologically, by Aurilia himself on his webpage on ResearchGate [192].

[64] The h-index has been introduced by Jorge E. Hirsch, a physicist at UC San Diego [193].

breaking papers before 1990s, data repositories could be incomplete. Another weakness of bibliometrics is connected to its basic essence. It is nothing more than blind counting. Such counting is easy to perform and quickly gives an aura of objectivity in evaluations. Unfortunately, an absolute faith in bibliometric figures is based on the false belief that science is democratic. Science is not an electoral system. It is not democratic. It is meritocratic. In science, the opinion of a single person may be worth more than everyone else's opinion. This is especially the case if such a person is Stephen Hawking!

For the above reasons, Aurilia's bibliometric data cannot represent his scientific profile nor the depths of his thoughts and the repercussions of his conjectures. As of today, according to INSPIRE-HEP, Aurilia's total citations are about 1,300, the average citations per paper are about 26 and the h-index is 17 [194]. ResearchGate reports about 1,500 citations overall [192], while SAO/NASA ADS [195] and Semantic Scholar [196] about 1,200 and 800 respectively. On the other hand, Hawking's citation [1] to the paper authored with Nicolai and Townsend [74] has definitely a huge weight in terms of recognition and reputation. The same can be said for the citations Aurilia received from two other highly decorated scientists: Alan Guth and Andrei Linde. Bibliometrics is simply not tailored to estimate the reputation of the citing authors.

Furthermore bibliometrics cannot even capture the quality of the collaborations that Aurilia established during his career. For example, Aurilia worked closely with the academic "family" of Nobel laureates from Japan. Umezawa, Takahashi and Kobayashi defended their Ph.D.s at the University of Nagoya under the guidance of the same advisor, Shoichi Sakata, in 1952, 1954 and 1972, respectively. Sakata, who was himself influenced by Nishina, is considered the father of modern physics in Japan.[65] In addition, Aurilia's time at CERN led to the collaboration with Nicolai [198] and Townsend [199], who eventually obtained leading positions at the world's top theoretical physics institutions.

It is therefore useful to analyze Aurilia's work and style in an alternative way. We can introduce a sort of uncertainty relation for scientific production. This is done by considering a pair of variables, namely the "generality" G and the "predictive power" P. The generality indicates how fundamental a certain formulation is, while the predictive power estimates the capacity to predict values of a set of observables. One can also introduce "uncertainties" Δ, that correspond to "intervals", in which G and/or P are distributed. In particular, we can interpret ΔG as the "loyalty" to a given physical principle, while ΔP corresponds to the accuracy of the predictions. Clearly, there has to be a bound for the product $\Delta G \Delta P$, namely

$$\Delta G \Delta P \geq \text{constant.} \tag{33}$$

---

[65] Sakata, who also became acquainted with Yukawa and Tomonaga, supervised Maskawa and mentored Nambu. While Yukawa, Tomonaga, Nambu, Kobayashi and Maskawa were all awarded the Nobel prize, Sakata slightly missed it in 1969 and in 1970, just before his death [197].

For instance, general relativity and the Standard Model of particle physics have great predictive power (e.g. period of binary pulsars, g-2). Therefore, they have a small ΔP. On the other hand, both general relativity and the Standard Model cannot be deemed as fundamental. Accordingly, their loyalty to something fundamental is "uncertain". General relativity and the Standard Model both have a large ΔG. From this viewpoint, Aurilia's scientific production is at the opposite end. He focused on the formulation of three paradigms, e.g. the spatial extension of quantum objects, the existence of a topological field in the Universe and geometric democracy. Aurilia's work, therefore, has a small ΔG, due to the closeness to the basic tenants of quantum gravity and the theory of everything.

I have also to add that the style of Aurilia's writings is unique. Although his papers have considerable mathematical content, his formalism is never sterile, confusing or redundant. For Aurilia, calculations are not the main goal. They are rather an instrument to understand and describe a given physical phenomenon. For his ability to draw scenarios, offer original interpretations and describe phenomena, Aurilia resembles Feynman. It is really fun to read Aurilia!

Such a style of writing implies that his scientific production is in the framework of "pure" theoretical physics, the science of Einstein and Fermi, that lies between phenomenology and mathematical physics. Accordingly, the uncertainty ΔP of Aurilia's production is large but not as large as in the case of formal, axiomatic theories that have been formulated over the years. Aurilia's papers always say something in a very concrete way! Aurilia, however, never fully explored the phenomenological repercussions of his findings. This can be explained also by the fact that phenomenology is generally a phase that follows the formulation of a paradigm. Phenomenology establishes a link between theory and observation and often leads to an explosion of citations. As a result, Aurilia's papers can be considered as "gemstones", which have not yet been observed in full light. The phenomenology has yet to come.

5.2 *The Trieste-Toronto-Pomona Triangle*

Aurilia's legacy is not only in terms of scientific publications. During his four decade career, Aurilia was influenced by a lot of great scientists, but, in turn, influenced just as many. In addition, one should not forget his commitment as a teacher and the large number of students he fascinated with his lectures.

At this point I could conclude with the summary that Aurilia was not only a brilliant researcher but also an engaging and enthusiastic teacher. Closing the section in such a way is, however, not possible. Things that work for others do not work for Aurilia. Indeed, a closer analysis of Aurilia's interaction with the world inside and outside academia reveals interesting and surprising aspects. Aurilia's yearning for scientific independence was the direct consequence of being a natural born leader. By leader, I mean a person that follows his/her principles and convictions, rather than other people's opinions or short term goals. This aspect of his character also concurred to develop the determination in pursuing career goals across two continents. On the side of the social life there are distinctive repercussions too. Throughout the years, Aurilia became increasingly selective. At a certain point, he reduced the contacts with the rest of the faculty to the minimum. He almost avoided any academic events and he seldom supervised students or junior scientists. He simply wanted to focus on his research and dedicate as much time as possible to his family. Aurilia was, however, never isolated. Despite not wanting to be in the limelight, he assumed representative positions. For instance, he served as a member of the Cal Poly Pomona Faculty Senate.

As a scientist, Aurilia developed his own research network. Such a network had three main nodes corresponding to the institutions where Aurilia had faculty positions, namely Trieste, Toronto and Pomona. Each node also had satellite centers. For example, in the vicinity of Trieste there were contacts and collaborators in Udine, Ljubljana, Bologna, Geneva, Frankfurt and London. In the vicinity of Toronto there were Waterloo, Edmonton and Vancouver. In the vicinity of Pomona there were Los Angeles, Stanford, Fresno and San Francisco. Therefore, it is probably more appropriate to speak of Italy-Ontario-California triangle or maybe Europe-Canada-United States, given the fact that the word "vicinity" becomes loose, in particular if used for Canadian towns. Whatever the definition one assumes, it is a fact that Aurilia's network is still active as of today. There exist both scientific collaborations and exchange programs. Interestingly, the contacts in the network are both direct and indirect. This means that there exist research activities between members that were never in direct contact with Aurilia or never met him. The network simply developed like a shower of particles. From this viewpoint, Aurilia was the "primary cosmic ray" impacting the Earth's atmosphere!

I myself have largely benefited from this network. Aurilia has been the forerunner of almost all my scientific collaborations. In this regard, I am not simply referring to the obvious connections to Spallucci and Balbinot. I am referring to other collaborators I engaged following Aurilia's suggestion. For instance, this is the case of Mann and Mureika, whom I met during one of my visits to the Perimeter Institute. Other times, new connections showed up because the network steadily developed by itself, due to Aurilia's reputation, talks and papers of its members. Over the course of time, Aurilia and I ended up having an array of mutual scientific acquaintances on a global scale. This actually had an interesting repercussion. In many places I have been for professional reasons, I could count on the support of a former Aurilia associate, e.g. Hermann Nicolai in Potsdam, Douglas Singleton in Fresno, William Pezzaglia in San Francisco, Matej Pavšič in Ljubjana.

As mentioned above, there exist curious "higher order effects" that do not directly stem from the initial cosmic ray. For instance, Mureika was under the guidance of Mann in Waterloo, and later became a Visiting Assistant Professor in Claremont, California, close to where Aurilia lived. Mureika published an initial paper with Spallucci about black holes due to a vector unparticle field [200]. This is probably a second order effect. My papers with Mureika are at the third order, but there exist, however, fourth order terms. For instance, through my contact with Mann, I supported the application of Antonia Frassino, a former Ph.D. student of mine in Frankfurt, to spend six months at the Perimeter Institute during her doctoral training. She ended up publishing extremely well received papers both with Mann and Mureika [201, 202]. Mureika himself expanded his collaborations at even higher orders through his papers with the gravity group in Bologna.[66]

In summary, it is possible to break up Aurilia's network in five distinct generations.[67]

First Generation
Umezawa, Takahashi, Rohrlich, Denardo and Salam

Second Generation
Aurilia, Legovini and Kobayashi

Third Generation
Spallucci, Balbinot, Smailagic, Gaete, Helayël-Neto, Castro, Martellini, Mann, Christodoulou, Nicolai and Townsend

---

Fourth Generation
Ansoldi, Vanzetta, Kissack, Palmer, Marinatto and Tung + members stemming from the third generation (currently mid career faculty members)

Fifth Generation
Members stemming mainly from fourth generation (currently fresh Ph.D.s, junior faculty members)

The first generation includes Aurilia's mentors and senior collaborators. The second generation includes Aurilia himself and collaborators of his own age, i.e. those born in the 1940s. The third generation includes scientists born in the 1950s, about ten years younger than Aurilia. The fourth generation includes scientists born between the 1960s and 1980s. It includes former students of Aurilia and the third generation scientists. The fifth generation is the last and includes scientists born in the 1980s or later.

---

[66] There is also a weak, Nth order effect. Aurilia and Murcika had a very early institutional connection: in 1971, Mureika was born at the University of Alberta Hospital when Aurilia was a postdoc there!.

[67] For brevity, the above list includes only the names of "first order" collaborators, scientists who had an official role during Aurilia's academic preparation (e.g. Ph.D. advisor or postdoctoral sponsor) and Aurilia's former students.

6 My Time with Aurilia

I first met Aurilia on October 19, 2008 near the train station on 1st Street in the City of
Claremont, California. It was late evening. A colleague gave me a ride from Carson,
CA on his way back to Fresno, CA. This was actually the second attempt to meet.
Initially, Aurilia and I planned to attend together the California-Nevada Section APS
Meeting at CSU Dominguez Hills. At that time, I had a one semester visiting post
in the group of Douglas Singleton at CSU Fresno. Due to an unexpected problem,
Aurilia was forced to cancel his participation in the APS Meeting. He therefore asked
me via email if I would be interested to visit him in Pomona, after the end of the
meeting in Carson.

Aurilia was already well known to me. In June 1995 I had my first meeting with
Spallucci during which we talked about some research themes in theoretical physics.
On that occasion, I heard the name of Aurilia for the first time. Spallucci spoke
very highly of him, like in the case of a superstar. I can still remember his words:
"Aurilia has been the only scientist in Trieste, that could actually engage Salam in
a discussion and turn it into a conversation between peers". Some years later in
Bologna, Balbinot recalled details of his collaboration with Aurilia and spoke about
him with great admiration. I gathered additional evidence about Aurilia's talent from
Pavšič in 2005, Nicolai in 2006 and Singleton in 2007.

Despite much praise, there were some conflicting aspects in Aurilia. For instance,
no one was able to explain to me why such a brilliant person was working at a teaching institution rather than at a research oriented university. My question remained
unanswered until I actually answered it for myself 20 years later by drafting the
present chapter. At the time, however, the anomaly of his career path made me think
of Aurilia as a fighter for a good cause. Furthermore, the fact that he was in a kind
of exile, namely in a place that seemed so remote in my eyes as a student, made his
character even more interesting.

Aurilia hosted me at his house at 2460 Beck Street, Upland for about a week.
During the day we would hang out in a cafe in the center of Claremont. Other times
we went to Cal Poly Pomona, where we could talk about physics and do some
calculations on the blackboard. At the university, I presented my results on noncommutative black holes in a series of private seminars [113, 114]. Nathan Tung joined
us for these. During our talks, Aurilia stressed limitations and risks of the use of natural units. "It seems trivial, but it is crucial to write $\hbar$, G and c explicitly"—he said.
He further insisted on the existence of some sort of symmetry between Newton's
constant and Planck's constant. As we noticed in previous sections, Aurilia's idea
has several features in common with Dvali's gravitational self-completeness [156,
157] and Carr's black hole uncertainty principle correspondence [159, 160]. Aurilia
also mentioned the concept of the Planckion and the fractal properties of a quantum spacetime. Interestingly, both paradigms, i.e. fractality and self-completeness,
opened the route to research lines I pursued in subsequent years [203, 204].

A second meeting took place again in Pomona one month later. This time I rented
a car and I drove more than 400 km (250 mi) to reach Upland from Fresno, via

Bakersfield and Pasadena. It was a nice ride, but largely delayed by wildfires in the San Fernando Valley. Traffic jams developed at the fork between the Interstate 5 (Golden State Freeway) and Interstate 210 (Foothill Freeway) near the Newhall Pass [205]. Our schedule followed more or less the same pattern: chats at the cafe in Claremont and seminars at the university. During that Fall, Aurilia held a lab course two times a week from 8 to 10:45 pm. At first, I did not understand why Aurilia agreed to work that late to cover a course, which was certainly not about his favorite subject. Then, he explained to me that he accepted this assignment because it was an easy course and the schedule allowed him to have free time during the day. Other times, Aurilia told me that the position he had in Pomona was not ideal, but it was the best possible compromise between several factors such as family life, job security, teaching schedule and research. In practice, what did not work in Trieste and in Toronto, was working in Pomona. "I do not want to be like Einstein"—he told me, alluding to Einstein's great focus on research at the price of a troubled family life.

When he was free from teaching, we spent the time after dinner discussing science, education, politics, the American society, the differences in the way of life between Italy and other countries. Our discussions were fueled by the events that occurred in the previous weeks or months, such as the beginning of operations at the Large Hadron Collider, the collapse of the "first" Alitalia, Italy's flag carrier airline,[68] the collapse of Lehman Brothers Holdings, the Russo-Georgian war and the presidential campaign of Barack Obama. The election actually took place on November 4, 2008, namely between the two visits of mine at Aurilia's place. Both in Fresno and in Pomona, there was a lot of enthusiasm for the outcome of the elections. In particular, there was a sort of new hope for a better society in America and in the world.[69] Aurilia was, on the other hand, moderately optimistic. He knew that some of Obama's promises would have been perceived as too European or maybe socialist. In practice, Aurilia doubted that even Obama himself could carry out the reforms that America needed at the time.

Concerning Italy, Aurilia was even less optimistic. He noted that Italy can only oscillate between anarchy and dictatorship. In Italy, individuals do not have the sense of community, namely the awareness and pride of being part of a greater system. The system is actually weak and inefficient, while individually Italians sometimes have a talent for something and become successful.[70] He further stressed that the world does not judge a country through isolated individuals, but through the quality of the system. For example, in America there is great consideration for car manufacturing and high tech products in general, while football, fashion, good food, art, singing are not considered as features qualifying a great nation.[71] For Aurilia, it was, therefore,

---

[68] After recapitalization, a "second" Alitalia ceased its operations in 2021.

[69] Obama's campaign was marked by the slogan "Yes We Can".

[70] Aurilia's remark has been confirmed by a recent analysis about the considerable number of Italian trained faculty at top US universities [206].

[71] Aurilia's considerations are very similar to FIAT-Chrysler former CEO Marchionne's criticism of Italy in one of his memorable key notes at Bocconi University in Milan [207].

not surprising that Italy is often perceived as one of the least advanced countries in Europe.[72]

According to Aurilia, the United States have a social structure that is the exact opposite of the Italian one. The system is strong in the US, while family bonds are not equally strong. One can see the strength of the American organization from many aspects at different scales, i.e., from large infrastructures to public libraries, which barely exist in Italy. Americans tend to move to pursue the best job prospects even at the cost of living far from their hometown. For Italians moving to another town is in general difficult, because for them the support of the family is often irreplaceable. In addition, for Italians, personal financial success is not always the main goal in life. Aurilia further noticed that the actual risk for Italy is the weakening of the value of the family. This is already gradually happening and is having negative impacts on the economy of the country. Miserable living conditions could trigger large scale emigration and exacerbate the existing demographic crisis. A default scenario cannot be ruled out in Italy.

Aurilia expressed the above considerations with mixed feelings for Italy, namely love, concern and indignation. He had equally mixed feelings for the United States too. In general, he was concerned about the future of mankind as a whole.

Aurilia was always wise and gentle. He was able to read into people's souls and leave indelible memories of himself in their minds. When I think about his persona, a flux of thoughts never stops and I can only write a list of adjectives... shy, ironic, kind, empathetic, sensitive, caring, silent, meditative, intuitive, selfless, elegant, honest, respectful, dreamer, embittered...

Our last meeting took place in the evening of January 29, 2016. We met in the same place of our first meeting, namely in the City of Claremont, California and we spent the time at "our" cafe. Jonas Mureika joined us and we had a very good time together. Aurilia talked about his ideas on physics at the Planck scale, and gave us a copy of a manuscript titled "Fundamentals of Planckian physics" [210], that was an extended version of a paper published three years earlier with Spallucci [148]. Aurilia wanted us to continue his work. He told us: "I do not have enough stamina to work on this now, but if you think the ideas of this manuscript are interesting, you can freely use them". We actually talked about a joint paper and we made plans for the future, including the organization of a conference.

At that time, Aurilia has already moved to Oregon. He flew back to Pomona for the last time for two main reasons. He wanted to seal his retirement from Cal Poly. This meant that he officially terminated any teaching activity in Pomona. He also had additional things to do, like leaving his car at a workshop in Claremont for good. The other reason was actually meeting with me. In Fall 2015 a faculty position at Cal Poly was advertised. After an initial online interview, I was invited for an on-campus interview on January 27–30, 2016. Aurilia was certainly contacted by the search committee and his support was probably the main reason why I was short listed for such a job opening. He actually told me that he would have liked me to

---

[72] At that time, a derogatory acronym, PIGS, was largely used to designate Portugal, Italy, Greece and Spain. On other occasions, Italy was termed "the real sick man of Europe" [208, 209].

continue the work he started thirty years earlier in Pomona. He also informed me that the department considered me to be a strong candidate. Some days prior the interview, my odds for the final victory had further increased. The candidate that was interviewed two weeks earlier was no longer interested in the post and had officially withdrawn from the competition.

Our meeting at the cafe actually took place that Friday, soon after the end of my 3-day visit to Cal Poly. During the conversation Aurilia was a little bit enigmatic. He was confident that I could get the job offer, but at the same time he was troubled by some doubts. For instance, Aurilia expressed some concerns in relation to the seminar I held at Cal Poly. It would have been better for me not to mention the difficulty in detecting Planck scale effects. He feared that my research line might have appeared "bizarre" to the search committee which was composed of experimentalists. He also warned me that the teaching load at Cal Poly would have been substantially heavier than in Frankfurt.[73] This would have compressed the time for my research.

In the last minutes of our meeting just before saying good bye, the three of us sat in the car chatting for a while. Aurilia seemed to have a lot to say and no intention to leave. He started speaking about the importance of his family during his career and he inquired about mine. When he learned that my family would have not followed me to California, he said that this job would cost me too much sacrifice. He told me: "If you get the offer, consider it very carefully. Having a family in Europe and a job in California can be hard. To be honest, I would not advise you to accept the job at Cal Poly".

From that moment on, our paths never crossed again. I did not get the job at Cal Poly and Aurilia retired in Oregon for good.

6.1 *Winter of Theoretical Physics*

I decided to be moderately optimistic after the first writing of this section. Indeed, the title was originally "Death of theoretical physics", and given the hard times we are living there would have been nothing wrong with it.

The content of this section is centered on a conversation I had with Aurilia in the dining room of his house in 2008. Aurilia made an analysis of the state of research in theoretical physics and drew a scenario about its future. With the hindsight one can say that Aurilia's vision is an anticipation of some of the current trends of the present day academia.

The conversation began almost by chance with a question that Aurilia asked me: "Why the hell did you study theoretical high energy physics?" Aurilia's tone was a mixture of concern, understanding and rhetoric. He also added: "Sorry for

---

[73] The reference teaching load at the CSU system is actually 12 weighted teaching units (WTU). In reality, with release time for various activities, hardly anyone teaches that much. The department informed me that the average classroom assignment was 5.2 WTU during that semester. At that time, my teaching assignment in Frankfurt was equivalent to about 2.4 WTU.

the question, but you chose the worst possible research field ... you know it, don't you?" Aurilia saw himself in myself because we both experienced struggles along the career path. He knew that theoretical physics is a life mission, a fate it is difficult to escape from. Due to very few job prospects, theoretical physics can even turn into a damnation. In practice, Aurilia knew I was facing a challenging time, but at the same time, he perceived my determination in overcoming any hurdle.

Aurilia continued by talking about the crisis of theoretical physics. Such a crisis is due to a superposition of an array of crises that involve education, science and its funding, the model of society and customs and habits of the present era. In other words, the crisis of theoretical physics is just the tip of an iceberg.

To better understand the point Aurilia made it is worthwhile to introduce a structure made of levels. The lowest level, i.e. the ground level, it the widest. It presents the most severe crises. In a nutshell, such a level has to do with the future of the mankind in the current post–Cold War era. The other three levels progressively narrow around Theoretical Physics according to the following scheme:

1. Society in the current post–Cold War era;
2. Education, scientific literacy and research, sociology of science;
3. Research in Physics, methods and practices;
4. State of Theoretical Physics, its disciplinary classification and future.

6.1.1 Global Challenges and the Risk of Societal Collapse

Following the above scheme 1–4., we start by discussing the crises of current society. Most of them are simply due to the inability of mankind to face global challenges, i.e., long standing questions of economical, environmental, demographic, social and cultural nature. World's crises can also be associated to some unpredictable threats such as natural disasters and uncontrolled technological developments.[74] In addition, there is the problem of the widespread political instability and the risk of large scale wars. For science and education, the transition from a two superpower system to a world led by a sole superpower has had some important consequences, such as a reduced competition and a standardization of knowledge, research and research methods. The global impoverishment of knowledge is exacerbated by the presence of two coexisting trends: a progressive cultural loss and an uncertain rise of potential superpowers like the European Union (EU) and BRICS[75] countries. In the absence of an actual competition, there is a permanent risk of a societal collapse. This is a scenario that is difficult to predict, but it can suddenly occur in case of a combination

---

[74] According to Hawking, two major risks of mankind extinction are asteroid impacts [211] and the artificial intelligence (AI) take over [212]. In his final book, he suggested space colonization as viable method to survive a doomsday event [213].

[75] BRICS is an acronym for Brazil, Russia, India, China and South Africa. BRICS are growing economies whose strength is due to two main factors, i.e. abundance of natural resource and cheap labor force.

of factors such as a loss of cultural identity, followed by a decline of civilization and the rise of violence.[76]

The fate of the EU is particularly uncertain for its inherent lack of cohesion: The EU, as of today, suffers from a general crisis of identity at all levels. The EU looks like an artificial union, with the currency as a sole common denominator. It is not only split due to language differences, but also due to differences in the structure of the society and the economy. This aspect is particularly evident if one compares northern and southern Europe or western and eastern Europe. Aurilia suggested that the stability of the EU can be evaluated by considering similar entities from the past. For instance, the Roman Empire, the Holy Roman Empire, the Austrian-Hungarian Empire, the Soviet Union broke up in smaller countries. They, however, all enjoyed two major cohesive factors that the EU misses: the presence of a dominant culture (language, traditions and institutions) and the presence of defense forces (army and navy). Therefore, for Aurilia the EU looked even more fragile: Without a radical change, one cannot rule out the possibility of a dissolution of the EU in the next few years.

The stability of the EU is called into question also for economic reasons. The EU is modeled along the principles of 1949 West Germany's constitution: The economy and the society should correspond to a socially corrected capitalistic system. The balance between the social correction and the principle of the free marked is, however, very complicated. It is like a walk along a very narrow bridge. Below the bridge there are two monsters, a wild, uncontrolled capitalism and a state supported, planned economy. In the absence of an efficient political leadership, there is a permanent risk of falling down from the bridge and be eaten by one of the two monsters. For now, such a risk has been avoided, but at the price of having a system that has never been pushed to its full regime. In other words, the EU has been kept in a semi-dormant state and never reached the prosperity it is capable of. This fact is connected to the tendency in the EU to a superposition of negative aspects from both the two antagonist models of society: The economy suffers from a certain rigidity of the job market, i.e., a portion of the population is super protected and another portion can access to the unregulated job market only; good welfare services are at the reach of an elite only, while most of the people can only afford inefficient and outdated public services. A symptom of the above structural problems of the EU is the demographic crisis, namely a low birth rate even in high income European countries like Germany.

For what concerns the academia and the academic job market, the EU has not led to those benefits one would have expected from a super-national entity. In particular, there is the problem of a general lack of academic standard in the EU. For example, titles (e.g. degrees, Ph.D., etc.) earned in one country of the EU (e.g. in Germany) are not automatically recognized in fellow countries of the EU (e.g. Italy). One has

---

[76] In 1972, MIT researchers predicted a societal collapse by 2040, due to a rapid and uncontrollable decline in population and industrial capacity [214]. Such conclusions have, however, obtained mixed reviews and are not universally recognized. Historically, examples of societies that suffered collapses are the Aztec, Inca and Mayan civilizations.

to go through a long, expensive, heavily bureaucratic process of title recognition.[77] This is in part due to a remnant of an old mentality of some politicians, that aim to create meaningless jobs to counteract large unemployment rates.[78] The net result is a depression of the job market, a general tendency to provincialism and brain drain to non-EU countries.

An additional negative factor that affects not only the EU but the world as a whole is connected to the technological evolution. In human history, the fact that some jobs became outdated is not a novelty. Technological progress led to the creation of new needs and new professions showed up to replace the old ones. The nasty aspect of the current time is that the rate of change is too rapid and the job supply can hardly cope with the demand of new professional profiles. This is the reason why phenomena, like automation and digitalization, are actually a threat for the job security in some sectors [215, 216].

6.1.2 Education and Science in a Changing World

The above crises affecting the world society have further repercussions on science and education. Aurilia quickly moved to the second point of the above list with the following remark: "Academia is progressively shifting its focus from its traditional goals, e.g. education, science and research, to new objectives such as marketing and funding. The human capital, i.e. students, researchers, faculty, is no longer considered of vital importance as in the past." Aurilia also pointed out: "As an educator, I aim to instill a love of objectivity in my students and prepare them to be able to chase and reach the truth. Unfortunately, in the present day society what one "sells" is becoming more important than what one is capable of doing."

Aurilia further specified his ideas by pointing out a general cultural impoverishment also in case of people having an academic degree. He feared that, following a wave of human work elimination via large scale automation, the destiny of mankind could be that of mere supervisors of the work done by machines. His apocalyptic view was motivated by a certain homogeneity of academic programs, whose boundary are becoming increasingly loose due to the workforce demand from the high tech and digital communication sector.[79] In particular, Aurilia feared the advent of

---

[77] By long and expensive I mean that the orders of magnitude are years and thousands euros.

[78] Bureaucratic hurdles are very common in the EU. This is reason why the European citizenship guarantees freedom of movement only on paper. Border controls between countries of the Union have de facto been reintroduced on permanent basis, namely beyond six months, that is the maximal duration for border controls in case exceptional events (e.g. war, terrorism, pandemic). In addition, there is no European residence. This means that each EU citizen is actually tied to a specific country for what concerns tax, social insurance and any sort of document, e.g. driving license, vehicle registration, etc. In case of change of residence from a member state to another, none of the above documents are recognized. Expensive fines are applied by the border police, that can also confiscate the vehicle and the license, if the necessary documents are not converted within a certain deadline after the change of residence.

[79] The phenomenon of academic homogeneity appears to be driven by "buzzwords", that decay rapidly with a half-life not exceeding seven years [217].

an era of "press-the-button-education", namely a system that aims to create professional figures with less competence, but with more adaptability to the automatized production [218, 219].

Aurilia expressed additional concerns about science and research in general. He was not convinced that the core of the mechanism governing academic research was actually working or could be beneficial for the actual progress of mankind. By "core", he meant the process of evaluation with peer review. Aurilia claimed, that even in case of ideal peer review (e.g. with unbiased, highly qualified reviewers), excellence is actually impossible to determine. He made his case stronger by citing notable rejections of papers that contained results for which the authors subsequently won the Nobel prize.[80] Aurilia also noticed that the problem is vaster than one can think: It is not limited to paper publication, but it has to do with grant applications, hiring process and career progress too.

According to Aurilia, at the crux of the issue there is the very nature of research, that consists in a challenge of the unknown. Therefore, the output of research can never be fully predictable. Major discoveries have often been attained either from unexpected experimental results or unconventional theoretical speculations. Today's evaluations can only be the consequence of current limited knowledge on a given topic. On the contrary, the only efficient way to evaluate the validity of a research proposal or research result is to observe its repercussions in the future: If something is really a major breakthrough in a given field, it will be freely used by future researchers in follow-up studies; Conversely, the research output will be simply forgotten.[81]

Aurilia was probably influenced by the thought of Michael Polanyi, a Hungarian-British polymath, who claimed that scientific knowledge must follow its spontaneous order resulting from the debate among specialists. The process is actually similar to what happens in a free market when consumers determine the prices of products according to the law of demand and supply [222, 223]. In conclusion, the essence of Aurilia's ideal of academia was aligned to the Humboldtian model of university, in which research and teaching are connected and governed by the principle of academic freedom. Such a freedom can be achieved if the university is free from governmental and economic constraints [224]: Financial resources should be equally distributed among scholars with no particular a priori selection.[82]

Any attempt to modify or alter the law of free market leads to distortions, unfair situations and an overall poorer academic performance. According to Aurilia, this is already happening within the "elite model" of academia, namely a model based

---

[80] Aurilia referred to Fermi and Gell-Mann [220].

[81] Aurilia's vision actually coincides with what expressed by Feynman about "honors", during one of his interviews: "I don't like honors. ... I've already got the prize: the prize is the pleasure of finding the thing out, the kick in the discovery, the observation that other people use it. Those are the real things." [221, p. 82].

[82] It has been shown that there exist benefits and no disadvantages from a funding distribution without peer review. In particular, the worry for an unsustainable dilution of resources in case of an egalitarian distribution of funds among all researchers is unjustified: Data from the UK, the USA and the Netherlands show that it would still be possible to maintain the current PhD and Postdoc employment levels and have enough budget for travel and equipment [225].

on concentration of resources on few excellent schools. Recent studies actually indicate that Aurilia was right and that the weakness of the elite system is not limited to a specific field, such as theoretical physics [226]. In particular, statistical analyses have determined the existence of two aspects of the methodology of academic evaluations: the so-called "excellence" is seldom excellent; researchers do not recognize excellence when they see it.[83] In addition, data suggests that excellence based funding schemes do not promote scientific and scholarly progress, but they are actually an obstacle to it [226]. This can be attributed to the fact that peer reviewed funding systems discourage intellectual risky activities and support run-of-the-mill research [228, 229].

To this purpose, Aurilia was even more radical. He pointed out that, for its intrinsic nature, funding via research grants can only lead to mediocre results. This is because a grant application can never contain a really outstanding research proposal: Nobody would share a revolutionary idea with a funding agency! In case of applied sciences, one would rather protect his/her intellectual property by a patent application with the aim of commercializing the results. In case of fundamental science, one would never run the risk of sharing the solution of a long standing open problem with an array of referees. From this perspective, science is light years away from engineering. In the industry business, there are strict rules about the non disclosure of trade secrets and any kind of information that potentially can lead to revenues.

The shortcomings of funding via research grants are also exacerbated by the phenomenon of "homophily": It is easier for applicants to be successful if their appearance, work patterns and research goals are aligned to those the reviewers have previous experience with [226]. Such a trend is one of the possible explanations for the "Matthew effect" in science funding, namely the tendency of researchers and institutions to accrue financial resources in proportion to their past success in funding history [230, 231].[84] In the long term, such an effect has determined a tendency to form academic oligarchies [232] and a drop of the success rate below 10% for most of the applicants. More importantly, applying for a grant has simply become too expensive and time consuming. In a study based on Natural Science and Engineering Research Council Canada (NSERC) statistics, it has been shown that the costs to prepare and review a grant application, i.e. $40, 000, exceeds the cost of giving every qualified investigator a baseline grant, i.e. $30, 000 in average [233].[85]

---

[83] Such conclusions, based on extensive studies on peer review and research funding [226], are motivated as follows. There is a general difficulty in defining what "excellence" actually is. In practice, "excellence" only has a linguistic function, but no actual meaning in academia [227]. Peer review does not guarantee the acceptance of strong research results and the rejection of poor ones: Data on scholarly publications show that the majority of papers rejected by a journal are subsequently published on journals of similar reputation.

[84] The Matthew effect is often summarized by the sentence "the rich get richer and the poor get poorer".

[85] Additional evidences has been provided by analyses of the Australian research system. Given an estimated yearly researchers' time of about 550 working years, the costs of grant applications in a country like Australia are equivalent to $68 million per year. This exceeds the budget of a major Australian research center, that produces hundreds of papers per year [234].

An additional drawback of the elite model is its poor overall performance: In countries like US and France, top schools are de facto not accessible to a large portion of the population, because social status and family background become more important than the bare performance of prospective students [235]. This is an aspect that is seldom captured by university rankings, but it has generated an intensive discussion about the proliferation of "excellence initiatives" in many countries, such as Germany, the UK, Russia, China and India [236, 237]. Unfortunately, the elite model based on peer reviewed funding schemes is not only inefficient, expensive and hardly fair. It can also be harmful, because it represents an incentive for an array of academic misconducts [226, 238, 239].

There exist many forms of academic misconduct that can increase a scientist's reputation, chance for publication and funding. One of the most common of such misconducts is probably the authorship and citation manipulation. For instance, "padding" a reference list is the intentional inclusion of superfluous citations. This typically occurs when authors aim to please decision makers (e.g. referee, senior faculty), by inflating their bibliometric indicators. If such a padding is the result of blackmail or demand, one rather speaks of "coercive citations" [240]. The emblematic example of such misbehavior is when editors demand authors to add spurious citations to enhance the impact factor of their journals. The "honorary authorship" is an additional widespread phenomenon.[86] It consists in including scholars as authors of manuscripts or research proposals, even if they did not contribute to the research.[87] It is estimated that such misconducts have been perpetrated with a percentage ranging from 14% to 40% of survey respondents in one academic study [242]. More recent analyses, however, tend to be more pessimistic, having reported a percentage of honorary authorship peaking 55% in the US and 70% among European based researchers [243].

---

[86] Until recently, this was the norm in Germany for heads of department or institution [241].

[87] There exists at least four situations in which honorary authorship is convenient. First, due to the pressure for publication and funding, early career researchers are induced to include more experienced scholars as authors in their publications and grant applications to acquire reputation and increase their success rate. Second, honorary authorship is attractive to senior scholars that can fictitiously inflate their bibliometric indicators and improve their chances of acquiring funds, also taking into account the tailwind of the aforementioned Matthew effect. Third, the misbehavior can become quadratically convenient for early career researchers. Having included senior scholars in author lists can pave the way to the acquisition of fellowships and research contracts, offered by the same senior scholars. A fourth benefit is when the honorary authorship becomes a reciprocal relation between scientists with similar seniority and reputation. The inclusion in the author list becomes a favor the beneficiaries will return in a future paper of their own. Along this process, both members of the relation can inflate their bibliometric indicators.

The spectrum of misconduct also includes more severe malpractices. Simultaneous and duplicate submissions, incremental publications ("salami slicing") and self-plagiarism aim to magnify the number of publications and citations of the authors. In case of conventional plagiarism, one reuses someone's else work without giving the proper credit. For example, some authors tend to excessively cite their own papers ("self-citation") and ignore seminal papers from other scholars.[88] It has been estimated that plagiarism (including self-plagiarism) is one of the most common forms of academic misconducts.[89] Other forms of serious misconduct include fabrication (i.e. invention of data or cases), falsification (e.g. image manipulation and re-use, data and result falsification), opportunistic interpretation of statistics, misrepresentation of research results and fake peer review [248]. Such extreme cases seem to be less common, but such a conclusion could be related to the difficulties in detecting them [249].[90]

In a nutshell, there is an increasing evidence of a progressive decline of peer reviewed based educational system across all disciplines. According to Aurilia, by 2008 a prompt action was imperative. He stressed the importance of "rescuing" the young generation. His concern was motivated by data showing a correlation between low income and vulnerability to forms of misconduct [242, 250]. Since 2008, many reforms have been proposed but, for now, none of them has actually been implemented [225–229, 251].[91]

---

[88] According to a study, a computer scientist has "received 94% of his citations from himself or his co-authors up to 2017 [244]". In case of simple self-citation, i.e., without omission of references to papers from other scholars, one rather speaks of a form of reference list padding.

[89] The Editor-in-Chief of Applied Surface Science has estimated that 10% of the submissions is affected by some misconduct, with an unacceptable overlap with previously published material being the most common issue [245]. A scrutiny of the bio-medical literature published in the year 2020 has revealed a plagiarism percentage around 30% [246, 247].

[90] Less than 2% of scientists have admitted to having fabricated or falsified data or results at least once, while about 34% of them admit other misconduct. In surveys about the behavior of colleagues, admissions were 14% for falsification and 72% for other misconduct. Considering the nature of these questions, it is reasonable to assume that the above percentages underestimate the actual occurrence of scientific misconduct [249].

[91] Among the alternatives to the elite model, we recall here the allocation of funds based on an assessment of researchers rather than a grant application [252–254], the distribution of resources among all qualified researchers without any assessment [233, 255], a combination of peer review and a lottery [256].

6.1.3 The Evolution of Physics in the Third Millennium

Aurilia was somehow less pessimistic for what concerned the Physics community. He expected a more frequent occurrence of misconduct in fields where financial interests are larger. Statistics confirms Aurilia's intuition [249], but Physics cannot be considered a happy island in the academic landscape. There have been major scandals[92] as well as questionable research practices.[93]

The ethical problem in Physics is probably the side effect of a bigger problem: Physics is no longer dominant. If one goes back to the 1960s, the situation was drastically different. The global competition between the two Cold War blocs led not only to military conflicts, but also to a hot rivalry in other fields, e.g., sport, education and technology. This explains why the Space Race was at its peak and provoked a growing interest for Physics. As an example, following the Sputnik mission, the NSF aimed to counteract the Soviet space supremacy by improving the quality of science teaching. As a result, the Physical Science Study Committee (PSSC) was founded by MIT professors and the first edition of their textbook was published in 1960 [262]. The Space Race also forged a Physics oriented culture by other means. For instance, space explorations became the main subject of science fiction, that boomed through the production of films, television series, cartoons, novels and music.[94] Kids dreamed of becoming astronauts and perceived life on another planet as a likely occurrence for their generation. Becoming a physicist was a sort of mandatory step to make such a dream come true. Society enjoyed a general climate of optimism and confidence in science. Everything seemed possible.

By the 1960s, physicists have already obtained a streak of triumphs that chronologically can be listed as classical mechanics, electrodynamics, statistical mechanics, special relativity, general relativity, quantum mechanics and quantum field theory. The formulation of the Standard Model in those days and the rise of QCD in the 1970s are probably the latest hits of Physics. From the beginning of the aforementioned Hawking revolution to the present time, progress in Physics has consisted in corroboration of the existing knowledge owing to improved experimental facilities. One might even dare say that Physics has become boring: It is a research field in which nothing new happens. This view is perhaps too extreme, but it contains a

---

[92] Among the most infamous cases, one should recall the Schön scandal (various claims on innovative materials including the discovery a transistor on molecular scale—data falsification [257]), the Ninov fraud (discovery of two new elements—data falsification [258]), the Bogdanov affair (proposal of a nonsensical solution of the cosmological singularity problem—fabrication [259]) and the El Naschie incident (proposal of a controversial E-infinity theory for the dimensions of the Universe—fake peer review [260]). Before the incident, El Naschie had legitimate contacts with Walter Greiner and Werner Martiennsen in Frankfurt. As editor-in-chief of Chaos, Solitons & Fractals, El Naschie invited Aurilia and Spallucci to submit a manuscript to one of the journal special issues [2].

[93] A 2020 survey of the American Physical Society confirmed that research misconducts and harassment are still serious problems in the physics community, despite campaigns for the good scientific practice have recently been intensified [261].

[94] "Space rock" is a genre that emerged in the late 1960s and is characterized by themes related to space explorations, extraterrestrial sounds and use of synthesizers (e.g. vocoder).

foundation of truth: The most important open questions in Physics have not yet been answered and we are not even close to addressing them.[95]

There exists an additional threat for Physics: the rise of competing disciplines that are backed by industry. The feverish interest in artificial intelligence (AI) is just the latest example of a field that has been eroding the dominance of Physics for the last thirty years. The virtue of AI is that it looks dynamic, modern and interdisciplinary, being at the confluence of multiple fields, such as engineering, applied mathematics, computer science and communications science.

The absence of results in combination with today's pressure for accruing financial resources has led to a potentially dangerous situation for the Physics community. In order to find a remedy, physicists have taken some initiatives. For instance, science communication has become increasingly important to connect science and society. If on the one hand popular science is per se a value, Aurilia, on the other hand, complained of an excessive focus on a marketing mentality in today's Physics community. There is also a disproportionate relation between the revenues from selling science and doing science.[96] This is particularly true by taking into account that faculty engage themselves in outreach activities as a service to the society. Conversely, the business of science communication is governed by non-experts, that mainly aim to maximize their profit. This fact implies a certain risk of conveying incorrect information. Furthermore, the overall picture is aggravated by the fact that science is not yet fully perceived as a profession but more like a hobby. Such a distorted perception can be one of the factors that determined precarious living conditions for many scientists having income below the poverty line [250].

The other feature that has emerged as an action to contain the risk of a loss of dominance is the shift of Physics towards engineering and technology. A possible way to evaluate such a tendency in quantitative terms is the analysis of the career path of Physics graduates. For instance, in Germany around 80% of all graduates end up in non-physics professions over the course of their professional career [264]. In total there are 120,000 active physicists in Germany, 15% of them are employed in universities and research institutions, 15% are school teachers, 70% work elsewhere. The unemployment rate is 2.5% [265].[97] The above figures vary from country to country and are not constant in time. Nevertheless, one can conclude that Physics graduates acquire a spectrum of skills and competences, that is wider than the traditional academic curriculum of a degree in science.

---

[95] Quantum gravity is probably the final chapter in Physics and the hardest to write. We do not have a universally recognized formulation for quantum gravity and we do not even have a single data point for it. For a list of unsolved problems in Physics see [263].

[96] While chatting with Mureika and myself during a recent meeting in Trieste, Carr pointed out a detail in relation to the film dedicated to his supervisor: "Eddie Redmayne earned more money for playing Stephen Hawking in the film "The Theory of Everything" than Stephen earned for being Stephen over his entire life!".

[97] There exist several methods to calculate the unemployment rate. Who is actively looking for a job is generally not listed as unemployed in Germany. For this reason, 2.5% of physicists without a job could be an optimistic estimate.

The employment percentages also show that there is a surplus of Physics researchers and faculty.[98] This is a problem for many prospective students. The basic idea behind studying Physics (e.g. rather than engineering) is to become a Physicist and stay in the academia. The very low success rate and demanding working conditions in the postdoc phase, however, determine a brain drain to other professions. A career change can be frustrating and difficult. For instance, further training could be necessary especially for physicists holding a Ph.D. Leaving academia can also determine the period of underemployment[99] or even unemployment. Germany is often considered a model because it is one of the few countries where a Ph.D. has become a required title for leadership positions, both within and outside academia. In most countries, however, science Ph.D. holders never fully exploit their qualifications and too often take jobs that do not require a doctorate [267]. The above situation has suggested to reorganize science PhD programs [268].

The career path of Physics graduates can also be seen as a problem for faculty, who invest funds, time and energy in the human capital. Such a waste of resources is generally accompanied by a slow down of the daily research activity. For instance, it is customary for Ph.D. candidates in Germany to attend courses where they can acquire transferable skills. This can be seen as a positive step, but there is, however, the risk that valuable time is subtracted to research, publication and thesis writing.

A third feature in the evolution of Physics is the tendency of getting bigger and bigger. Aurilia noticed that everything proliferates, e.g., journals,[100] publications, authors. He feared that physicists have become too concerned with the image of their research rather than the results of their research. As an example, in recent years some discoveries have been announced in events or press releases, even before having been published in a journal.[101] This is probably the consequence of the "hyperauthorship",[102] namely the publication of papers with more than fifty coauthors, that has become a routine in some research fields.

---

[98] Official data does not capture this phenomenon. Physicists are not listed among the professions with largest supply on the European job market. It is also likely that figures simply consider Physics as a degree rather than a profession [266].

[99] It is the condition of workers that do not work full time or take a job that does not reflect their actual training and financial needs.

[100] A notable side effect of such a numerical growth is the phenomenon of "predatory publishing" [269].

[101] The detection of superluminal neutrinos [270] and the observation of a primordial gravitational wave [271] are among the discoveries that have been announced and later retracted.

[102] The term has originally been introduced by Cronin in 2001 [272]. For a recent discussion of this phenomenon see [273–275].

Aurilia was aware that changes in Physics were unavoidable. He, however, feared that many of the measures adopted to address the loss of scholarly supremacy were ineffective and maybe counterproductive. In the third millennium, there could be a general paucity of research results. In addition, physicists could run the risk of losing focus on one of their social mission, namely the scientific literacy of society.[103] Such a mission has become more important than ever, as an increasing number of political decisions on scientifically sensitive issues will have to be made in the near future [277].

6.1.4 Will Theoretical Physics Celebrate Its 200th Birthday?

If we stop for a moment to reflect on the objects around us, on the things we use in our daily lives, or on the goods we would like to have, we soon realize that each item is ultimately a byproduct of the work of some theoretical physicist of the past. This is true also for many things we experience, such as watching TV, having a phone call, flying in a plane. In some cases, the technological development can be difficult to track back to the initial scientific discovery. Nevertheless, one could say that virtually everything descends from the "tip of the diamond".[104]

The character, which makes theoretical physics unique is its predictive power, which is unmatched in science. A crystal clear example is the kinetic theory of gases, in which a mechanism at the microscopic level, i.e. molecular dynamics, explains the macroscopic behavior of the system. Theoretical physics has been so successful that the meaning of the word "theory" has probably been forged after it. In a non-scientific context, theory is synonym of guesswork, speculation, presumption, supposition. In a scientific context, theory means an experimentally confirmed mathematical formulation, or simply a scientific truth. Therefore theory drastically differs from "model", a phenomenological formulation that describes something without explaining it. Models are generally the result of "bottom-up" approaches, e.g. data interpolation. Sometimes, there exists a "top-down" reasoning behind a model, namely some basic principle. For a given model, however, the connection between the principle and

---

[103] It has been estimated that roughly 90% of the Americans are scientific illiterate [276].

[104] Being the hardest of natural sciences, theoretical physics can metaphorically be compared to a diamond. The term hard science is used to qualify rigorous, objective, exact sciences. The hardness determines the so called "hierarchy of the sciences" [278–280].

actual predictions is loose or poorly understood. Therefore, in contrast to theories, models are never fundamental.[105]

Despite such a premise, Aurilia noticed that theoretical physics is in a state of despair. First, there is a stalemate situation concerning the solution of long standing issues. As previously mentioned, quantum gravity posits a series of insurmountable problems. The situation is similar to the crisis of classical physics, with string theory playing the role similar to that of quantum mechanics in the 1920s. Quantum mechanics had, however, considerable experimental corroboration, in contrast to what happens with string theory. The other problem of string theory is the absence of competitors. There have not been new paradigms for the last forty years, or at least, they have not yet reached the maturity to challenge string theory on the ground of becoming the theory of everything.[106] Second, none of the current predictions for the physics beyond the Standard Model have been corroborated by experimental data. This fact is worsened by the intertwining with the problem of dark matter. The Standard Model above the TeV and Einstein gravity beyond the kiloparsec cannot be both correct, given the observed galaxy rotation curves. In addition, there is a conflict with the existence of hypothetical fields describing dark energy as quintessence.

A further problem of theoretical physics is of sociological nature. There is a general subalternity of theoretical physics to experimental physics. Research funding is dominated by large collaborations, while theorists struggle, unless they are aligned to the research topics under investigation at some big experiment. Theorists predominantly belong to an isolated minority in almost every department in the world. It is interesting to note that being the weakest link in the chain is nothing new for theorists. Before 1850 there were no theoretical physicists. There were professorships only for mathematicians and experimental physicists. It was unthinkable that research about a natural science such as physics could be conducted without laboratory activity.[107]

The role of the theoretical physicist emerged almost by chance in Germany in the second half of the 19th century. At that time, physics professors preferred not to hold lectures that required the presentation of mathematical material. The simple reason for this was that such lectures were demanding and far from the research focus of the department. Another factor was the academic prestige. Experiments could generate more visibility, recognition and larger funding, while the abstract reasoning was deemed speculative and unimportant. As a result, mathematically oriented courses

---

[105] This is the main difficulty with theoretical formulations (e.g. theoretical chemistry, mathematical biology and applied mathematics), that can only rely on models. Typical examples are weather forecasting or earthquake predictions. For each of them, there exist governing principles, i.e. fluid-dynamics and rock mechanics, but their mathematical connection to some macroscopic variable is virtually impossible to obtain.

[106] For a discussion about the ongoing "quantum gravity war" see a review written by Nicolai [281].

[107] For this reason, Lagrange, Maxwell and Hamilton were officially labeled as mathematicians having interests for equations emerging from a physical context.

were customarily assigned to researchers associated to the chair of (experimental) physics or to lower ranked faculty.[108] The work of a "calculating physicist that can never experiment"[109] became acknowledged only after 1850. This is the time in which von Helmholtz, Clausius and Kirchhoff obtained professorships in physics, but they were were "allowed" to conduct theoretical research in parallel with the "mandatory" laboratory activity.[110] Actually, the first chair of theoretical physics, i.e., with no laboratory duties, was assigned to Kirchhoff in Berlin in 1875. To stress that the position was laboratory free, the initial title was "ordinary professor of mathematical physics". We must recall here that at that time theoretical and mathematical physics had become synonyms.[111]

Another turning point is represented by the hiring of Planck in Berlin as Kirchhoff's successor in 1889.[112] He became the first director of a theoretical physics institute, without having any prior experience in experimental physics [282]. This feat can be considered a sort of declaration of independence of theoretical physics. The process was, however, not complete and another separation occurred. In 1895, Boltzmann, who himself previously held a chair in mathematics, clarified that theoretical physicists are not mathematical physicists. The main focus of theoretical physics is on the physical content and the predictions that can be tested in experiments. Conversely, for mathematical physicists, the goal of the investigation lies within the mathematical framework.

We note that the separation between mathematical and theoretical physics did not happen in all countries simultaneously. For instance, in Italy the first professors of theoretical physics were appointed in 1926.[113] In the British system, theoretical physics is still considered part of mathematics. Nevertheless, it is clear that the schism with mathematicians further weakened the community of theorists. As a worsening factor, there was the progressive evolution of mathematics towards purists positions, a fact that is currently at the center of a hot debate.[114] In practice, rather than

---

[108] Due to such a hierarchy, proto-theoretical physicists have recently been termed as "second physicists" [282].

[109] Sentence attributed to Boltzmann [282].

[110] Kirchhoff obtained the chair in Heidelberg in 1854, while Clausius was appointed "professor of physics, preferably for mathematical and technical physics..." at the Zurich Polytechnic in 1855. Von Helmholtz obtained a chair in physics in Berlin in 1870, after having been professor physiology since 1848. The case of Meyer is opposite and complementary, since he initially obtained the chair for mathematics and theoretical physics in 1865 in Breslau, but he was forced to conduct "experimental" investigations, after becoming director of the physics institute in 1867 [282].

[111] Until 1870 theoretical physics indicated advanced physics lectures without laboratory. Experimental physics covered introductory laboratory courses. Mathematical physics concerned mathematical method for physics and was taught either by mathematicians or physicists [282].

[112] Planck was ranked third, after Boltzmann and Hertz, who both declined the offer. Boltzmann, who was already professor in Graz, opted for another offer from the Ludwig Maximilian University of Munich, while Hertz preferred to stay at the technical university in Karlsruhe, where he had obtained the chair in 1885 [282].

[113] The holders of the newly created chairs were Fermi, Persico and Pontremoli [283].

[114] Arnold has been one of the harshest critics of the trend of mathematics towards abstraction, that started in France with the Bourbaki school in the 1930s and now has spread worldwide.

mathematical physics, one should better speak of "physical mathematics", namely a field with just a nominal connection to theoretical physics [284, 285].

It is interesting to notice that many of the problems related to the divisions between experimental and theoretical physics and between theoretical and mathematical physics were already a source of concern at the end of the 19th century. In a letter to Sommerfeld in 1898, Wien pointed out:

> Theoretical physics in Germany lies as good as completely fallow. That ought to be a reason to help revive it, but the situation has already gotten to the point that even the need for theoretical physics disappears more and more. The reasons for this are, first, that physicists do almost nothing but pure experiments and have hardly any interest in theory and, second, that most mathematicians have turned to entirely abstract areas and do not concern themselves with applications. This reveals itself externally in that pure theoretical physics is taught from only two chairs (Berlin and Göttingen), and such an important chair as Munich has come entirely to an end. Theoretical physics currently finds no takers. Later it will all be different again, indeed, because otherwise it would go completely to ruin; but I must make some allowance for trend and thoroughly busy myself with purely experimental researchers as long as I still have to work for an external position.[115]

Given the above disciplinary fractures, theoretical physics can maybe appear closer to some aspects of applied mathematics, than to mathematical physics and experimental physics.[116] The bottom line result is, however, that theoretical physicists never reached a critical mass to be properly represented in physics (or mathematics) departments. Incidentally, applied mathematics has a similar problem of recognition. In very few institutions, there are departments of applied mathematics.[117] This has contributed to the present state of suffering on the research, teaching and funding side.

At some point in our discussion, Aurilia pointed out that mathematical physicists generally have a happier life than theoretical physicists. In fact, mathematical physicists have a far quicker career progress, as a result of a certain academic policy. For instance, courses that theoretical physicists routinely teach in the UK are reserved to mathematics Ph.D. holders in most countries worldwide, creating a de facto a shortage of mathematics faculty positions. Unfortunately, the converse is not true: There is no shortage of theoretical physics faculty. This is in part due to the reduction or cancellation of advanced physics courses in some programs, such as mathematics and engineering, and the contextual assignment of introductory physics courses to experimentalists. Another advantage of mathematical physics is being less exposed to the pressure of bibliometrics. Due to the absence of hyperauthorship among mathematicians, citations records tend to be homogeneous between different fields of mathematics. This fact has also contributed to create the aforementioned homophily

---

[115] Wien to Sommerfeld, 11 June 1898, Sommerfeld correspondence, Ms. Coll., DM [282].

[116] The relation between mathematics and physics is still at the center of many discussions among scholars [286, 287].

[117] Two notable exceptions are DAMTP at Cambridge (where Carr got his Ph.D.) and the Department of Applied Mathematics at Waterloo (where Mann is cross-appointed faculty).

between mathematical physicists and mathematicians with other research interests. On the other hand, in physics departments bibliometrics can be a threat for theorists, that must compare their citation records with that of faculty working in large experimental collaborations.

According to Aurilia, there are additional phenomena, signaling that theorists may be in the wrong department. He touched this point, while giving me tips about how to reply to questions during my job interview at Cal Poly. First, he started by saying that the number of vacancies offered to theorists is a small portion of those offered to experimentalists. Second, experimentalists have the tendency to forge the profile of the theorist to hire. He stressed that, in many physics departments worldwide, theorists have been hired only because they are instrumental to the needs of experimental physics. Conversely, the needs of theoretical physics are seldom taken into account. In practice, after almost 200 years, the situation is similar to what happened to Clausius in Zurich: He was a theorist, but his background work in experimental physics turned to be decisive to get the job. According to Aurilia, by 2008 there was a risk for theoretical physics of being progressively substituted by more experimental physics akin topics, such as data analysis, computational and numerical physics.[118]

In general, theoretical physics is still culturally ostracized. Mathematics is deemed as necessary for every science or engineering sector. Experimental physics is perceived as the science at the forefront of technological revolutions. On the other hand, the work of a theoretical physicist is often debated with questions like "What is your research for?". Indeed, it is a widespread opinion that every penny spent on theoretical physics is a waste of money.[119]

Such a reputation is certainly a symptom of poor communication, but it is likely to be part of a bigger problem: Theoretical physics, for its very essence, does not get along with modernity. By modernity, I mean all the rapidly evolving features of the current (technological) society, e.g., virtual learning, popularization, artificial intelligence. Such an opposition to evolve can be explained for the presence of at least three features. First, theoretical physics is not predictable. The major scientific discovery in theoretical physics were found by chance, i.e., without a research plan. This is in marked contrast to the current practice in any experimental science, for which research phases, such as apparatus development, data collection, data analysis, can be easily put in a Gantt chart with tasks and milestones. In general, there is not even a guarantee that research efforts in theoretical physics can lead to any good. This is connected to the second point. It can take decades to understand if research

---

[118] According to Hossenfelder, the turn of theoretical physics towards numerics is unavoidable [288].

[119] This explains the fact that theoretical physics is underfunded. For instance, in 1909 the budget of Planck's theoretical physics institute was 700 marks, while the physics institute received 26,174 marks [282].

results are good or bad. Sometimes formal results in theoretical physics become technological applications 50 years after the initial discovery.[120] The third feature is that the work of a theorist is obscure, namely difficult to convey to a broad public. This fact does not come as a surprise. There are limits to popular science. The idea that everybody can understand everything is a myth. This is very true especially for abstract research fields, that being immaterial, have nothing concrete to show, i.e., no lab, no device, no picture. In a nutshell, theoretical physics is deep and slow, the contrary of the modern tendency of exchanging information very rapidly although superficially.[121]

Aurilia pointed out that the problems of theoretical physics should not be attributed only to conflicts against external forces. The community of theoretical physicists is torn among a multitude of research groups that fight for control of a relatively modest amount of funds. The separation is not only due to differences between research areas (e.g. particle physics, astrophysics, solid state physics), but also due to antithetical approaches and methods to tackle the same problem. The typical example is the aforementioned "quantum gravity war". There exists an orthodoxy, namely a dominant proposal to address the issue of gravity quantization and heresies.[122] The bottom line result is that there is no fraternity among theoretical physicists. Their fragmentation in minuscule parishes is particularly harmful in the light of previous considerations about the homophily and the numerical growth of other scientific communities. An additional negative consequence is that evaluations with peer-review become ineffective. There is always a conflict of interest because theoretical physicists are just a bunch of people worldwide and they know each other in part or fully. Aurilia stressed that there is virtually no bias free report when a research proposal or a manuscript is evaluated. A further disadvantage is the absence of dialogue and debate. According to Aurilia, theoretical physicists tend to be selective, rigid and narrow minded: They focus on a topic and they defend their ideas with intolerance. Aurilia also recalled a sentence by Planck [292]:

> A new scientific truth does not triumph by convincing its opponents and making them see the light, but rather because its opponents eventually die, and a new generation grows up that is familiar with it.

At this point, Aurilia posed a question about the fate of theoretical physics: "Will it survive in the long term future?" Of course, predictions are always difficult. Nevertheless, Aurilia was moderately optimistic. Even if there are long standing problems and

---

[120] A notable example is the work of Bethe dated back in 1930 [289]. His equation for the passage of charged particles through matter is nowadays used in radiation therapy.

[121] There is probably just one advantage connected to having an old-fashioned character. It is likely that theoretical physics will be not (or to a lesser extent) affected by AI. In some sectors, such as engineering, experimental physics, jobs will be soon at risk. For instance, it is expected that the building of particle accelerator will be designed and coordinated by AI [290].

[122] Here we mean alternative paradigms that can co-exist with the main theoretical proposal when there is no full scientific consensus. It is useful to recall that Asimov introduced the distinction between endoheresies and exoheresies [291]. In contrast to an exoheresy, an endoheresy forms within the scientific community.

threats, he was confident that the kernel of theoretical physics is somehow "immortal". There will always be people like Clausius, Kirchhoff and Planck having the drive for theoretical speculations, despite the most grueling working and personal conditions. In other words, theoretical physics will overcome the present crises and resuscitate at a higher level. The timescale of the process is, however, uncertain and there is no guarantee that the rebirth will occur during the present civilization.

Unfortunately, Aurilia missed the development of AI, during the last five years. It is possible that the transition from current era of AI to a large scale implementation of AGI (artificial general intelligence) will occur "soon", i.e. within a few generations. By AGI, one means a type of super-intelligent agent, with consciousness and better than human ability to accomplish intellectual tasks. The question arises about the possibility that such a super-intelligent agent can solve problems in theoretical physics, whose solution humans cannot even conceive. If this occurs, it will be probably coincide with the end of theoretical physics as we know it now. It is also very likely that such an end is accompanied by the so call AGI "take over". This would correspond to the end of human civilization and the beginning of an era dominated by computers and robots.

Alternatively, there is the possibility that theoretical physics, as we currently know it, turns to be the decisive factor to rescue mankind. If the Earth becomes inhabitable due to AGI take over or any other threat, space colonization may result the only possible way to prevent human extinction. Finding a planet similar to Earth is not an easy task. Current spacecraft have a limited reach and cannot cover interstellar distances. On the other hand, one might be led to consider a short cut in space. According to general relativity, it is possible to cross the Universe through special space-time configurations, called traversable wormholes [293]. Of course, there exists a myriad of scientific and technological questions about the feasibility of a travel of this kind. The issue is, however, not just science fiction. Theoretical physicists currently run programs aiming to create a space-time wormhole in the laboratory [294]. Even if we will not know the outcome of such a research before decades, such a line research must continue, given what is at stake.

In conclusion, theoretical physics may be the science destined to mark the end of one civilization and the rebirth of a new one. Perhaps the next civilization will start from scratch and there will be a new Clausius, a new Planck, a new Boltzmann... and also a new Aurilia. Despite many unknowns, we have learned at least one thing: theoretical physics is essentially the pursuit of infinity. This is the truth that Aurilia leaves us all.

7 Timeline

- 1942-May-14, born in Naples
- 1956–1961, student at Liceo "Gian Battista Vico", Naples
- 1961–1966, undergraduate studies at the University of Naples Federico II
- 1966, Laurea degree in Physics with Umezawa
- 1966, beginning of the Scuola di Pefezionamento
- 1967, start of the PhD program at the University of Wisconsin-Milwaukee
- 1970-Aug-15, PhD in Physics with Umezawa
- 1970–1972, postdoc in Edmonton with Takahashi
- 1972-Dec-01, married Elizabeth Adams
- 1972–1974, postdoc in Syracuse with Rohrlich
- 1974–1975, researcher at the ICTP, Trieste
  ↪ start of the collaboration with Christodoulou and Legovini (7 papers)
- 1975, the son Darius was born
- 1975–1986, permanent researcher at INFN, Trieste
- 1977, visiting scientist at Imperial College, London
- 1977, the daughter Alex was born
- 1978, visiting scientist at the Chalmers Institute of Science and Technology, Gothenburg
- 1978–1979, visiting scientist at Imperial College, London
- 1979, visiting scientist at the University of Sussex
- 1980, visitor at CERN theory division
  ↪ collaboration with Kobayashi, Nicolai and Townsend (3 papers)
- 1981–1983, return to Takahashi's group in Edmonton, Canada
  ↪ beginning of the collaboration with Spallucci (30 papers)
- 1983–1986, assistant professor at the University of Toronto in Mississauga
- 1986–1989, assistant professor at California State Polytechnic University, Pomona
- 1987–1989, visiting scholar at SLAC, Stanford University, Stanford
- 1989, tenured professorship position at California State Polytechnic University, Pomona
- 1991, honorable mention at the Gravity Research Foundation essay competition
- 2010-Oct, last visit at the University of Trieste and ICTP
- 2012, professor emeritus at California State Polytechnic University, Pomona
- 2015-Jul, retirement to Milwaukie, Oregon
- 2016-Jan, last visit at Department of Physics and Astronomy, California State Polytechnic University, Pomona
- 2017-May, final peer review publication
- 2018-Sep, return to the Gulf of Naples for good (see Fig. 15).

Fig. 15 View of the Gulf of Naples. Courtesy of Salvatore Samuele Sirletti. All rights reserved

Acknowledgements For their help, support and source of information, PN is grateful to the family of Antonio Aurilia. In particular PN thanks Antonio's wife Elizabeth, his daughter Alex as well as his sisters Anna Maria, Elena and Liliana. PN thanks Euro Spallucci, Aurilia's closest collaborator, for the time dedicated in describing the atmosphere he lived while working with Aurilia. PN is grateful to the late Franco Legovini for details about his collaborations with Aurilia and Christodoulou. PN is grateful to Roberto Balbinot for information about the INFN in Bologna and his relation with Aurilia. PN is grateful to William Pezzaglia for information about his discussions with Aurilia about Clifford algebra. PN is grateful to Nathan Tung for sharing his experience as Aurilia's research student at Cal Poly. PN is grateful to Athanasios Tzikas and Salvatore S. Sirletti for the pictures in this chapter. PN is grateful to Jonas Mureika and Douglas Singleton for proofreading the text. PN is grateful to Springer-Nature and the other contributors of the present book dedicated to Aurilia for their patience and long wait. The work of PN was partially supported by GNFM, Italy's National Group for Mathematical Physics.

References

1. S. W. Hawking, Phys. Lett. B **134**, 403 (1984). https://doi.org/10.1016/0370-2693(84)91370-4
2. S. Ansoldi, A. Aurilia, E. Spallucci, Chaos Solitons Fractals **10**, 197 (1999). https://doi.org/10.1016/S0960-0779(98)00115-5
3. A. Aurilia, Theory of high spin fields. Laurea thesis, Università degli Studi di Napoli Federico II, Naples, Italy, 1966
4. A. Aurilia, H. Umezawa, Nuovo Cimento A Serie **51**(1), 14 (1967). https://doi.org/10.1007/BF02739983
5. A. Aurilia, H. Umezawa, Phys. Rev. **182**, 1682 (1969). https://doi.org/10.1103/PhysRev.182.1682

6. Fulbright Scholar Directory, https://cies.org/fulbright-scholar-directory
7. Graduate Programs in Physics, University of Wisconsin–Milwaukee, https://uwm.edu/physics/graduate/academic-program/
8. S. Ansoldi, A. Aurilia, R. Balbinot, E. Spallucci, Phys. Essays **9**, 556 (1996). https://doi.org/10.4006/1.3029270
9. A. Aurilia, Y. Takahashi, H. Umezawa, Lett. Nuovo Cim. **1S2**, 603 (1971). https://doi.org/10.1007/BF02770237
10. A. Aurilia, Y. Takahashi, H. Umezawa, Prog. Theor. Phys. **48**, 290 (1972). https://doi.org/10.1143/PTP.48.290
11. A. Aurilia, Self consistency in the theory of quantified fields. PhD thesis, University of Wisconsin–Milwaukee, Milwaukee, Wisconsin, USA, 1970
12. H. S. Leff, in *Touring the Planck Scale: Antonio Aurilia Memorial Volume*, ed. by P. Nicolini. Fundamental Theories of Physics vol. 219 (Springer-Nature, 2025)
13. K. Lam, in *Touring the Planck Scale: Antonio Aurilia Memorial Volume*, ed. by P. Nicolini. Fundamental Theories of Physics vol. 219 (Springer-Nature, 2025)
14. A. Aurilia, Y. Takahashi, H. Umezawa, Phys. Rev. D **5**, 851 (1972). https://doi.org/10.1103/PhysRevD.5.851
15. A. Aurilia, Y. Takahashi, N.J. Papastamatiou, H. Umezawa, Phys. Rev. D **5**, 3066 (1972). https://doi.org/10.1103/PhysRevD.5.3066
16. A. Aurilia, Y. Takahashi, Phys. Rev. D **6**, 2294 (1972). https://doi.org/10.1103/PhysRevD.6.2294
17. A. Aurilia, F. Rohrlich, Am. J Phys. **43**, 261 (1975). https://doi.org/10.1119/1.10066
18. A. Aurilia, F. Rohrlich, Action at a distance quantum field theory (1974), Unpublished Paper
19. A. Aurilia, F. Rohrlich, Model for unified strong and electromagnetic interactions (1974). Internal Report, University of Syracuse
20. Abdus Salam International Centre for Theoretical Physics (ICTP), https://www.ictp.it/about-ictp/mission-history.aspx
21. Theoretical Physics Group at Imperial College London, https://www.imperial.ac.uk/theoretical-physics/about-us/history/
22. J. Strathdee, Y. Takahashi, Nucl. Phys. **8**, 113 (1958). https://doi.org/10.1016/0029-5582(58)90138-X
23. J. Strathdee, Y. Takahashi, Nucl. Phys. **9**(3), 558 (1958). https://doi.org/10.1016/0029-5582(58)90389-4
24. W. C. Eells, A.C. Cleveland, J. High. Educ. **6**, 261 (1935). https://doi.org/10.1119/1.10066
25. Istituto Nazionale di Fisica Nucleare, https://en.wikipedia.org/wiki/Istituto_Nazionale_di_Fisica_Nucleare
26. Cinquant'anni di ricerca e sviluppo, https://agenda.infn.it/event/30154/timetable/?view=standard
27. A. Aurilia, Nucl. Phys. B **92**, 241 (1975). https://doi.org/10.1016/0550-3213(75)90181-9
28. W. K. H. Panofsky, SLAC Beam Line **27N1**, 36 (1999)
29. D. Rickles, *A Brief History of String Theory*. (Springer, Berlin Heidelberg, 2014). https://doi.org/10.1007/978-3-642-45128-7
30. M. Gell-Mann, Phys. Lett. **8**, 214 (1964). https://doi.org/10.1016/S0031-9163(64)92001-3
31. G. Zweig, An SU(3) model for strong interaction symmetry and its breaking. Version 1 (1964). CERN-TH-401, CERN Preprint
32. G. Zweig, in *Developments in the quark theory of Hardons*, ed. by D.B. Lichtenberg, S.P. Rosen (Hadronic Press, Nonantum, Massachusetts, USA, 1964), pp. 22–101
33. E. Spallucci, Introduction to string theory: a guided tour from elementary particles to fundamental strings (2020). Unpublished paper
34. G. F. Chew, S.C. Frautschi, Phys. Rev. Lett. **7**, 394 (1961). https://doi.org/10.1103/PhysRevLett.7.394
35. R. Dolen, D. Horn, C. Schmid, Phys. Rev. Lett. **19**, 402 (1967). https://doi.org/10.1103/PhysRevLett.19.402

67. F. Lund, T. Regge, Phys. Rev. D **14**, 1524 (1976). https://doi.org/10.1103/PhysRevD.14.1524
68. J. S. Schwinger, Phys. Rev. **125**, 397 (1962). https://doi.org/10.1103/PhysRev.125.397
69. A. Aurilia, G. Denardo, F. Legovini, E. Spallucci, Phys. Lett. B **147**, 258 (1984). https://doi.org/10.1016/0370-2693(84)90112-6
70. Department of Physics and Astronomy, California State Polytechnic University, Pomona, https://www.cpp.edu/~aaurilia/site/euro.html
71. A. Aurilia, M. Kobayashi, Y. Takahashi, Phys. Rev. D **22**, 1368 (1980). https://doi.org/10.1103/PhysRevD.22.1368
72. D. Z. Freedman, P. van Nieuwenhuizen, S. Ferrara, Phys. Rev. D **13**, 3214 (1976). https://doi.org/10.1103/PhysRevD.13.3214
73. G. Velo, D. Zwanziger, Phys. Rev. **186**, 1337 (1969). https://doi.org/10.1103/PhysRev.186.1337
74. A. Aurilia, H. Nicolai, P.K. Townsend, Nucl. Phys. B **176**, 509 (1980). https://doi.org/10.1016/0550-3213(80)90466-6
75. M. Dine, in *in Proceedings of the Theoretical Advanced Study Institute in Elementary Particle Physics (TASI 2000): Flavor Physics for the Millennium* ed. by J. L. Rosner (World Scientific, 2001), pp. 349–369
76. P. K. in *Touring the Planck Scale: Antonio Aurilia Memorial Volume*, ed. by P. Nicolini. Fundamental Theories of Physics vol. 219 (Springer-Nature, 2025)
77. S. W. Hawking, *A Brief History of Time: From the Big Bang to Black Holes* (Bantam Press, London, UK, 1988)
78. E. Cremmer, B. Julia, J. Scherk, Phys. Lett. B **76**, 409 (1978). https://doi.org/10.1016/0370-2693(78)90894-8
79. E. Witten, Nucl. Phys. B **186**, 412 (1981). https://doi.org/10.1016/0550-3213(81)90021-3
80. B. de Wit, in *Unity from Duality: Gravity, Gauge Theory and Strings*, ed. by C. Bachas, A. Bilal, M. Douglas, N. Nekrasov, F. David. Les Houches - Ecole d'Ete de Physique Theorique vol.76 (EDP Sciences; Springer-Verlag, 2002)
81. A. Aurilia, Y. Takahashi, P.K. Townsend, Phys. Lett. B **95**, 265 (1980). https://doi.org/10.1016/0370-2693(80)90484-0
82. G. 't Hooft, Phys. Rept. **142**, 357 (1986). https://doi.org/10.1016/0370-1573(86)90117-1
83. European Commission, Remuneration of Researchers in the Public and Private Commercial Sectors (2007). Final Report
84. A. Aurilia, Y. Takahashi, Phys. Rev. D **23**, 1752 (1981). https://doi.org/10.1103/PhysRevD.23.1752
85. A. Aurilia, Y. Takahashi, Prog. Theor. Phys. **66**, 693 (1981). https://doi.org/10.1143/PTP.66.693
86. A. Aurilia, Y. Takahashi, Bosonization and confinement: from two-dimensions to four-dimensions (1981). Unpublished Paper
87. S. Deser, P.K. Townsend, W. Siegel, Nucl. Phys. B **184**, 333 (1981). https://doi.org/10.1016/0550-3213(81)90222-4
88. A. Aurilia, M. Martellini, Phys. Lett. B **123**, 303 (1983). https://doi.org/10.1016/0370-2693(83)91205-4
89. https://www.uninsubria.it/hpp/maurizio.martellini
90. G. 't Hooft, in *Conference on Highlights of Particle and Condensed Matter Physics (SALAMFEST)*. Conf. Proc. C930308 (World Scientific, 1993), pp. 284–296
91. J. Mureika, in *Touring the Planck Scale: Antonio Aurilia Memorial Volume*, ed. by P. Nicolini. Fundamental Theories of Physics vol. 219 (Springer-Nature, 2025)
92. S. Weinberg, in *Understanding the Fundamental Constituents of Matter*, ed. by A. Zichichi. The Subnuclear Series vol. 14, (Springer New York, NY, 1978), pp. 1–52
93. S. Weinberg, in *General Relativity: An Einstein centenary survey*, ed. by S.W. Hawking, W. Israel (Cambridge University Press, 1979), pp. 790–831
94. S. W. MacDowell, F. Mansouri, Phys. Rev. Lett. **38**, 739 (1977). https://doi.org/10.1103/PhysRevLett.38.739. [Erratum: Phys. Rev. Lett. **38**, 1376 (1977)]

36. R. Dolen, D. Horn, C. Schmid, Phys. Rev. **166**, 1768 (1968). https://doi.org/10.1103/Phys 166.1768
37. G. Veneziano, Nuovo Cim. A **57**, 190 (1968). https://doi.org/10.1007/BF02824451
38. Y. Nambu, in *Proceedings of the International Conference on Symmetries and Quark Mode* ed. by R. Chand (Gordon and Breach, New York, NY, USA, 1969)
39. H. Nielsen, An almost physical interpretation of the dual N point function (1969). Nordi Preprint
40. L. Susskind, Phys. Rev. D **1**, 1182 (1970). https://doi.org/10.1103/PhysRevD.1.1182
41. E. D. Bloom, D. H. Coward, H. Destaebler, J. Drees, G. Miller, L. W. Mo, R. E. Taylor, M. Breidenbach, J. I. Friedman, G. C. Hartmann, H. W. Kendall, Phys. Rev. Lett. **23**(16), 930 (1969). https://doi.org/10.1103/PhysRevLett.23.930
42. M. Breidenbach, J. I. Friedman, H. W. Kendall, E. D. Bloom, D. H. Coward, H. Destaebler, J. Drees, L. W. Mo, R. E. Taylor, Phys. Rev. Lett. **23**(16), 935 (1969). https://doi.org/10.1103/PhysRevLett.23.935
43. R. P. Feynman, Phys. Rev. Lett. **23**, 1415 (1969). https://doi.org/10.1103/PhysRevLett.23.1415
44. D. J. Gross, F. Wilczek, Phys. Rev. Lett. **30**, 1343 (1973). https://doi.org/10.1103/PhysRevLett.30.1343
45. H. Fritzsch, M. Gell-Mann, H. Leutwyler, Phys. Lett. B **47**, 365 (1973). https://doi.org/10.1016/0370-2693(73)90625-4
46. J. J. Aubert et al., Phys. Rev. Lett. **33**, 1404 (1974). https://doi.org/10.1103/PhysRevLett.33.1404
47. J. E. Augustin et al., Phys. Rev. Lett. **33**, 1406 (1974). https://doi.org/10.1103/PhysRevLett.33.1406
48. S. W. Hawking, Nature **248**, 30 (1974). https://doi.org/10.1038/248030a0
49. S. W. Hawking, Comm. Math. Phys. **43**, 199 (1975). https://doi.org/10.1007/BF02345020
50. J. D. Bekenstein, Phys. Rev. D **7**, 2333 (1973). https://doi.org/10.1103/PhysRevD.7.2333
51. B. J. Carr, S. W. Hawking, Mon. Not. Roy. Astron. Soc. **168**, 399 (1974)
52. B. J. Carr, G. F. R. Ellis, G. W. Gibbons, J. B. Hartle, T. Hertog, R. Penrose, M. J. Perry, K. S. Thorne, Biogr. Mems Fell. R. Soc. **66**, 267 (2019). https://doi.org/10.1098/rsbm.2019.0001
53. Marcel Grossmann Meetings on General Relativity, http://www.icra.it/MG/
54. A. Salam, J. A. Strathdee, Phys. Rev. D **18**, 4596 (1978). https://doi.org/10.1103/PhysRevD.18.4596
55. Y. Nambu, Phys. Rev. D **10**, 4262 (1974). https://doi.org/10.1103/PhysRevD.10.4262
56. H. B. Nielsen, P. Olesen, Nucl. Phys. B **61**, 45 (1973). https://doi.org/10.1016/0550-3213(73)90350-7
57. P. Nicolini. Instabilità del vuoto e creazione di coppie in un mezzo dissipativo. Laurea thesis, Università degli Studi di Trieste, Trieste, Italy, 1997
58. M. Kalb, P. Ramond, Phys. Rev. D **9**, 2273 (1974). https://doi.org/10.1103/PhysRevD.9.2273
59. A. Chodos, R.L. Jaffe, K. Johnson, C.B. Thorn, V.F. Weisskopf, Phys. Rev. D **9**, 3471 (1974). https://doi.org/10.1103/PhysRevD.9.3471
60. A. Aurilia, F. Legovini, Phys. Lett. B **67**, 299 (1977). https://doi.org/10.1016/0370-2693(77)90376-8
61. A. Aurilia, D. Christodoulou, Phys. Lett. B **71**, 90 (1977). https://doi.org/10.1016/0370-2693(77)90747-X
62. A. Aurilia, D. Christodoulou, F. Legovini, Phys. Lett. B **73**, 429 (1978). https://doi.org/10.1016/0370-2693(78)90757-8
63. A. Aurilia, D. Christodoulou, Phys. Lett. B **78**, 589 (1978). https://doi.org/10.1016/0370-2693(78)90646-9
64. A. Aurilia, Phys. Lett. B **81**, 203 (1979). https://doi.org/10.1016/0370-2693(79)90524-0
65. A. Aurilia, D. Christodoulou, J. Math. Phys. **20**, 1446 (1979). https://doi.org/10.1063/1.524228
66. A. Aurilia, D. Christodoulou, J. Math. Phys. **20**, 1692 (1979). https://doi.org/10.1063/1.524251

95. G. Denardo, R. Ruffini, Phys. Lett. B **45**, 259 (1973). https://doi.org/10.1016/0370-2693(73)90198-6

96. G. Denardo, L. Hively, R. Ruffini, Phys. Lett. B **50**, 270 (1974). https://doi.org/10.1016/0370-2693(74)90557-7

97. A. Aurilia, G. Denardo, F. Legovini, E. Spallucci, Nucl. Phys. B **252**, 523 (1985). https://doi.org/10.1016/0550-3213(85)90460-2

98. S. R. Coleman, F. De Luccia, Phys. Rev. D **21**, 3305 (1980). https://doi.org/10.1103/PhysRevD.21.3305

99. A. Vilenkin, Phys. Rev. D **27**, 2848 (1983). https://doi.org/10.1103/PhysRevD.27.2848

100. C. Sivaram, K. P. Sinha, Phys. Rept. **51**, 111 (1979). https://doi.org/10.1016/0370-1573(79)90037-1

101. S. M. Carroll, How to get tenure at a major research university (2012), https://www.discovermagazine.com/the-sciences/how-to-get-tenure-at-a-major-research-university

102. History of the Department of Physics, University of Toronto, https://www.physics.utoronto.ca/physics-at-uoft/history/

103. The Adjunct Crisis, https://adjunctcrisis.com/

104. J. C. Young, R. B. Townsend, The adjunct problem is a data problem (2021), https://www.amacad.org/news/adjunct-problem-data-problem

105. H. S. Leff, https://en.wikipedia.org/wiki/Harvey_S._Leff

106. R. S. Kissack, Heat flow and collisional energy exchange in Farley-Buneman waves. PhD thesis, University of Western Ontario, London, Ontario, Canada, 1993

107. N. Altamirano, E. Gould, N. Afshordi, R. B. Mann, in *in Touring the Planck Scale: Antonio Aurilia Memorial Volume*, ed. by P. Nicolini. Fundamental Theories of Physics vol. 219 (Springer-Nature, 2025)

108. R. B. Mann, Investigations of an alternative theory of gravitation. PhD thesis, University of Toronto, Toronto, Ontario, Canada, 1982

109. A. Aurilia, R.S. Kissack, R.B. Mann, E. Spallucci, Phys. Rev. D **35**, 2961 (1987). https://doi.org/10.1103/PhysRevD.35.2961

110. A. Aurilia, M. Palmer, E. Spallucci, Phys. Rev. D **40**, 2511 (1989). https://doi.org/10.1103/PhysRevD.40.2511

111. A. Aurilia, E. Spallucci, Phys. Rev. D **42**, 464 (1990). https://doi.org/10.1103/PhysRevD.42.464

112. A. Aurilia, E. Spallucci, Phys. Lett. B **251**, 39 (1990). https://doi.org/10.1016/0370-2693(90)90228-X

113. P. Nicolini, A. Smailagic, E. Spallucci, Phys. Lett. B **632**, 547 (2006). https://doi.org/10.1016/j.physletb.2005.11.004

114. P. Nicolini, Int. J. Mod. Phys. A **24**, 1229 (2009). https://doi.org/10.1142/S0217751X09043353

115. P. Nicolini, E. Spallucci, Class. Quant. Grav. **27**, 015010 (2010). https://doi.org/10.1088/0264-9381/27/1/015010

116. O. Klein, Z. Phys. **53**, 157 (1929). https://doi.org/10.1007/BF01339716

117. E. Farhi, A.H. Guth, J. Guven, Nucl. Phys. B **339**, 417 (1990). https://doi.org/10.1016/0550-3213(90)90357-J

118. R. Balbinot, A. Fabbri, in *Touring the Planck Scale: Antonio Aurilia Memorial Volume*, ed. by P. Nicolini. Fundamental Theories of Physics vol. 219 (Springer-Nature, 2025)

119. A. Aurilia, R. Balbinot, E. Spallucci, Phys. Lett. B **262**, 222 (1991). https://doi.org/10.1016/0370-2693(91)91558-D

120. A. Aurilia, F. Legovini, E. Spallucci, Phys. Lett. B **264**, 69 (1991). https://doi.org/10.1016/0370-2693(91)90705-U

121. A. Aurilia, E. Spallucci, The Dual Higgs mechanism and the origin of mass in the universe (1991). Unpublished essay, awarded with an "honorable mention" by the Gravity Research Foundation in the 1991 competition, where the authors first suggested a possible connection between Cosmological Constant and Dark Matter

122. A. Aurilia, E. Spallucci, Phys. Lett. B **282**, 50 (1992). https://doi.org/10.1016/0370-2693(92)90478-M

123. E. Spallucci, A. Smailagic, in *Touring the Planck Scale: Antonio Aurilia Memorial Volume*, ed. by P. Nicolini. Fundamental Theories of Physics vol. 219 (Springer-Nature, 2025)

124. A. Aurilia, A. Smailagic, E. Spallucci, Class. Quant. Grav. **9**, 1883 (1992). https://doi.org/10.1088/0264-9381/9/8/010

125. Y. B. Zeldovich, JETP Lett. **6**, 316 (1967)

126. A. D. Sakharov, Sov. Phys. Dokl. **12**, 1040 (1968)

127. A. Aurilia, A. Smailagic, E. Spallucci, Phys. Rev. D **47**, 2536 (1993). https://doi.org/10.1103/PhysRevD.47.2536

128. A. Aurilia, E. Spallucci, Class. Quant. Grav. **10**, 1217 (1993). https://doi.org/10.1088/0264-9381/10/7/004

129. A. Aurilia, E. Spallucci, I. Vanzetta, Phys. Rev. D **50**, 6490 (1994). https://doi.org/10.1103/PhysRevD.50.6490

130. E. Heller, S. Tomsovic, Phys. Today **48**(7), 38 (1993)

131. A. Aurilia, A. Smailagic, E. Spallucci, Phys. Rev. D **51**, 4410 (1995). https://doi.org/10.1103/PhysRevD.51.4410

132. S. Ansoldi, A. Aurilia, E. Spallucci, Phys. Rev. D **53**, 870 (1996). https://doi.org/10.1103/PhysRevD.53.870

133. T. Eguchi, Phys. Rev. Lett. **44**, 126 (1980). https://doi.org/10.1103/PhysRevLett.44.126

134. S. Ansoldi, A. Aurilia, E. Spallucci, Int. J. Mod. Phys. B **10**, 1695 (1996). https://doi.org/10.1142/S0217979296000775

135. S. Ansoldi, A. Aurilia, E. Spallucci, Phys. Rev. D **56**, 2352 (1997). https://doi.org/10.1103/PhysRevD.56.2352

136. L. F. Abbott, M. B. Wise, Am. J. Phys. **49**, 37 (1981). https://doi.org/10.1119/1.12657

137. S. Ansoldi, A. Aurilia, R. Balbinot, E. Spallucci, Class. Quant. Grav. **14**, 2727 (1997). https://doi.org/10.1088/0264-9381/14/10/004

138. S. Ansoldi, A. Aurilia, E. Spallucci, Eur. J. Phys. **21**, 1 (2000). https://doi.org/10.1088/0143-0807/21/1/301

139. S. Ansoldi, A. Aurilia, A. Smailagic, E. Spallucci, Phys. Lett. B **471**, 133 (1999). https://doi.org/10.1016/S0370-2693(99)01323-4

140. S. Ansoldi, A. Aurilia, L. Marinatto, E. Spallucci, Prog. Theor. Phys. **103**, 1021 (2000). https://doi.org/10.1143/PTP.103.1021

141. S. Ansoldi, A. Aurilia, E. Spallucci, Phys. Rev. D **64**, 025008 (2001). https://doi.org/10.1103/PhysRevD.64.025008

142. S. Ansoldi, A. Aurilia, C. Castro, E. Spallucci, Phys. Rev. D **64**, 026003 (2001). https://doi.org/10.1103/PhysRevD.64.026003

143. W. M. Pezzaglia, Jr., A Clifford algebra multivector reformulation of field theory. PhD thesis, University of California, Davis, Davis, California, USA, 1983

144. W. M. Pezzaglia, Jr., in *Clifford Algebras and their Applications in Mathematical Physics, Volume 1: Algebra and Physics*, ed. by R. Ablamowicz, B. Fauser (Birkhäuser, Boston, Massachusetts, USA, 2000)

145. M. Pavsic, *The Landscape of Theoretical Physics: A Global view. From Point Particles to the Brane World and Beyond, in Search of a Unifying Principle* (Kluwer, Dordrecht, Netherlands, 2001)

146. A. Aurilia, S. Ansoldi, E. Spallucci, Class. Quant. Grav. **19**, 3207 (2002). https://doi.org/10.1088/0264-9381/19/12/307

147. A. Aurilia, E. Spallucci, Planck's uncertainty principle and the saturation of Lorentz boosts by Planckian black holes (2013). Unpublished essay submitted to the Gravity Research Foundation for the 2002–03 competition. arXiv:1309.7186 [gr-qc]

148. A. Aurilia, E. Spallucci, Adv. High Energy Phys. **2013**, 531696 (2013). https://doi.org/10.1155/2013/531696

149. H. J. Treder, Found. Phys. **15**(2), 161 (1985). https://doi.org/10.1007/bf00735287

150. M. Maggiore, Phys. Lett. B **304**, 65 (1993). https://doi.org/10.1016/0370-2693(93)91401-8

151. A. Kempf, G. Mangano, R.B. Mann, Phys. Rev. D **52**, 1108 (1995). https://doi.org/10.1103/PhysRevD.52.1108
152. L. J. Garay, Int. J. Mod. Phys. A **10**, 145 (1995). https://doi.org/10.1142/S0217751X95000085
153. D. Amati, M. Ciafaloni, G. Veneziano, Phys. Lett. B **197**, 81 (1987). https://doi.org/10.1016/0370-2693(87)90346-7
154. D. Amati, M. Ciafaloni, G. Veneziano, Int. J. Mod. Phys. A **3**, 1615 (1988). https://doi.org/10.1142/S0217751X88000710
155. D. Amati, M. Ciafaloni, G. Veneziano, Phys. Lett. B **216**, 41 (1989). https://doi.org/10.1016/0370-2693(89)91366-X
156. G. Dvali, C. Gomez, Self-Completeness of Einstein Gravity (2010). Unpublished paper. arXiv:1005.3497 [hep-th]
157. G. Dvali, S. Folkerts, C. Germani, Phys. Rev. D **84**, 024039 (2011). https://doi.org/10.1103/PhysRevD.84.024039
158. G. Dvali, in *Touring the Planck Scale: Antonio Aurilia Memorial Volume*, ed. by P. Nicolini. Fundamental Theories of Physics vol. 219 (Springer-Nature, 2025)
159. B. Carr, in *1st Karl Schwarzschild Meeting on Gravitational Physics*, ed. by P. Nicolini, M. Kaminski, J. Mureika, M. Bleicher. Springer Proceedings in Physics, vol. 170 (Springer International Publishing, 2016), pp. 159–167
160. B. J. Carr, J. Mureika, P. Nicolini, JHEP **07**, 052 (2015). https://doi.org/10.1007/JHEP07(2015)052
161. B. J. Carr, in *Touring the Planck Scale: Antonio Aurilia Memorial Volume*, ed. by P. Nicolini. Fundamental Theories of Physics vol. 219 (Springer-Nature, 2025)
162. L. Susskind, Phys. Rev. D **49**, 6606 (1994). https://doi.org/10.1103/PhysRevD.49.6606
163. G. Amelino-Camelia, Nature **418**, 34 (2002). https://doi.org/10.1038/418034a
164. J. Magueijo, L. Smolin, Phys. Rev. Lett. **88**, 190403 (2002). https://doi.org/10.1103/PhysRevLett.88.190403
165. J. Magueijo, L. Smolin, Phys. Rev. D **67**, 044017 (2003). https://doi.org/10.1103/PhysRevD.67.044017
166. E. Aiken, M. Bishop, D. Singleton, in *Touring the Planck Scale: Antonio Aurilia Memorial Volume*, ed. by P. Nicolini. Fundamental Theories of Physics vol. 219 (Springer-Nature, 2025)
167. A. Aurilia, E. Spallucci, Phys. Rev. D **69**, 105004 (2004). https://doi.org/10.1103/PhysRevD.69.105004
168. A. Aurilia, E. Spallucci, Phys. Rev. D **69**, 105005 (2004). https://doi.org/10.1103/PhysRevD.69.105005
169. Department of Physics and Astronomy, California State Polytechnic University, Pomona, https://www.cpp.edu/sci/physics-astronomy/people/alumni/nathan-tung.shtml
170. Department of Physics and Astronomy, California State Polytechnic University, Pomona, https://www.cpp.edu/sci/physics-astronomy/people/index.shtml
171. UCLA, Physics and Astronomy, https://www.pa.ucla.edu/researchers.html
172. Uloop Cal Poly Pomona, https://csupomona.uloop.com/professors/view.php/41710/Antonio-Aurilia
173. Rate my professors, https://www.ratemyprofessors.com/professor?tid=1324215
174. Docsity, https://www.docsity.com/en/professors/antonio-aurilia/reviews/
175. A. Aurilia, P. Gaete, E. Spallucci, Chern-Simons-Schwinger model of confinement in QCD (2015). Unpublished paper. arXiv:1504.05810 [hep-th]
176. A. Aurilia, P. Gaete, J.A. Helayël-Neto, E. Spallucci, EPL **117**(6), 61001 (2017). https://doi.org/10.1209/0295-5075/117/61001
177. P. Gaete, J.A. Helayël-Neto, in *Touring the Planck Scale: Antonio Aurilia Memorial Volume*, ed. by P. Nicolini. Fundamental Theories of Physics vol. 219 (Springer-Nature, 2025)
178. G. Gabadadze, Phys. Rev. D **58**, 094015 (1998). https://doi.org/10.1103/PhysRevD.58.094015
179. A. G. Tzikas, P. Nicolini, J. Mureika, B. Carr, JCAP **1812**(12), 033 (2018). https://doi.org/10.1088/1475-7516/2018/12/033

180. K. F. Dialektopoulos, P. Nicolini, A.G. Tzikas, JCAP **10**, 008 (2020). https://doi.org/10.1088/1475-7516/2020/10/008
181. A. Aurilia, H. Nicolai, P.K. Townsend, in *Nuffield Workshop on Superspace and Supergravity*, ed. by S. Hawking, M. Roček (Cambridge University Press, Cambridge, UK, 1981), pp. 403–412. Invited talk
182. A. Aurilia, M. Martellini, in *10th International Conference on General Relativity and Gravitation (Vol. 2)*, ed. by B. Bertotti, F. de Felice, A. Pascolini (Consiglio Nazionale delle Ricerche, Rome, Italy, 1984)
183. A. Aurilia, G. Denardo, F. Legovini, E. Spallucci, in *Advanced Series in Astrophysics and Cosmology: Galaxies, Quasars and Cosmology*, vol. 2, ed. by L.Z. Fang, R. Ruffini (World Scientific, Singapore, 1985), pp. 181–200. https://doi.org/10.1142/9789814415354_0010. Invited talk presented at the "First Equatorial School on Relativistic Astrophysics and the Large Scale of the Universe", Bogotá, 12–24 February 1984
184. A. Aurilia, R.S. Kissack, G. Denardo, E. Spallucci, in *Proceedings of the International Symposium in Honour of Hiroomi Umezawa, Positano, Salerno, Italy, 5–7 June 1985*, ed. by F. Mancini (North-Holland, Amsterdam, Netherlands, 1986). Invited paper
185. A. Aurilia, E. Spallucci, in *Trieste Conference on Supermembranes and Physics in 2+1 Dimensions.* ed. by M.J. Duff, C.N. Pope, E. Sezgin (World Scientific, Singapore, 1990), pp. 29–39
186. A. Aurilia, F. Legovini, E. Spallucci, in *Particles & Fields 91: Meeting of the Division of Particles & Fields of the APS*, ed. by D. Axen, D. Bryman, M. Comyn (World Scientific, Singapore, 1992)
187. A. Aurilia, Phys. Persp. **6**, 478 (2003). https://doi.org/10.1007/s00016-004-0230-2. Review of the book "Volta: Science and culture in the age of Enlightenment", authored by Giuliano Pancaldi, Princeton University Press, 2003
188. A. Aurilia, Equivalent field theories and the bound state problem (1974). Lecture Notes, Syracuse University
189. A. Aurilia, Dynamical symmetry breaking in gauge theories: superconductivity and particle physics (1974). Internal Report, University of Syracuse
190. S. Ansoldi, A. Aurilia, E. Spallucci, Fluctuating quantum dimensions and p-brane unification (2001). Unpublished essay submitted to the Gravity Research Foundation for the 2001 competition
191. A. Aurilia, Dynamical symmetry breaking and manifest covariance in gauge theories (1974). Unpublished paper
192. ResearchGate, https://www.researchgate.net/profile/Antonio-Aurilia
193. J. E. Hirsch, Proc. Nat. Acad. Sci. **102**(46), 16569 (2005). https://doi.org/10.1073/pnas.0507655102
194. AAIH. INSPIRE HEP, https://inspirehep.net/authors/1017871
195. AAADS. SAO/NASA Astrophysics Data System (ADS), https://ui.adsabs.harvard.edu/
196. AAADS. Semantic Scholar, https://www.semanticscholar.org/author/A.-Aurilia/66580287
197. Shoichi Sakata, https://en.wikipedia.org/wiki/Shoichi_Sakata
198. Hermann Nicolai, https://en.wikipedia.org/wiki/Hermann_Nicolai
199. Paul Townsend, https://en.wikipedia.org/wiki/Paul_Townsend
200. J. R. Mureika, E. Spallucci, Phys. Lett. B **693**, 129 (2010). https://doi.org/10.1016/j.physletb.2010.08.025
201. A. M. Frassino, D. Kubiznak, R. B. Mann, F. Simovic, JHEP **09**, 080 (2014). https://doi.org/10.1007/JHEP09(2014)080
202. A. M. Frassino, R. B. Mann, J. R. Mureika, Phys. Rev. D **92**(12), 124069 (2015). https://doi.org/10.1103/PhysRevD.92.124069
203. L. Modesto, P. Nicolini, Phys. Rev. D **81**, 104040 (2010). https://doi.org/10.1103/PhysRevD.81.104040
204. P. Nicolini, E. Spallucci, Adv. High Energy Phys. **2014**, 805684 (2014). https://doi.org/10.1155/2014/805684
205. Sayre Fire, https://en.wikipedia.org/wiki/Sayre_Fire

206. C. A. La Porta, S. Zapperi, Nature Italy (2022). https://doi.org/10.1038/d43978-022-00163-5, https://www.nature.com/articles/d43978-022-00163-5
207. Marchionne e il suo discorso a due facce all'università Bocconi (2012), https://www.milanofinanza.it/news/marchionne-e-il-suo-discorso-a-due-facce-all-universita-bocconi
208. The real sick man of Europe (2005), https://www.economist.com/leaders/2005/05/19/the-real-sick-man-of-europe
209. Italy: The sick man of Europe (2008), https://web.archive.org/web/20080416074329/, http://www.telegraph.co.uk/opinion/main.jhtml?xml=/opinion/2008/04/15/dl1502.xml
210. A. Aurilia, E. Spallucci, Fundamentals of Planckian physics (2001). Unpublished Paper
211. I. Stanley-Becker. Stephen Hawking feared race of 'superhumans' able to manipulate their own DNA (2018), https://www.washingtonpost.com/news/morning-mix/wp/2018/10/15/stephen-hawking-feared-race-of-superhumans-able-to-manipulate-their-own-dna/
212. S. Clark. Artificial intelligence could spell end of human race—Stephen Hawking (2014), https://www.theguardian.com/science/2014/dec/02/stephen-hawking-intel-communication-system-astrophysicist-software-predictive-text-type
213. S. Hawking, *Brief Answers to the Big Questions* (Bantam, London, UK, 2018)
214. D. H. Meadows, D. L. Meadows, J. Randers, W. W. Behrens III., *The Limits to Growth; A Report for the Club of Rome's Project on the Predicament of Mankind* (Universe Books, New York, 1972)
215. T. D. Team, A study finds nearly half of jobs are vulnerable to automation (2018), https://www.economist.com/graphic-detail/2018/04/24/a-study-finds-nearly-half-of-jobs-are-vulnerable-to-automation
216. R. Srinivasan, Opinion: automation is likely to eliminate nearly half our jobs in the next 25 years. Here's what to do (2019), https://www.latimes.com/opinion/story/2019-10-29/opinion-automation-is-likely-to-eliminate-40-of-jobs-in-the-next-25-years-heres-what-we-can-do-a
217. G. Press, Data science: what's the half-life of a buzzword? (2013), https://www.forbes.com/sites/gilpress/2013/08/19/data-science-whats-the-half-life-of-a-buzzword/?sh=1953b4c87bfd
218. Foghlam Consulting LLC, A push button education: the role of AI in higher education (2018), https://www.foghlamconsulting.com/research/2018/7/23/a-push-button-education-the-role-of-ai-in-higher-education
219. L. Jong-Wha, Education in the age of automation (2018), https://www.japantimes.co.jp/opinion/2018/09/17/commentary/world-commentary/education-age-automation/
220. F. MacDonald, 8 scientific papers that were rejected before going on to win a Nobel prize (2016), https://www.sciencealert.com/these-8-papers-were-rejected-before-going-on-to-win-the-nobel-prize
221. R. P. Feynman, J. Robbins, F. J. Dyson, *The Pleasure of Finding Things Out: The Best Short Works of Richard P Feynman* (Perseus Books, USA, 1999)
222. M. Polanyi, *The Logic of Liberty: Reflections and Rejoinders* (Routledge and K, Paul, 1951)
223. W. McGucken, Minerva **16**, 42 (1978). https://doi.org/10.1007/BF01102181, https://www.studentenwerke.de/
224. K. H. Günther, Prospects **18**, 127 (1988). https://doi.org/10.1007/BF02192965, https://link.springer.com/article/10.1007/BF02192965
225. K. Vaesen1, J. Katzav, PLoS One **12**, e0183967 (2017). https://doi.org/10.1371/journal.pone.0183967
226. S. Moore, C. Neylon, M.P. Eve, D.P. O'Donnell, D. Pattinson, Palg. Comm. (2017) https://doi.org/10.1057/palcomms.2016.105, https://www.nature.com/articles/palcomms2016105
227. Science needs to redefine excellence (2018). https://doi.org/10.1038/d41586-018-02183-y, https://www.nature.com/articles/d41586-018-02183-y
228. W. G. G. Benda, T. C. E. Engels, Int. J. Forecast. **27**, 166 (2011). https://doi.org/10.1016/j.ijforecast.2010.03.003, https://www.sciencedirect.com/science/article/abs/pii/S0169207010000439

229. J. M. Nicholson, J. P. A. Ioannidis, Nature **492**, 34 (2012). https://doi.org/10.1038/492034a, https://www.nature.com/articles/492034a

230. R. K. Merton, Science **159**, 56 (1968). https://doi.org/10.1126/science.159.3810.56, https://www.science.org/doi/10.1126/science.159.3810.56

231. T. Bol, M. de Vaan, A. van de Rijt, PNAS **115**, 4887 (2018). https://doi.org/10.1073/pnas.1719557115, https://www.pnas.org/doi/full/10.1073/pnas.1719557115

232. H. Horta, Minerva **60**, 593 (2022). https://doi.org/10.1007/s11024-022-09469-6, https://link.springer.com/article/10.1007/s11024-022-09469-6

233. R. Gordon, B. J. Poulin, Account. Res. **16**, 13 (2009). https://doi.org/10.1080/08989620802689821

234. D. L. Herbert, A. G. Barnett, N. Graves, Nature **495**, 314 (2013). https://doi.org/10.1038/495314d, https://www.nature.com/articles/495314d

235. M. Hartmann, DSW-Journal **3** (2019), https://www.studentenwerke.de/

236. D. Matthews, German excellence strategy 'risks creating closed social elite' (2019), https://www.timeshighereducation.com/news/german-excellence-strategy-risks-creating-closed-social-elite

237. S. Baker, Do university excellence initiatives work? (2020), https://www.timeshighereducation.com/features/do-university-excellence-initiatives-work-node-comments

238. M. Anderson, E. Ronning, R. De Vries et al., Sci. Eng. Ethics **13**, 437 (2007). https://doi.org/10.1007/s11948-007-9042-5

239. B. R. Martin, Res. Pol. **42**, 1005 (2013). https://doi.org/10.1016/j.respol.2013.03.011

240. A. W. Wilhite, E. A. Fong, Science **335**(6068), 542 (2012). https://doi.org/10.1126/science.1212540, https://www.science.org/doi/abs/10.1126/science.1212540

241. Credit where credit's due (2006). https://doi.org/10.1038/440591a, https://www.nature.com/articles/440591a

242. E. A. Fong, A. W. Wilhite, PLoS One **12**, e0187394 (2017). https://doi.org/10.1371/journal.pone.0187394

243. D. S. Chawla, Nature (2023). https://doi.org/10.1038/d41586-023-00016-1, https://www.nature.com/articles/d41586-023-00016-1

244. J. P. A. Ioannidis, J. Baas, R. Klavans, K. W. Boyack, PLoS Biol. **17**, e3000384 (2019). https://doi.org/10.1371/journal.pbio.3000384

245. H. Rudolph, U. Brandt, B. Martin. Research misconduct (2013), https://www.elsevier.com/connect/editors-update/research-misconduct

246. J. Brainard, Fake scientific papers are alarmingly common (2023), https://www.science.org/content/article/fake-scientific-papers-are-alarmingly-common

247. B. A. Sabel, E. Knaack, G. Gigerenzer, M. Bilc, medRxiv (2023), https://doi.org/10.1101/2023.05.06.23289563, https://www.medrxiv.org/content/early/2023/05/08/2023.05.06.23289563

248. C. Ferguson, A. Marcus, I. Oransky, Nature **515**, 480 (2014). https://doi.org/10.1038/515480a

249. D. Fanelli, PLoS One **4**, e5738 (2009). https://doi.org/10.1371/journal.pone.0005738

250. The scandal of researchers paid less than a living wage (2022). https://doi.org/10.1038/d41586-022-03472-3, https://www.nature.com/articles/d41586-022-03472-3

251. M. Spooner, The ugly side of performance-based funding for universities (2021), https://academicmatters.ca/the-ugly-side-of-performance-based-funding-for-universities-2/

252. J. Bollen, D. Crandall, D. Junk, Y. Ding, K. Börner, EMBO Rep. **15**, 131 (2014). https://doi.org/10.1002/embr.201338068

253. R. Roy, Science. Technol. Hum. Values **10**, 73 (1985). https://doi.org/10.1177/016224398501000309

254. R. N. Germain, Cell **161**, 1485 (2015). https://doi.org/10.1016/j.cell.2015.05.052

255. J. P. A. Ioannidis, Nature **477**, 529 (2011). https://doi.org/10.1038/477529a

256. F. Fang, A. Casadevall, mBio **7**, e00422 (2016). https://doi.org/10.1128/mBio.00422-16

257. R. F. Service, Science **299**, 31 (2003). https://doi.org/10.1126/science.299.5603.31b

258. C. Seife, Science **297**, 313 (2002). https://doi.org/10.1126/science.297.5580.313

259. D. Butler, Nature **420**, 5 (2002). https://doi.org/10.1038/420005a
260. Q. Schiermeier, Nature **456**, 432 (2008). https://doi.org/10.1038/456432a
261. F. A. Houle, K. P. Kirby, M. P. Marder, Phys. Today **76**, 28 (2023). https://doi.org/10.1063/PT.3.5156
262. Physical Science Study Committee, *Physics* (D.C, Heath and Company, 1960)
263. List of unsolved problems in physics, https://en.wikipedia.org/wiki/List_of_unsolved_problems_in_physics
264. F. Heil, Was Physiker verdienen: Gehalt und Einflussfaktoren (2020), https://www.academics.de/ratgeber/gehalt-physiker
265. F. Heil, Von Beruf Physiker: Berufsbild, Voraussetzungen, Jobaussichten (2022), https://www.academics.de/ratgeber/physik-berufsaussichten
266. J. McGrath, *Report on Labour Shortages and Surpluses* (2021), https://www.ela.europa.eu/en/media/725
267. D. Cyranoski, N. Gilbert, H. Ledford, A. Nayar, M. Yahia, Nature **472**, 276 (2011). https://doi.org/10.1038/472276a, https://www.nature.com/articles/472276a
268. A. McCook, Nature **472**, 280 (2011). https://doi.org/10.1038/472280a. https://www.nature.com/articles/472280a
269. J. Beall, Nature **489**, 179 (2012). https://doi.org/10.1038/489179a
270. G. Brumfiel, Nature (2001). https://doi.org/10.1038/news.2011.554
271. C. Moskowitz, Gravity waves from big bang detected (2014), https://www.scientificamerican.com/article/gravity-waves-cmb-b-mode-polarization/
272. B. Cronin, J. Am. Soc. Inf. Sci. Technol. **52**, 558 (2001). https://doi.org/10.1002/asi.1097
273. D. Castelvecchi, Nature (2015). https://doi.org/10.1038/nature.2015.17567, https://www.nature.com/articles/nature.2015.17567
274. D. S. Chawla, Nature (2019). https://doi.org/10.1038/d41586-019-03862-0, https://www.nature.com/articles/d41586-019-03862-0
275. B. Nogrady, Nature (2023). https://doi.org/10.1038/d41586-023-00575-3, https://www.nature.com/articles/d41586-023-00575-3
276. J. Maienschein, Science **281**, 917 (1998). https://doi.org/10.1126/science.281.5379.917
277. N. Augustine, Science **279**, 1640 (1998). https://doi.org/10.1126/science.279.5357.1640
278. A. Comte, Cours de philosophie positive bachelier (1830)
279. S. Cole, Am. J. Sociol. **89**, 111 (1983). https://doi.org/10.1126/science.281.5379.917
280. D. K. Simonton, Rev. Gen. Psychol. **10**(2), 98 (2006)
281. H. Nicolai, in *General Relativity, Cosmology and Astrophysics*, ed. by J. Bičák, J., T. Ledvinka. Fundamental Theories of Physics, vol. 177 (Springer Cham, 2014). https://doi.org/10.1007/978-3-319-06349-2_18
282. C. Jungnickel, R. McCormmach, *The Second Physicist: On the History of Theoretical Physics in Germany* (Springer International Publishing, Switzerland, 2017)
283. E. Amaldi. I miei giorni con Fermi (1986)
284. E. Zaslow, Physmatics (2005). Unpublished Paper
285. G. W. Moore, Physical mathematics and the future (2014), https://www.physics.rutgers.edu/gmoore/PhysicalMathematicsAndFuture.pdf
286. A. Jaffe, F. Quinn, Bull. Am. Math. Soc. **29**(1), 1 (1993)
287. M. Atiyah, A. Borel, G.J. Chaitin, D. Friedan, J. Glimm, J.J. Gray, M.W. Hirsch, S. Mac Lane, B.B. Mandelbrot, D. Ruelle, et al., Bull. Am. Math. Soc. **30**(2), 178 (1994)
288. S. Hossenfelder, Quanta Mag. (2018), https://www.quantamagazine.org/the-end-of-theoretical-physics-as-we-know-it-20180827/
289. H. Bethe, Annalen der Physik **397**(3), 325 (1930)
290. M. Gnida, AI learns physics to optimize particle accelerator performance (2021), https://www6.slac.stanford.edu/news/2021-07-29-ai-learns-physics-optimize-particle-accelerator-performance
291. I. Asimov, in *Scientists Confront Velikovsky*, ed. by D. Goldsmith (Cornell University Press, Ithaca, NY, USA, 1977), pp. 7–18

292. M. Planck, in *Scientific Autobiography and Other Papers* (Greenwood Press Publishers, West-port, Connecticut, 1968). Translated from German by Frank Gaynor, Section: A Scientific Autobiography, Start Page 13, Quote Page 33 and 34
293. M. S. Morris, K. S. Thorne, Am. J. Phys. **56**, 395 (1988). https://doi.org/10.1119/1.15620
294. R. Garattini, Eur. Phys. J. C **79**(11), 951 (2019). https://doi.org/10.1140/epjc/s10052-019-7468-y

Stringy Matter and Planckian Black Holes

The Compton-Schwarzschild Correspondence and Higher Dimensional Black Holes: Linking Quantum Theory and General Relativity

Bernard J. Carr

Abstract Black holes are important in linking microphysics with macrophysics, and those of the Planck mass must play a fundamental role in any marriage of quantum theory and general relativity. In particular, the Black Hole Uncertainty Principle correspondence posits a smooth transition between the Compton and Schwarzschild scales as a function of mass. This suggests that there is a duality between elementary particles below the Planck mass and black holes above it, with elementary particles being interpreted as sub-Planckian black holes. Higher dimensions may also elucidate the connection between quantum and classical physics, with higher-dimensional black holes relating to the birth of the Universe and the embedding space of general relativity providing a quasi-classical interpretation of some features of quantum theory. This may also elucidate the problem of the passage of time.

1 Black Holes in Microphysics and Macrophysics

The triumph of physics in understanding the multitude of structures in microphysics and macrophysics is beautifully summarised in Fig. 1. This shows the Cosmic Uroborus (the snake eating its own tail), with the various scales of structure in the Universe indicated along the side. It can be regarded as a sort of "clock" in which the scale changes by a factor of 10 for each minute—from the Planck scale at the top left to the scale of the observable universe at the top right. The head meets the tail at the Big Bang because at the horizon distance one is peering back to an epoch when the universe was very small, so the very large meets the very small there.

The Cosmic Uroborus can also be used to illustrate the pervasive role of black holes in physics. These exist over a huge range of masses and have a characteristic radius ($R = 2GM/c^2$), so can also be placed around the body of the snake. Those larger than several solar masses form at the endpoint of evolution of ordinary stars

B. J. Carr (✉)
School of Physics and Astronomy, Queen Mary University of London, Mile End Road, London E1 4NS, UK
e-mail: B.J.Carr@qmul.ac.uk

© The Author(s), under exclusive license to Springer Nature Switzerland AG 2025
P. Nicolini (ed.), *Touring the Planck Scale*, Fundamental Theories of Physics 219,
https://doi.org/10.1007/978-3-031-76066-2_4

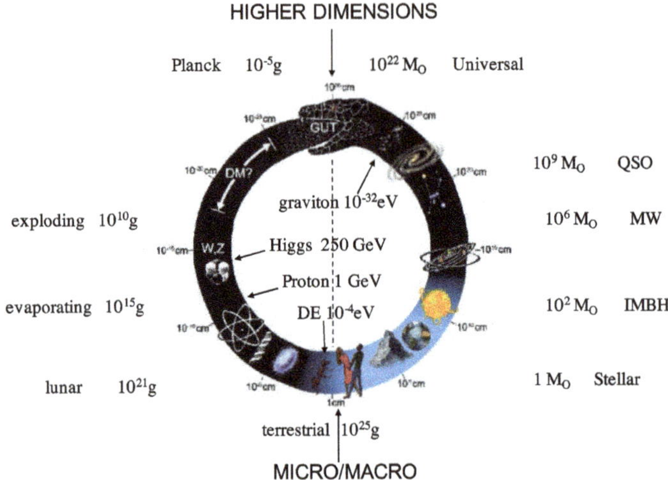

Fig. 1 The Cosmic Uroboros, indicating lengths and masses of various types of black holes and elementary particles. The divison between micro and macro domains is signified by the vertical line. QSO stands for "Quasi-Stellar Object", MW for "Milky Way", IMBH for "Intermediate Mass Black Hole", and "DE" for "Dark Energy" [1]. Reprinted from Carr, B. "Quantum Black Holes as the Link Between Microphysics and Macrophysics" 2nd Karl Schwarzschild Meeting on Gravitational Physics, eds. P. Nicolini et al. pp 159–167 (Springer 2018). Central image from "The New Universe and the Human Future" (Yale University Press, 2011) by Nancy Ellen Abrams and Joel R. Primack with permission, ©The Authors. All rights reserved

and there should be billions of these even in our own galaxy. "Intermediate Mass Black Holes" (IMBHs) might derive from stars bigger than 100 $M_\odot$, these being radiation-dominated and collapsing due to an instability during oxygen-burning, or from the merging of smaller holes. "Supermassive Black Holes" (SMBHs), with masses from $10^6\,M_\odot$ to $10^{10}\,M_\odot$, are thought to reside in galactic nuclei, with our own galaxy harbouring one of $4 \times 10^6\,M_\odot$ and quasars typically being powered by ones of around $10^8\,M_\odot$. In some sense, to be explained later, the whole Universe might be regarded as a black hole of $10^{22}\,M_\odot$. All these black holes might be described as "macroscopic" since they are larger than a kilometre in size.

Black holes smaller than a solar mass could only have formed in the early Universe, the density being $\rho \sim 1/(Gt^2)$ at a time t after the Big Bang. Since a region of mass M requires a density $\rho \sim c^6/(G^3 M^2)$ to form an event horizon, such "Primordial Black Holes" (PBHs) would initially have of order the horizon mass $M_H \sim c^3 t/G$. They could therefore span an enormous mass range. The smallest would form at the Planck time ($t_P \sim 10^{-43}$s) and have the Planck mass ($M_P \sim 10^{-5}$g). Those forming at the Quantum Chromodynamics (QCD) epoch at $t \sim 10^{-5}$s would have a mass of $\sim 1\,M_\odot$ and might explain the dark matter or some of the LIGO/Virgo gravitational wave events. Those forming at $t \sim 1$s would have a mass of $\sim 10^5 M_\odot$ and might be seeds for the SMBHs in galactic nuclei. There might even be "Stupendously Large Black Holes" (SLABs) larger than $10^{11}\,M_\odot$ of primordial origin.

PBHs initially lighter than $M_* \sim 10^{15}$ g would be smaller than a proton and have evaporated by now due to Hawking radiation, the temperature and evaporation time of a black hole of mass M being $T \sim 10^{12}(M/10^{15}g)^{-1}$K and $\tau \sim 10^{10}(M/10^{15}g)^3$y, respectively. Those smaller than 10^{26}g (i.e. the mass of the Earth) would be hotter than the Cosmic Microwave Background (CMB) temperature and might be classified as "quantum" rather than "clasical" since their evaporation would not be suppressed by accretion of the CMB. However, I will argue later that *all* black holes are in a sense quantum. The dividing scale is around a micron and corresponds to the bottom of the Cosmic Uroborus. A theory of quantum gravity would be required to understand the evaporation of a black hole with the Planck scale ($R_{\mathrm{P}} \sim 10^{-33}$ cm) at the top of the Uroborus and this might even allow stable relics. So the vertical line connecting the bottom and the top provides a convenient division between the microphysical and macrophysical domains.

Underlying Fig. 1 are the twin pillars of current physics: quantum theory in the microscopic domain and general relativity in the macroscopic domain. It is well known that these are incompatible and that a final theory of quantum gravity must somehow amalgamate them. However, whether the final paradigm will be more like quantum theory (as in M-theory) or general relativity (as in loop quantum gravity) is contentious. Whatever the solution, it will be argued in this chapter that it must involve two features: (1) what is termed the Black Hole Uncertainty Principle (BHUP) or Compton-Schwarzschild correspondence; and (2) the existence of higher dimensions beyond the four macroscopic ones of ordinary spacetime on sufficiently small scales. Both features are expected to become important on the Planck scale, $R_{\mathrm{P}} \sim 10^{-33}$ cm, although the scale of the extra dimensions could be larger than that—in string theory.

What is striking is that black holes are involved in both these features and indeed provide a link between them. For example, as regards feature (1), the key motivation is the duality under the transformation $M \rightarrow M_{\mathrm{P}}^2/M$ between the Compton wavelength for a *particle* of mass M, $R_{\mathrm{C}} = \hbar/(Mc)$, and the Schwarzschild radius for a *black hole* of the same mass, $R_{\mathrm{S}} = 2GM/c^2$. This suggests a unified Compton-Schwarzschild expression with a smooth minimum in the (M, R) plane, in which the Schwarzschild scale merges into the Compton scale. This implies that elementary particles may in some sense be sub-Planckian black holes. This proposal goes back to the 1970s, when it was motivated in the context of strong gravity theories by the link between Regge trajectories and extreme Kerr solutions. Figure 1 can also be used to represent elementary particles (with sub-Planckian mass) by locating their Compton wavelength on the *inside* of the Uroborus. For example, it shows the Higgs boson (250 GeV) and proton (1 GeV) on the left, the dark energy mass-scale (10^{-4} eV) at the bottom, and the mass (limit) on the gravitino (10^{-32} eV) at the top. Note that the inner scale also gives the temperature of a black hole with mass indicated by the outer scale.

As regards feature (2), the existence of extra spatial dimensions would clearly modify the formation and evaporation of sufficiently small black holes. These dimensions are usually assumed to be compactified on the Planck length but they could be much larger than this in some models, in which case gravity would grow more strongly at short distances than implied by the inverse-square law. If there are n extra spatial dimensions compactified on scale R_E, the Schwarschild radius then scales as $R^{1/(1+n)}$ for $R < R_E$, leading to the possibility of TeV quantum gravity and black hole production at accelerators if the Compton wavelength still scales as M^{-1} for $R < R_E$. Currently there is no evidence for this. However, the effective higher-dimensional Compton wavelength depends on the form of the $(3 + n)$-dimensional wavefunction. If this is spherically symmetric, then one indeed has $R_C \propto M^{-1}$. However, if the wave function is pancaked in the extra dimensions and maximally asymmetric, then $R_C \propto M^{-1/(1+n)}$ for $R < R_E$. This preserves the duality between R_C and R_S but TeV quantum gravity is precluded, so black holes cannot be generated in collider experiments. Nevertheless, the extra dimensions could still have consequences for the detectability of black hole evaporations and the enhancement of pair-production at accelerators on scales below R_E.

This shows that black holes provide a subtle link between the existence of higher dimensions and the Compton-Schwarzschild correspondence. Indeed, the main message of this chapter is that higher-dimensional black holes may amalgamate quantum theory and general relativity in a rather surprising way. This is because some of the counter-intuitive features of quantum theory may be explained if the embedding space of curved spacetime is associated with by the higher dimensional space of M-theory. One the other hand, it is curious that the Compton-Schwarzschild correspondence may also support the notion that the world is lower-dimensional on the Planck scale and it is not clear how one reconciles these two perspectives.[1]

The plan of this chapter is as follows. Section 2 discusses the BHUP correspondence in general terms. The next three sections discuss three applications of this: loop black holes in Sect. 3, the '$M + 1/M$' Schwarzschild model and its extension to the charged and rotating cases in Sect. 4, and the extended de Broglie interpretation in Sect. 5. The next two sections discuss higher-dimensional black holes: the implications of T-duality in Sect. 6 and the relevance to the origin of the Universe and higher-dimensional embeddings in Sect. 7. Some final conclusions are given in Sect. 8.

2 The Black Hole Uncertainty Principle Correspondence

A key feature of the microscopic domain is the (reduced) Compton wavelength for a particle of rest mass M, which is $R_C = \hbar/(Mc)$. In the (M, R) diagram of Fig. 2, the region corresponding to $R < R_C$ might be regarded as the "quantum domain" in the sense that the classical description breaks down there. A key feature of the macroscopic domain is the Schwarzschild radius for a body of mass M, $R_S = 2GM/c^2$, which corresponds to the size of the event horizon. The region $R < R_S$ might be regarded as the "relativistic domain" in the sense that there is no stable classical configuration in this part of Fig. 2.

---

[1] The lower-dimensional proposal is discussed in this volume by Jonas Mureika [2], his title nicely complementing my own!

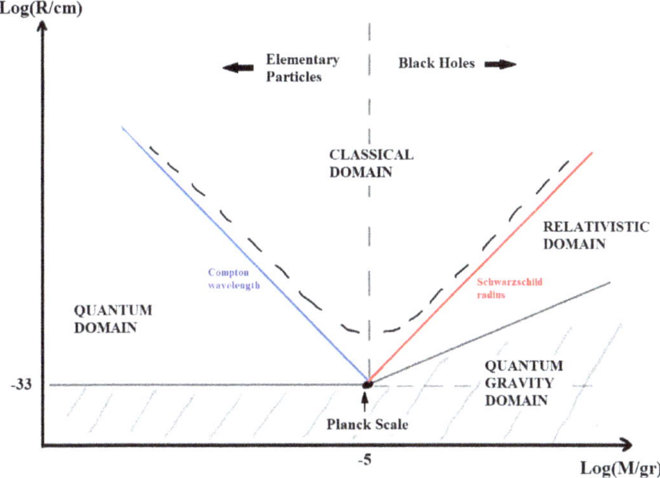

Fig. 2 Division of (M, R) diagram into classical, quantum, relativistic and quantum gravity domains. The boundaries are specified by the Planck density, Compton wavelength and Schwarzschild radius [1]

The Compton and Schwarzschild lines intersect at around the Planck scales,

$$R_{\mathrm{P}} = \sqrt{\hbar G/c^3} \sim 10^{-33}\mathrm{cm}, \quad M_{\mathrm{P}} = \sqrt{\hbar c/G} \sim 10^{-5}\mathrm{g}, \tag{1}$$

and divide the (M, R) diagram in Fig. 2 into three regimes, which we label quantum, relativistic and classical. There are several other interesting lines in the figure. The vertical line $M = M_{\mathrm{P}}$ marks the division between elementary particles ($M < M_{\mathrm{P}}$) and black holes ($M > M_{\mathrm{P}}$), since the size of a black hole is usually required to be larger than the Compton wavelength associated with its mass. The horizontal line $R = R_{\mathrm{P}}$ is significant because quantum fluctuations in the metric should become important below this [3]. Quantum gravity effects should also be important whenever the density exceeds the Planck value, $\rho_{\mathrm{P}} = c^5/(G^2\hbar) \sim 10^{94}\mathrm{g\,cm}^{-3}$, corresponding to the sorts of curvature singularities associated with the big bang or the centres of black holes. This implies $R < R_{\mathrm{P}}(M/M_{\mathrm{P}})^{1/3}$, which is well above the $R = R_{\mathrm{P}}$ line in Fig. 2 for $M \gg M_{\mathrm{P}}$, so one might regard the shaded region as specifying the 'quantum gravity' domain. This point has recently been invoked to support the notion of Planck stars [4] and could have important implications for the detection of evaporating black holes [5].

The Compton and Schwarzschild lines transform into one another under the transformation $M \to M_{\mathrm{P}}^2/M$, corresponding to a reflection in the line $M = M_{\mathrm{P}}$ in Fig. 2. This interchanges sub-Planckian and super-Planckian mass scales and suggests some connection between elementary particles and black holes. The lines also transform into each other under the transformation $R \to R_{\mathrm{P}}^2/R$, corresponding to a reflection in the line $R = R_{\mathrm{P}}$. This turns super-Planckian length scales into sub-Planckian ones.

Each line maps into itself under the combined T-duality transformation

$$M \to M_{\mathrm{P}}^2/M, \quad R \to R_{\mathrm{P}}^2. \tag{2}$$

T-dualities arise naturally in string theory and are known to map momentum-carrying string states to winding states and vice-versa [6].

Although the Compton and Schwarzschild boundaries correspond to straight lines in the logarithmic plot of Fig. 2, this form presumably breaks down near the Planck point due to quantum gravity effects. One might envisage two possibilities: either there is a smooth minimum, as indicated by the broken line in Fig. 2, so that the Compton and Schwarzschild lines in some sense merge, or there is some form of phase transition or critical point at the Planck scale, so that the separation between particles and black holes is maintained. Which alternative applies has important implications for the relationship between elementary particles and black holes [7]. This may relate to the issue of T-duality since this also purports to play some role in linking point particles and black holes.[2]

One way of smoothing the transition between the Compton and Schwarzschild lines is to invoke some form of unified expression which asymptotes to the Compton wavelength and Schwarzschild radius in the appropriate regimes. The simplest such expression would be

$$R_{\mathrm{CS}} = \frac{\beta \hbar}{Mc} + \frac{2GM}{c^2}, \tag{3}$$

where β is a free constant, associated with the first term since the factor of 2 in the expression for the Schwarzschild radius is precise, whereas the coefficient associated with the Compton term is somewhat arbitrary. In the sub-Planckian regime, this can be written as

$$R_{\mathrm{C}}' = \frac{\beta \hbar}{Mc} \left[1 + \frac{2}{\beta} \left(\frac{M}{M_{\mathrm{P}}} \right)^2 \right] \quad (M \ll M_{\mathrm{P}}), \tag{4}$$

with the second term corresponding to a small correction to the uual expression for the Compton wavelength. In the super-Planckian regime, it becomes

$$R_{\mathrm{S}}' = \frac{2GM}{c^2} \left[1 + \frac{\beta}{2} \left(\frac{M_{\mathrm{P}}}{M} \right)^2 \right] \quad (M \gg M_{\mathrm{P}}). \tag{5}$$

This is termed the Generalised Event Horizon (GEH) [8, 9], with the second term corresponding to a small correction to the usual Schwarzschild expression. More generally, one might consider any unified expression $R_{\mathrm{C}}'(M) \equiv R_{\mathrm{S}}'(M)$ which has the asymptotic behaviour $\beta \hbar/(Mc)$ for $M \ll M_{\mathrm{P}}$ and $2GM/c^2$ for $M \gg M_{\mathrm{P}}$.

---

[2] Antonio Aurilia may have been the first person to present a version of Fig. 2. This motivated him to assign a Schwarzschild radius to an elementary a particle and a Compton wavelength to a black hole [10]. With Euro Spallucci he also placed this in the context of string theory [11].

An expression of the form (5) arises in the quantum N-portrait model of Dvali et al. [12–17], which regards a black hole as a weakly-coupled Bose-Einstein condensate of gravitons. From holographic considerations, the number of gravitons in the black hole is

$$N \approx \frac{A_{\mathrm{BH}}}{R_{\mathrm{P}}^2} \approx \frac{M^2}{M_{\mathrm{P}}^2} \,, \tag{6}$$

where A_{BH} is the black hole area. As noted in Ref. [18], one can then argue that the black hole radius is

$$R_{\mathrm{CS}} \approx \frac{2M}{M_{\mathrm{P}}^2} \left(1 + \frac{\beta}{2N} \right) \quad (M > M_{\mathrm{P}}) \,, \tag{7}$$

which is equivalent to Eq. (4).

An expression of the form (4) arises in the context of the Generalized Uncertainty Principle (GUP). For example, Adler and colleagues [19–22] have given a series of heuristic arguments for why the Uncertainty Principle should take the GUP form

$$\Delta x = \frac{\hbar}{\Delta p} + \alpha \, \frac{R_{\mathrm{P}}^2 \, \Delta p}{\hbar} \,, \tag{8}$$

where α is a dimensionless constant and the usual factor of 2 in the first term on the right has been dropped. This term represents the uncertainty in the position due to the momentum of the probing photon and leads to the expression for the Compton wavelength if one substitutes $\Delta x \to R$ and $\Delta p \to cM$. The second term on the right represents the gravitational effect of the probing photon and is much smaller than the first term for $\Delta p \ll M_{\mathrm{P}}c$. Variants of Eq. (8) can be found in other approaches to quantum gravity, such as non-commutative quantum mechanics or general minimum length considerations [23–25], polymer corrections in the structure of spacetime in LQG [26, 27], some approaches to the problem of quantum decoherence [28, 29] and in string theory [30–36].

If we rewrite Eq. (8) using the substitution $\Delta x \to R$ and $\Delta p \to cM$, we obtain the revised Compton wavelength

$$R_{\mathrm{CS}} = \frac{\hbar}{Mc} + \alpha \, \frac{GM}{c^2} \,. \tag{9}$$

This resembles Eq. (3) except that the constant is associated with the second term rather than the first. Since R_{CS} asymptotes to the Schwarzschild form, apart from a numerical factor, for $M \gg M_{\mathrm{P}}$, this suggests that there is a different kind of positional uncertainty for an object larger than the Planck mass, related to the size of a black hole. This is not unreasonable since the Compton wavelength is below the Planck scale (and hence meaningless) here and also an outside observer cannot localize an object on a scale smaller than its Schwarzschild radius. This is termed the Black Hole Uncertainty Principle correspondence and also the Compton-Schwarzschild correspondence when discussing an interpretation in terms of extended de Broglie relations [37].

An important caveat is that Eq. (8) assumes that the two uncertainties add linearly. On the other hand, since they are independent, it might be more natural to assume that they add quadratically:

$$\Delta x > \sqrt{(\hbar/\Delta p)^2 + (\alpha R_{\rm P}^2 \Delta p/\hbar)^2}\,. \tag{10}$$

While the heuristic arguments indicate the form of the two uncertainty terms, they do not specify how one combines them. We refer to Eqs. (8) and (10) as the *linear* and *quadratic* forms of the GUP, respectively. The latter corresponds to a generalized Compton wavelength

$$R_{\rm C}' = \sqrt{(\hbar/Mc)^2 + (\alpha GM/c^2)^2}\,. \tag{11}$$

As in the linear case, it might be more natural to use the unified expression

$$R_{\rm CS} = \sqrt{(\beta\hbar/Mc)^2 + (2GM/c^2)^2}\,, \tag{12}$$

leading to the approximations

$$R_{\rm C}' \approx \frac{\beta\hbar}{Mc}\left[1 + \frac{2}{\beta^2}\left(\frac{M}{M_{\rm P}}\right)^4\right] \quad (M \ll M_{\rm P}) \tag{13}$$

and

$$R_{\rm S}' \approx \frac{2GM}{c^2}\left[1 + \frac{\beta^2}{8}\left(\frac{M_{\rm P}}{M}\right)^4\right] \quad (M \gg M_{\rm P})\,. \tag{14}$$

These might be compared to the *exact* expressions in the linear case, given by Eqs. (4) and (5). We will see later that a model inspired by LQG permits the existence of a type of black hole whose horizon size has precisely the form (12).

In the standard picture, one can calculate the black hole temperature from the HUP by identifying the Schwarzschild radius with Δx and the black hole temperature with a multiple η of Δp. This gives [38]

$$kT = \eta c\,\Delta p = \frac{\eta\hbar c}{\Delta x} = \frac{\eta M_{\rm P}^2}{2M}\,, \tag{15}$$

which is precisely the Hawking temperature if we take $\eta = 1/(4\pi)$. This approach can also be used to derive the black hole temperature for a model in which one adopts the GUP but assumes that the expression for the black hole size is unchanged. From Eq. (10), one then obtains [19]

$$kT = \frac{\eta M c^2}{\alpha} \left(1 \pm \sqrt{1 - \frac{\alpha M_{\mathrm{P}}^2}{M^2}} \right).$$ (16)

The negative sign just gives a small perturbation to the standard Hawking temperature in the super-Planckian regime:

$$kT \approx \frac{\eta M_{\mathrm{P}}^2 c^2}{2M} \left[1 - \frac{\alpha M_{\mathrm{P}}^2}{4M^2} \right] \quad (M \gg M_{\mathrm{P}}).$$ (17)

However, the solution becomes complex when M falls below $\sqrt{\alpha}\, M_{\mathrm{P}}$, corresponding to a minimum mass, and it then connects to the positive branch of Eq. (16). This asymptotes to $2\eta M c^2/\alpha$, which is presumably unphysical since it exceeds the Planck temperature. This form is indicated by the curve on the right of Fig. 3.

The BHUP correspondence goes beyond the GUP because it also modifies the relationship between the black hole radius Δx and M (i.e. it involves the parameter β in the expression for the generalised event horizon). Although we have argued that the parameters α and β are equivalent [cf. Eqs. (3) and (9)], they could in principle be unrelated since there are independent arguments for the GUP and GEH expressions. In this case, one never reaches the limiting Adler mass of $\sqrt{\alpha}\, M_{\mathrm{P}}$ for $\alpha < 2\beta$, so the temperature reaches a maximum and then decreases rather than going complex. The dependence of T on M in the asymptotic limits can then be approximated by

$$kT \approx \begin{cases} \frac{\eta M_{\mathrm{P}}^2 c^2}{2M} \left[1 - \left(\frac{2\beta - \alpha}{4} \right) \left(\frac{M_{\mathrm{P}}}{M} \right)^2 \right] & (M \gg M_{\mathrm{P}}) \\[2ex] \frac{\eta M c^2}{\beta} \left[1 - \left(\frac{2\beta - \alpha}{\beta^2} \right) \left(\frac{M}{M_{\mathrm{P}}} \right)^2 \right] & (M \ll M_{\mathrm{P}}). \end{cases}$$ (18)

The overall behaviour of T is shown by the lowest curve in Fig. 3. For $\alpha > 2\beta$, it has the same qualitative form as in the Adler model. In the special case $\alpha = 2\beta$, the effects of the α and β terms cancel and one obtains the simple solution

$$kT = \min \left[\frac{\eta M_{\mathrm{P}}^2 c^2}{2M}, \frac{2\eta M c^2}{\alpha} \right].$$ (19)

This is indicated by the middle curve in Fig. 3, although this is not required for the BHUP correspondence. The first expression is the *exact* Hawking temperature, but one must cross over to the second expression below $M = \sqrt{\alpha/4}\, M_{\mathrm{P}}$ to avoid the temperature going above the Planck value. However, it seems more natural to have a single-parameter expression which applies in both the sub- and super-Planckian regimes and this is what we assume henceforth.

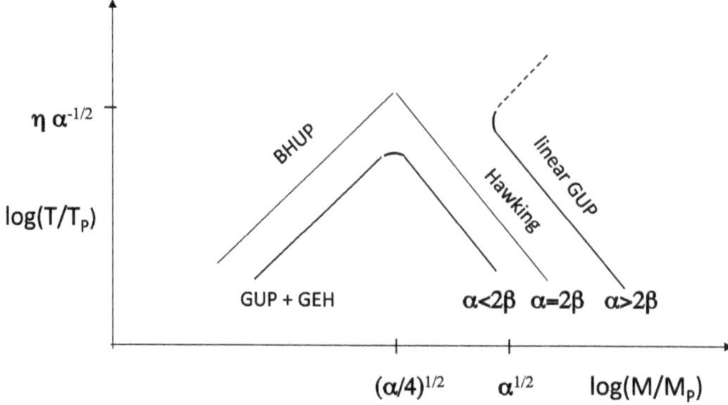

Fig. 3 Temperature in general case with both α and β for $\alpha < 2\beta$ (cf. the '$M + 1/M$' model), $\alpha = 2\beta$ (cf. the Hawking solution) and $\alpha > 2\beta$ (cf. Adler's solution) [7]. Reprinted from Carr, B., Mureika, J. & Nicolini, P. "Sub-Planckian black holes and the Generalized Uncertainty Principle". J. High Energ. Phys. 07, 52 (2016). https://doi.org/10.1007/JHEP07(2015)052 under CC BY, ©The Authors

3 Loop Black Holes

This section shows that Loop Quantum Gravity (LQG) leads to a quadratic version of the BHUP correspondence.[3] LQG is based on a canonical quantization of the Einstein equations written in terms of the Ashtekar variables [40–45]. One consequence of this is that the area is quantized, with its smallest possible value being

$$A_{\min} = 4\pi \sqrt{3}\, \gamma\, R_{\mathrm{P}}^2 , \tag{20}$$

where γ is the Immirzi parameter. The exact value of the coefficient on the right is not certain, so we parametrize our ignorance with another constant ζ and use

$$a_o = A_{\min}/8\pi = \sqrt{3}\,\gamma\,\zeta R_{\mathrm{P}}^2/2 . \tag{21}$$

The expected values of γ and ζ are of order 1 but the precise choice is not crucial. Another relevant constant is the dimensionless polymeric parameter δ. Together with a_0, this determines the deviation from classical theory.

One version of LQG, using the mini-superspace approximation, gives rise to cosmological solutions which resolve the initial singularity problem [46–48]. Another

---

[3] The work with Modesto and Prémont-Schwarz [39] described here arose as a result of a visit to the Perimeter Institute in 2011. I was tremendously excited to realise that the BHUP correspondence could be given a geometrical interpretation in the context of their work on loop black holes. Unfortunately, our paper has only appeared on the arxiv. The referee for the journal where it was first submitted rejected it on the grounds that no theory which disagrees with both general relativity and quantum theory could possibly be correct. The referee for the second journal accepted the paper but the editor rejected it because he did not believe in loop black holes.

version gives the loop black hole (LBH) solution [49–57] and this has a self-duality property that removes the singularity and replaces it with another asymptotically flat region. The metric in this solution depends only on the combined dimensionless parameter $\varepsilon \equiv \delta\gamma$, which must be small if quantum gravitational corrections are relevant only when the curvature is in the Planckian regime. It can be expressed as

$$ds^2 = -G(r)c^2 dt^2 + \frac{dr^2}{F(r)} + H(r)d\Omega, \tag{22}$$

with $d\Omega \equiv d\theta^2 + \sin^2\theta d\phi^2$ and

$$G(r) = \frac{(r - r_+)(r - r_-)(r + r_*)^2}{r^4 + a_o^2},$$

$$F(r) = \frac{(r - r_+)(r - r_-)r^4}{(r + r_*)^2(r^4 + a_o^2)},$$

$$H(r) = r^2 + \frac{a_o^2}{r^2}. \tag{23}$$

Here $r_+ = 2Gm/c^2$ and $r_- = 2GmP^2/c^2$ are the outer and inner horizons, respectively, and $r_* \equiv \sqrt{r_+ r_-} = 2GmP/c^2$, where m is the black hole mass and

$$P \equiv \frac{\sqrt{1 + \varepsilon^2} - 1}{\sqrt{1 + \varepsilon^2} + 1} \tag{24}$$

is called the polymeric function. For $\varepsilon \ll 1$, we have $P \approx \varepsilon^2/4 \ll 1$, so $r_- \ll r_* \ll r_+$. Since $g_{\theta\theta}$ is not exactly r^2 in the above metric, r is only the usual radial coordinate asymptotically. In the limit $r \to \infty$ one has

$$G(r) \to 1 - \frac{2GM}{c^2 r}(1 - \varepsilon^2), \quad F(r) \to 1 - \frac{2GM}{c^2 r}, \quad H(r) \to r^2, \tag{25}$$

so the deviations from the Schwarzschild solution are of order $GM\varepsilon^2/(c^2 r)$. Here

$$M = m(1 + P)^2 \tag{26}$$

is the ADM mass, which is determined solely by the metric at flat asymptotic infinity and might be associated with the quantity M appearing in our earlier discussion.

The expression for $H(r)$ shows that the physical radial coordinate is

$$R \equiv \sqrt{r^2 + \frac{a_o^2}{r^2}}, \tag{27}$$

in the sense that this measures the proper circumferential distance. As r decreases from infinity to zero, R first decreases from infinity to a minimum value of $\sqrt{2a_0}$ at

$r = \sqrt{a_0}$ and then increases again to infinity. In particular, the value of R associated with the event horizon is

$$R_{EH} = \sqrt{H(r_+)} = \sqrt{\left(\frac{2Gm}{c^2}\right)^2 + \left(\frac{a_0 c^2}{2Gm}\right)^2} \, . \tag{28}$$

This corresponds to Eq. (12) if

$$\beta = \frac{a_0 c^2}{2G} = \frac{\sqrt{3}\gamma}{4} \, , \tag{29}$$

tending to the Schwarzschild radius for $m \gg M_{\mathrm{P}}$ and the Compton wavelength for $m \ll M_{\mathrm{P}}$.

The important physical implication of Eq. (27) is that central singularity of the Schwarzschild solution is replaced with another asymptotic region, so the collapsing matter bounces and the black hole becomes part of a wormhole. Equation (27) has three important consequences: it removes the singularity; it permits the existence of black holes with $m \ll M_{\mathrm{P}}$; and it allows a unified expression for the Compton and Schwarzschild scales. The fact that a purely geometrical condition in LQG implies the quadratic version of the GUP suggests some deep connection between general relativity and quantum theory.

The temperature is determined by the black hole's surface gravity [38]:

$$T \propto \frac{GM}{R_{EH}^2} \propto \begin{cases} M^{-1} & (M \gg M_{\mathrm{P}}) \\ M^3 & (M \ll M_{\mathrm{P}}) \, , \end{cases} \tag{30}$$

so this scales as M^3 rather than M for $M \ll M_{\mathrm{P}}$ [cf. Eq. (17)]. More precisely, the surface gravity at the outer horizon of the LBH solution is

$$\kappa_+ = \frac{4G^3 m^3 c^4 (1 - P^2)}{16G^4 m^4 + a_0^2 c^8} \, . \tag{31}$$

The relationship between m and M given by Eq. (26) involves another factor of $(1 + P)^2$ but this is close to 1. The mass dependence of the temperature $\hbar \kappa_+ / (2\pi k c)$ is indicated in Fig. 4.

If one calculates the temperature using Adler's heuristic argument, in which one associates the positional uncertainty $(\Delta x)_{BH}$ with the black hole radius and the momentum uncertainity $(\Delta p)_{BH}$ with the black hole temperature, one obtains

$$\left[\left(\frac{\hbar \eta c}{kT}\right)^2 + \left(\frac{\alpha R_{\mathrm{P}}^2 kT}{\hbar \eta c}\right)^2\right]^{1/2} = \left[\left(\frac{\hbar \beta}{Mc}\right)^2 + \left(\frac{2GM}{c^2}\right)^2\right]^{1/2} \, , \tag{32}$$

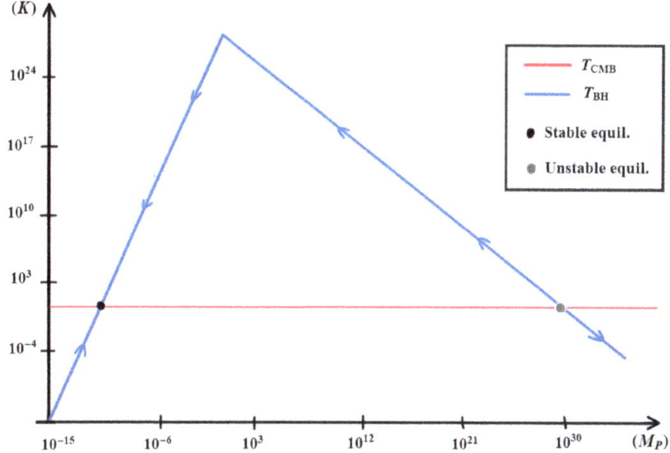

Fig. 4 Comparing black hole temperature predicted by GUP with CMB temperature [39]

where the parameters α and β are taken to be independent. This leads to

$$kT = \frac{\sqrt{2}\,\eta Mc^2}{\alpha}\left[1 + \frac{\beta^2 M_{\rm P}^4}{4M^4} - \sqrt{1 + \frac{(2\beta^2 - \alpha^2)}{4}\left(\frac{M_{\rm P}}{M}\right)^4 + \frac{\beta^4}{16}\left(\frac{M_{\rm P}}{M}\right)^8}\,\right]^{1/2} \tag{33}$$

which is real for all M providing $\alpha < 2\beta$ and implies

$$kT \approx \begin{cases} \frac{\eta\hbar c^3}{2GM}\left[1 + \left(\frac{\alpha^2 - 4\beta^2}{32}\right)\left(\frac{M_{\rm P}}{M}\right)^4\right] & (M \gg M_{\rm P}) \\[2ex] \frac{\eta Mc^2}{\beta}\left[1 + \left(\frac{\alpha^2 - 4\beta^2}{2\beta^4}\right)\left(\frac{M}{M_{\rm P}}\right)^4\right] & (M \ll M_{\rm P}). \end{cases} \tag{34}$$

This is similar to Eq. (18) but is inconsistent with Eq. (30). Both equations predict that the temperature deviates from the Hawking expression when M falls below $M_{\rm P}$ and that it never goes above $T_{\rm P}$, so which prediction is correct?

The source of this discrepancy is that there are two asymptotic spaces in the LBH solution, described by the limits $r \to \infty$ and $r \to 0$. Since the mass is different in these two spaces, so is the temperature and it would seem natural to assume that the GUP argument selects the asymptotic space which is on the same side of the wormhole throat as the black hole event horizon. In order to analyse the situation more carefully, we examine the behaviour of the surface gravity in the two asymptotic spaces.[4] This can be found using the expression

$$\kappa^2 = -g^{\mu\nu} g_{\rho\sigma} \nabla_\mu \chi^\rho \nabla_\nu \chi^\sigma / 2, \tag{35}$$

---

[4] This argument was not presented in the original paper.

where the metric is given by Eq. (22) and χ^μ is the timelike Killing vector with unit length in the relevant asymptotic space. Using the symmetry of the metric and the fact that χ^μ only has a 0 component gives

$$\kappa = \sqrt{g^{rr} g_{00}} \, \Gamma^0_{r0} \chi^0 = \sqrt{\frac{F}{G} \frac{\partial G}{\partial r}} \chi^0 , \tag{36}$$

where the first term is just $(1 + r_*/r)^{-2}$ and the expression must be evaluated at the appropriate event horizon. In the limit $r \to \infty$, $g_{00} = G(r) \to 1$, so $\chi^\mu = (1, 0, 0, 0)$ and we obtain the temperature given by Eq. (31). In the limit $r \to 0$, Eq. (23) implies

$$\chi^\mu = P^{-2}(m/M_{\rm P})^{-2}(\beta/2)(1, 0, 0, 0) , \tag{37}$$

where the factor $(m/M_{\rm P})^{-2}$ arises because the mass in the other space is the dual of that in our space. Another difference is that r_- rather than r_+ must be regarded as the outer horizon from the perspective of an observer at $r = 0$, so the appropriate temperature is

$$T = \frac{\hbar \kappa_-}{2\pi kc} = \frac{\hbar G m (1 - P^2)}{2\pi k c P^4 R_{\rm EH}^2} . \tag{38}$$

In calculating the temperatures in the two asymptotic regimes, we assume that an observer at our infinity can always see emission from r_+ and that an observer at the other other infinity can always see emission from r_- but that an observer at our infinity cannot see radiation from r_-, since this is always in their causal future. One then has three possible situations. (i) For $M > P^{-2}\sqrt{\beta/2}\, M_{\rm P}$, one has $\sqrt{a_0} < r_- < r_+$, so the black hole temperature has the standard form: $T \propto M^{-1}$ in our space but $T \propto P^{-6} M^{-3}$ in the other space. The latter result assumes that the radiation from r_- can propagate through the wormhole throat into the other space since the wormhole throat ($r \sim R_{\rm P}$) in this case is timelike and traversible. (ii) For $M < \sqrt{\beta/2}\, M_{\rm P}$, one has $r_- < r_+ < \sqrt{a_0}$, so the black hole has $T \propto M^3$ in our space and $T \propto P^2 M$ in the other space. The former result assumes that the radiation from r_+ can propagate through the wormhole throat into our space since the wormhole throat is again timelike and traversible. (iii) For $\sqrt{\beta/2}\, M_{\rm P} < M < P^{-2}\sqrt{\beta/2}\, M_{\rm P}$, one has $r_- < \sqrt{a_0} < r_+$, so the black hole has $T \propto M^{-1}$ in our space and $T \propto P^2 M$ in the other space. In this case, the radiation cannot cross the wormhole throat since it is spacelike. The full dependence of the black hole temperature on mass in both spaces is illustrated in Fig. 4 for $\beta = 2$.

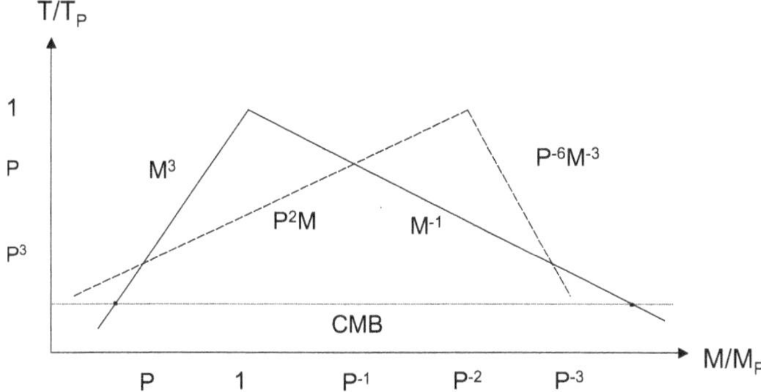

Fig. 5 The dependence of temperature on the black hole mass predicted by the surface gravity argument for our space (solid line) and the other space (broken line), with evaporation being suppressed below the CMB temperature (horizontal line). Reprinted from "Quantum Black Holes" by Xavier Calmet, Bernard Carr, Elizabeth Winstanley with permission, ©Springer Nature BV. All rights reserved

4 Carr-Mureika-Nicolini-Mentzer Approach

This section describes a particular interpretation of the linear version of the BHUP correspondence, which was first described in my work with Mureika and Nicolini [7]. In the standard picture, the mass in the Schwarzschild solution is obtained by matching the metric coefficients with the Newtonian potential and this gives the Komar integral

$$M \equiv \frac{1}{4\pi G} \int_{\partial \Sigma} d^2 x \sqrt{\gamma^{(2)}} \, n_\mu \sigma_\nu \nabla^\mu K^\nu \,, \tag{39}$$

where K^ν is a timelike vector, Σ is a spacelike surface with unit normal n^μ, and $\partial \Sigma$ is the boundary of Σ (typically a 2-sphere at spatial infinity) with metric $\gamma^{(2)ij}$ and outward normal σ^μ. For $M \gg M_P$, quantum effects are negligible and one finds the usual Schwarzschild mass. For $M < M_P$, however, the expression can simultaneously refer to a particle and a black hole. One usually considers the particle case and writes Eq. (39) as

$$M \equiv \int_\Sigma d^3 x \sqrt{\gamma} \, n_\mu K_\nu T^{\mu\nu} \simeq -4\pi \int_0^{R_C} dr \, r^2 T_0{}^0 \,, \tag{40}$$

where γ is the determinant of the spatially induced metric γ^{ij}, $T^{\mu\nu}$ is the stress-energy tensor and $T_0{}^0$ accounts for the particle distribution on a scale of order R_C. This corresponds to the mass appearing in the expression for the Compton wavelength. When the black hole reaches the final stages of evaporation, the major contribution to integral (39) becomes

$$M = -4\pi \int_0^{R_P} dr \, r^2 T_0{}^0 \,, \tag{41}$$

where $T_0{}^0$ accounts for an unspecified quantum-mechanical distribution of matter and energy. Integral (41) is then unknown and might lead to a completely different definition of the Komar energy.

Inspired by the dual role of M in the GUP, we explored a variant of the last scenario, based on the existence of sub-Planckian black holes, i.e. quantum mechanical objects that are simultaneously black holes and elementary particles. In this context, we suggested that the Arnowitt-Deser-Misner (ADM) mass, which coincides with the Komar mass in the stationary case, should be

$$M_{\mathrm{ADM}} = M \left(1 + \frac{\beta}{2} \frac{M_{\mathrm{P}}^2}{M^2} \right) , \tag{42}$$

which is equivalent to Eq. (5). We thus posited a quantum-corrected Schwarzschild metric, like the usual one but with M replaced by M_{ADM} and noted a possible connection with the energy-dependent metric proposed in the framework of "gravity's rainbow" [58]. It may also relate to the distinction between the bare and renormalized mass in QFT in the presence of stochastic metric fluctuations [59].

The horizon size for the modified metric is given by

$$R'_{\mathrm{S}} = \frac{2M_{\mathrm{ADM}}}{M_{\mathrm{P}}^2} \approx \begin{cases} 2M/M_{\mathrm{P}}^2 & (M \gg M_{\mathrm{P}}) \\ (2+\beta)/M_{\mathrm{P}} & (M \approx M_{\mathrm{P}}) \\ \beta/M & (M \ll M_{\mathrm{P}}), \end{cases} \tag{43}$$

where in this section we use units with $\hbar = c = 1$. The first expression is the standard Schwarzschild radius. The intermediate expression gives a minimum of order R_{P}, so the Planck scale is never actually reached for $\beta > 0$ and the singularity remains inaccessible. The last expression resembles the Compton wavelength. If the temperature is determined by the black hole's surface gravity [38], one has

$$T = \frac{M_{\mathrm{P}}^2}{8\pi M_{\mathrm{ADM}}} \approx \begin{cases} M_{\mathrm{P}}^2/(8\pi M)[1 - \beta(M_{\mathrm{P}}/M)^2] & (M \gg M_{\mathrm{P}}) \\ M_{\mathrm{P}}/(8\pi(1 + \beta/2)) & (M \approx M_{\mathrm{P}}) \\ M/(4\pi\beta)[1 - (M/M_{\mathrm{P}})^2/\beta] & (M \ll M_{\mathrm{P}}). \end{cases} \tag{44}$$

This temperature is plotted in Fig. 6, which might be contrasted with Fig. 4. The large M limit is the usual Hawking temperature with a small correction. However, as the black hole evaporates, the temperature reaches a maximum at around T_{P} and then decreases to zero as $M \to 0$.

As discussed in Mureika's chapter, a possible explanation for the $M \ll M_{Pl}$ behaviour is that a decaying black hole makes a temporary transition to a (1+1)-D dilaton black hole when approaching the Planck scale, since this naturally encodes a $1/M$ term in its gravitational radius. According to t'Hooft [60], gravity might experience a (1+1)-D phase at the Planck scale due to spontaneous dimensional reduction, such a conjecture being further supported by studies of the fractal properties of a quantum spacetime at the Planck scale. At this point the Komar mass can be defined as for dilaton black holes by [61]

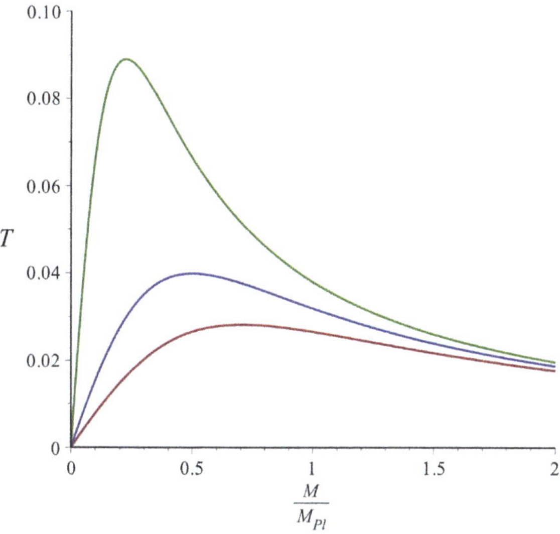

Fig. 6 Hawking temperature from surface gravity argument as a function of M/M_P for $\beta = 1$ (bottom), $\beta = 0.5$ (middle) and $\beta = 0.1$ (top). As M decreases, T reaches a maximum below T_P and then falls to zero [7]. Reprinted from Carr, B., Mureika, J. & Nicolini, P. Sub-Planckian black holes and the Generalized Uncertainty Principle. J. High Energ. Phys. 2015, 52 (2015). https://doi.org/10.1007/JHEP07(2015)052 under CC BY, ©The Authors

$$M \sim \int dx \sqrt{g^{(1)}}\, n_i^{(2)} T_0^{\ i}\,, \tag{45}$$

where $g^{(1)}$ is the determinant of the spatial section of g^{ij}, the effective 2D quantum spacetime metric, and $^{(2)}T_0^{\ i}$ is the dimensionally reduced energy-momentum tensor.[5]

The black hole luminosity in this model is $L = \eta^{-1} M_{\mathrm{ADM}}^{-2}$ where $\eta \sim t_{\mathrm{Pl}}/M_\mathrm{P}^3$. Although the black hole loses mass on a timescale

$$\tau \sim M/L \sim \eta M^3 (1 + \beta M_\mathrm{P}^2/2M^2)^2\,, \tag{46}$$

it never evaporates entirely because the mass loss rate decreases when M falls below M_P. There are two values of M for which τ is comparable to the age of the Universe ($t_0 \sim 10^{17}$s). One is super-Planckian,

$$M_* \sim (t_0/\eta)^{1/3} \sim (t_0/t_\mathrm{P})^{1/3} M_\mathrm{P} \sim 10^{15}\mathrm{g}\,, \tag{47}$$

this being the standard expression for the mass of a PBH evaporating at the present epoch, and the other is sub-Planckian,

---

[5] An interesting application of the (1+1)-D approach to PBH formation was studied in collaboration with Athanasios Tzikas [62] and is discussed in Mureika's chapter.

$$M_{**} \sim \beta^2 (t_{\mathrm{P}}/t_o) M_{\mathrm{P}} \sim 10^{-65}\mathrm{g}\,. \tag{48}$$

The usual Hawking lifetime ($\tau \propto M^3$) gives the time for the mass to decrease to M_{P}, after which it quickly falls to the value M_{**}. Although this mass-scale is very tiny, it arises naturally in some estimates for the photon or graviton mass [63]. Note that the PBH mass cannot actually reach M_{**} at the present epoch because the black hole temperature is less than the CMB temperature below $M_{\mathrm{CMB}} \sim 10^{-36}\mathrm{g}$, leading to *effectively* stable relics which might provide the dark matter.

To summarise the advantages of this proposal: it encodes the GUP duality in the expression for the mass; it smooths the $M(R)$ curve, so that there is no critical point; it cures the thermodynamic instability of evaporating black holes; it exhibits dimensional reduction in the sub-Planckian regime; and it gives a consistent theory of gravity in different spacetime dimensions without needing two regimes governed by different theories (GR and QM). Indeed, in some sense, the BHUP correspondence implies that all black holes are quantum and that the Uncertainty Principle has a gravitational explanation.

Recently we have extended the CMN work, together with Heather Mentzer, to charged and rotating black holes [64], since this is clearly relevant to elementary particles.[6] The standard Reissner-Nordström (RN) and Kerr solutions already exhibit features of the GUP-modified Schwarzschild solution. However, interesting new features arise if the charged and rotating solutions are themselves GUP-modified by replacing M with the expression M_{ADM} given by Eq. (42). In particular, there is an interesting transition below some value of β from the GUP solutions (spanning both super-Planckian and sub-Planckian regimes) to separated super-Planckian and sub-Planckian solutions. Equivalently, for a given value of β, there is a critical value of the charge and spin above which the solutions bifurcate and become separated by a mass gap.

Here we examine the characteristics and thermodynamics of the RN solutions which are self-complete, in the sense that the Compton line intersects the outer black hole horizon. The metric has the well-known form

$$ds^2 = f(r)dt^2 - \frac{dr^2}{f(r)} - r^2 d\Omega^2\ , \tag{49}$$

with

$$f(r) = 1 - \frac{2M}{M_{\mathrm{P}}^2 r} + \frac{\alpha_e n^2}{M_{\mathrm{P}}^2 r^2} \tag{50}$$

where ne is the charge.

The outer ($+$) and inner ($-$) horizons are then given by

---

[6] The results of this work are described on more detail in Mureika's chapter, which covers the GUP-modified RN solution. This chapter just focuses on the RN solution itself.

$$f(r_\pm) = 0 \implies r_\pm = \frac{M}{M_\mathrm{P}^2}\left(1 \pm \sqrt{1 - \frac{\alpha_e M_\mathrm{P}^2 n^2}{M^2}}\right) \tag{51}$$

where $\alpha_e \, 10^{-2}$ is the electric fine structure constant.

For a black hole which is far from extremal ($M \gg \sqrt{\alpha_e}\, n M_\mathrm{P}$), this can be written as

$$r_\pm \approx \begin{cases} \frac{2M}{M_\mathrm{P}^2}\left(1 - \frac{\gamma M_\mathrm{P}^2}{M^2}\right) \ (+) \\ \frac{2\gamma}{M}\left(1 + \frac{\gamma M_\mathrm{P}^2}{M^2}\right) \ (-) \end{cases} \tag{52}$$

where $\gamma \equiv \alpha_e n^2/4$. The form of the outer and inner horizons for different values of n are shown by the upper and lower parts of the solid curves in Fig. 7, respectively. The outer horizon asymptotes to $2M/M_\mathrm{P}^2$ (upper dotted curve) at large r and the inner horizon to $2\gamma/M$ at low r.

For each n, the two horizons merge on the line $r = GM$ (lower dotted curve) at the minimum value of M and have an infinite gradient (dr/dM) there. This corresponds to a sequence of "extremal" solutions (shown by the dots in Fig. 7) with a spectrum of masses given by

$$1 - \frac{\alpha_e M_\mathrm{P}^2 n^2}{M^2} = 0 \implies M_n = \sqrt{\alpha_e}\, n M_\mathrm{P}\,. \tag{53}$$

For given n, there are no solutions with M less than this since these would correspond to naked singularities. In particular, n could be at most the integer part of $1/\sqrt{\alpha_e}$ (i.e. 11) for a Planck-mass black hole. It is interesting that Eq. (52) has two asymptotic behaviors in the $M \gg M_\mathrm{P}$ regime: the outer horizon correponds to Eq. (5) but with a negative value of β; the inner horizon corresponds to Eq. (9) but with a positive value of α and it asymptotes to the Compton wavelength for $n = 16$, this being the integer part of $\sqrt{2/\alpha_e}$.

The (standard) Compton line intersects the outer black hole horizon, as required by the self-completeness condition, at the mass-scale

$$M = \frac{M_\mathrm{P}}{\sqrt{2 - \alpha_e n^2}} \approx \frac{M_\mathrm{P}}{\sqrt{2 - n^2/137}}\,. \tag{54}$$

The relation between the Compton and outer horizon scales is shown in Fig. 7. For $n = 0$, the intersect mass is $M_\mathrm{P}/\sqrt{2}$ but it increases with n and tends to M_P as $n \to \sqrt{137}$ (middle curve). This implies a constraint $n \leq 11$ on the charge of a self-complete RN black hole. The Compton line still intersects the *inner* horizon for $\sqrt{137} < n < \sqrt{274}$, with $n = 16$ (right curve) being the last solution which allows this. However, these solutions do not exhibit self-completeness since they penetrate the $r < R_\mathrm{P}$ region where quantum gravity applies. For $n > \sqrt{274}$, not even the inner horizon intersects the Compton line, ensuring a clear distinction between particles and black holes.

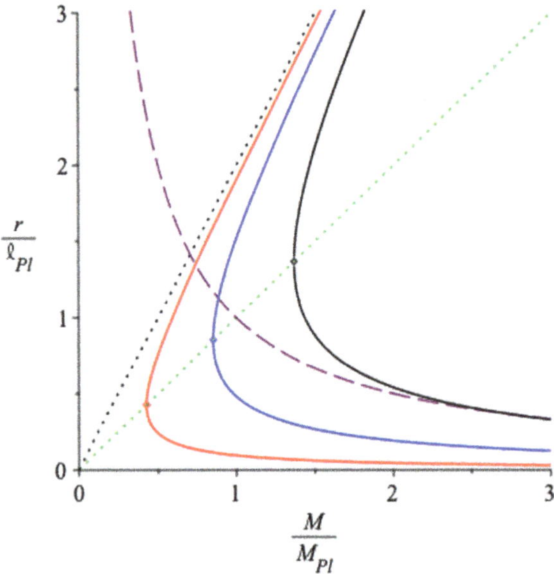

Fig. 7 The solid curves show the outer and inner horizons for a standard RN black hole with $n = 5, 11, 16$ (left to right). For each n, the horizons meet at the extremal mass on the line $r = M/M_{\mathrm{P}}^2$ (green dotted) and are bounded from above by the Schwarzschild radius $r_{\mathrm{S}} = 2M/M_{\mathrm{P}}^2$ (black dotted line). The Compton curve is shown by the dashed line and the inner horizon asymptotes to this for $n = 16$. Solutions with $n < 11$ penetrate the sub-Planckian RN regime [64]. Reprinted from Carr, B., Mentzer, H., Mureika, J. et al. Self-complete and GUP-modified charged and spinning black holes. Eur. Phys. J. C 80, 1166 (2020). https://doi.org/10.1140/epjc/s10052-020-08706-0 under CC BY, ©The Authors

The temperature of the RN solution for quantized charge $Q = n\sqrt{\alpha_e}$ is calculated from the surface gravity as

$$kT = \frac{1}{4\pi}\frac{df}{dr}\bigg|_{r_+} = \frac{M_{\mathrm{P}}^2\sqrt{M^2 - \alpha_e n^2 M_{\mathrm{P}}^2}}{2\pi\,(M + \sqrt{M^2 - \alpha_e n^2 M_{\mathrm{P}}^2}\,)^2}\,. \tag{55}$$

Figure 8 shows the function $T(M)$. It asymptotes to the Hawking expression for $M \gg M_{\mathrm{P}}$ but, as M decreases, it reaches a maximum and then goes to zero as M tends to the minimum mass $\sqrt{\alpha_e}\,n M_{\mathrm{P}}$. Hawking evaporation stops at this mass, as in the Adler model. Such a feature is common to many quantum-gravity-corrected black hole models, for example, the non-commutative geometry inspired models [65, 66], the string T-duality corrected black holes [67], the Hayward model [68], the holographic screen model [69] and a class of GUP modified metrics [70, 71].

The extension of this model to the GUP-modified RN solutions is discussed in Mureika's chapter. Providing $\beta > n^2\alpha_e/2$, the outer horizon behaves as in the GUP-Schwarzschild case [7], with a continuous transition between the gravitational ($r_{\mathrm{CS}} \propto$

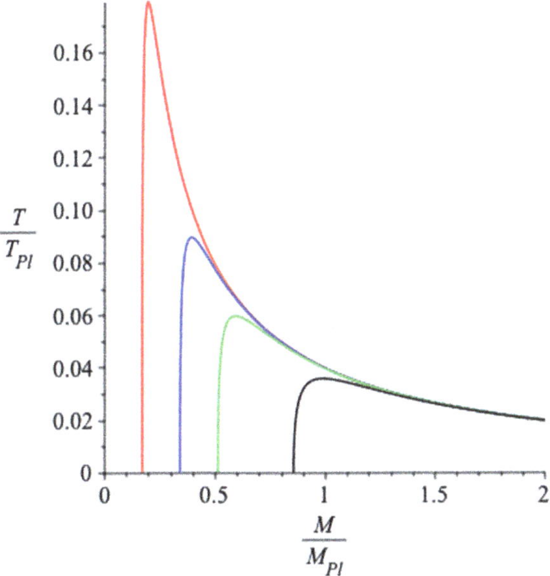

Fig. 8 The function $T(M)$ for a standard RN black hole with $n = 2, 4, 6, 10$ (left to right). The curves end at $M = n\sqrt{\alpha_e}M_{\rm P}$, corresponding to a charged remnant with $T = 0$ [64]. Reprinted from Carr, B., Mentzer, H., Mureika, J. et al. Self-complete and GUP-modified charged and spinning black holes. Eur. Phys. J. C 80, 1166 (2020). https://doi.org/10.1140/epjc/s10052-020-08706-0 under CC BY, ©The Authors

M) and Compton ($r_{\rm CS} \propto M^{-1}$) scaling. For given β, as n increases, the smoothness of the minimum sharpens as the charge reaches a maximum value $n_{max} = \sqrt{2\beta/\alpha_e}$, which differs from the maximum of 11 required for self-completeness in the RN case. Above this value, there is a mass gap in the (r, M) diagram, which replicates the behaviour of the standard RN inner and outer horizons. We therefore speculate that these solutions represent super-Planckian black holes on the right and Compton-like objects (*i.e.* particles) on the left. The GUP-RN black hole has a temperature similar to the GUP-Schwarzschild one, with a zero-temperature remnant for some range of charge up to the maximum value of n. The GUP-Kerr metric exhibits similar behaviour but there is a critical spin instead of a critical charge.

These considerations suggest that there is a fundamental link between elementary particles and black holes. This proposal goes back to the 1970s, when it was motivated in the context of strong gravity theories by the link between Regge trajectories and extreme Kerr solutions. In our case, it is prompted by the Compton-Schwarzschild correspondence, which is based on the $M \to 1/M$ duality. Indeed, this suggests that elementary particles could be black holes with sub-Planckian mass. However, this duality no longer applies in the charged and rotating cases since the Compton wavelength is independent of Q and J for a particle.

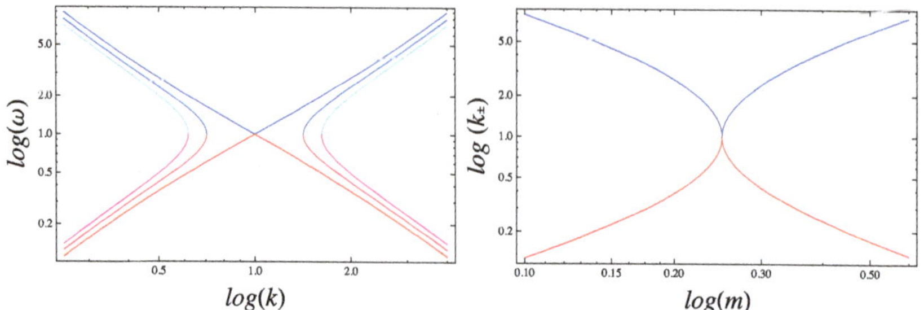

Fig. 9 Illustrating how the dispersion relations $\omega_\pm(m, k)$ for three values of m (left) and the limiting wavenumbers $k_\pm(m)$ are changed in the proposed model (right) [37]. Reprinted from Lake, M.J., Carr, B. The Compton-Schwarzschild correspondence from extended de Broglie relations. J. High Energ. Phys. 2015, 105 (2015). https://doi.org/10.1007/JHEP11(2015)105 with permission, ©Springer Nature BV. All rights reserved

5 Compton-Schwarzschild Correspondence From Extended de Broglie Relations

Canonical (non-gravitational) quantum mechanics is based on the concept of wave-particle duality, encapsulated in the de Broglie relations $E = \hbar\omega$ and $p = \hbar k$. When combined with the energy-momentum relation for a non-relativistic point particle, these lead to the dispersion relation $\omega = (\hbar/2m)k^2$. However, these relations break down above the Planck energy, since they correspond to wavelengths $\lambda \ll R_P$ or periods $t \ll t_P$. Ref. [37] therefore proposes modified forms for the de Broglie relations which may be applied even for $E \gg M_P c^2$, with the additional terms being interpreted as representing the self-gravitation of the wave packet.

The simplest such relations are $E = \hbar\Omega$ and $p = \hbar\kappa$ with

$$
\Omega = \begin{cases} \omega_P^2 \left(\omega + \omega_P^2/\omega\right)^{-1} & (m < M_P) \\ \beta \left(\omega + \omega_P^2/\omega\right) & (m > M_P) \end{cases}, \quad \kappa = \begin{cases} k_P^2 \left(k + k_P^2/k\right)^{-1} & (m < M_P) \\ \beta \left(k + k_P^2/k\right) & (m > M_P). \end{cases}
\tag{56}
$$

Continuity of E, p, $dE/d\omega$ and dp/dk at $\omega = \omega_P$ and $k = k_P$ is ensured by setting $\beta = 1/4$. The relation $\Omega = (\hbar/2m)\kappa^2$ then leads to new dispersion relations, quadratic in ω, which can be solved for both $E \ll M_P c^2$ and $E \gg M_P c^2$. The two solution branches, $\omega_\pm(k, m)$, are shown as functions of k for the three values of m in Fig. 9 (left). The solutions are dual under the transformation $m \to M_P'^2/m$ where $M_P' \equiv (\pi/2)M_P$. Canonical non-relativistic quantum mechanics is recovered in the bottom left region, where $\omega_- \approx (\hbar/2m)k^2$. The branches meet at $\omega_\pm(k_P) = \omega_P$ for the critical case $m = M_P'$ but there is a gap in the allowed values of k for $m \neq M_P'$. The limiting values for a given mass, $k_\pm(m)$, are shown in Fig. 9 (right) and these also exhibit duality. These values correspond to the Schwarzschild formula for $E \gg M_P c^2$ and the Compton formula for $E \ll M_P c^2$, so this is another interpretation of the BHUP correspondence.

6   Higher-Dimensional Black Holes

The black hole boundary in Fig. 2 assumes there are three spatial dimensions but many theories suggest that the dimensionality could increase on small scales. Although the extra dimensions are often assumed to be compactified on the Planck length, there are also models [72] in which they are extended on a scale much larger than R_P. For example, in the Randall-Sundrum (RS) picture [73, 74], the universe corresponds to a D_3 brane in a 5-dimensional bulk. The bulk dimension can be infinitely extended in this case, although it looks locally compactified since the Anti-de Sitter space is warped, so that the D_3-brane has some finite thickness. One could also consider models with more than one large dimension. The model of Arkani-Hamed et al. [75] has n extra spatial dimensions, all compactified on the same scale, and one could also consider models with a hierarchy of compacitifed dimensions, so that the dimensionality increases as one goes to smaller scales.

Let us now consider the behavior of black holes and quantum mechanical particles in spacetimes with extra directions. For simplicity, we first assume that all the extra dimensions are compactified on a single length scale R_E and that the standard expression for the Compton wavelength ($R_C \propto M^{-1}$) applies even in the higher-dimensional case. There are then two interesting mass scales: the mass whose Compton wavelength is R_E,

$$M_E \equiv \frac{\hbar}{cR_E} \simeq M_P \frac{R_P}{R_E} , \tag{57}$$

and the mass whose Schwarzschild radius is R_E,

$$M'_E \equiv \frac{c^2 R_E}{G} \simeq M_P \frac{R_E}{R_P} . \tag{58}$$

These mass scales are reflections of each other in the line $M = M_P$, so that $M'_E = M_P^2 / M_E$. We now examine why the existence of large extra dimensions could lead to TeV quantum gravity and the production of black holes at accelerators.

If the number of extra dimensions is n, then in the Newtonian approximation the gravitational potential generated by a mass M is [76, 77]

$$V_{\text{grav}} = \frac{G_D M}{R^{1+n}} \quad (R < R_E) , \tag{59}$$

where G_D is the higher-dimensional gravitational constant and $D = 4 + n$ is the number of spacetime dimensions. This also applies in general relativity since for $R < R_E$ we may take the weak field limit of Einstein's field equations in $4 + n$ dimensions. For $R > R_E$, the factor R^{1+n} is replaced by $R R_E^n$, where R is the 3-dimensional radius, so one recovers the usual form:

126   B. J. Carr

$$V_{\text{grav}} = \frac{GM}{R} \quad \text{with} \quad G = \left(\frac{G_D}{R_E^n}\right) \quad (R > R_E). \tag{60}$$

Thus the higher-dimensional nature of the gravitational force is only manifest for $R < R_E$. In the Newtonian limit, the effective gravitational constants at large and small scales are different because of the dilution effect of the extra dimensions.

The implication of Eq. (59) is that the usual expression for the Schwarzschild radius no longer applies for masses below M_E'. If the black hole is assumed to be spherically symmetric in the higher-dimensional space, the expression for R_S must be replaced with [78]

$$R_S \simeq R_E \left(\frac{M}{M_E'}\right)^{1/(n+1)} \simeq R_* \left(\frac{M}{M_P}\right)^{1/(1+n)}, \tag{61}$$

where

$$R_* \simeq (R_P R_E^n)^{1/(1+n)}. \tag{62}$$

Therefore the slope of the black hole boundary in Fig. 2 becomes shallower for $M < M_E'$. This is indicated in Fig. 10a for various values of n. The intersect with the (usual) Compton line then becomes

$$R_P' \simeq (R_P^2 R_E^n)^{1/(2+n)}, \quad M_P' \simeq (M_P^2 M_E^n)^{1/(2+n)}. \tag{63}$$

This gives $M_P' \simeq M_P$ and $R_P' \simeq R_P$ for $R_E \simeq R_P$ but $M_P' \ll M_P$ and $R_P' \gg R_P$ for $R_E \gg R_P$. The higher-dimensional Planck mass therefore decreases (allowing the possibility of TeV quantum gravity) and the higher-dimensional Planck length increases. The relationship between the various key scales in the above analysis (R_E, R_E', R_P, R_P', M_P, M_P', R_*) is illustrated in Fig. 11 for the case of one extra spatial dimension ($n = 1$).

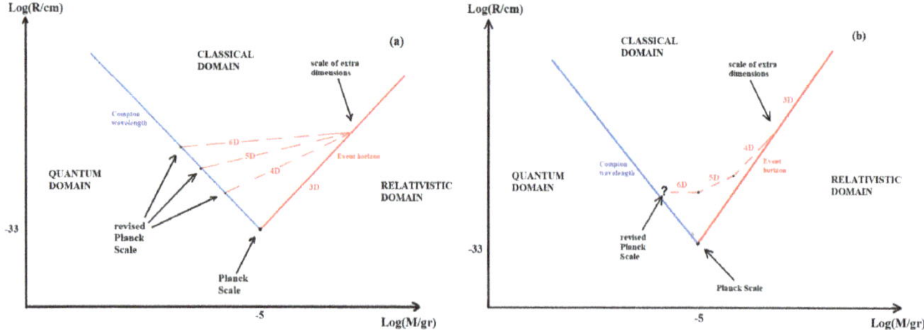

Fig. 10 Modification of the Schwarzschild line in the (M, R) diagram for extra compact dimensions associated with a single length scale (a) or a hierarchy of length scales (b). If the Compton scale preserves its usual form, the Planck scales are shifted as indicated [86]

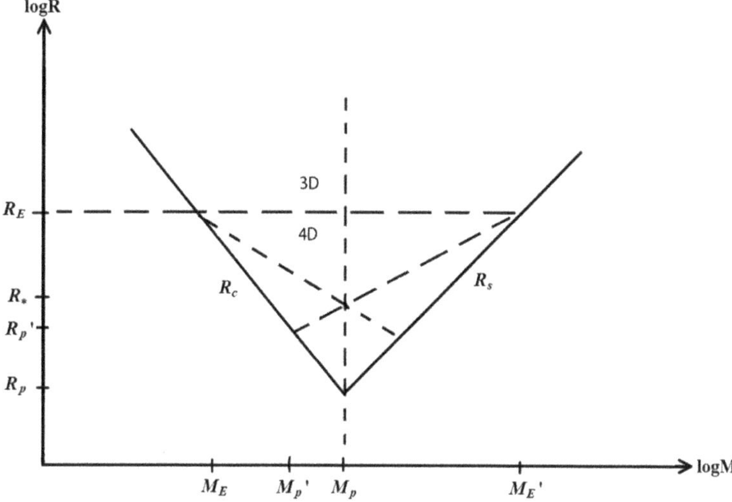

Fig. 11 Key mass and length scales in the 3D case (solid lines) and 4D cases (dotted lines) if the extra dimension is compactified on a scale R_E. The associated Compton and Schwarzschild masses are M_E and M_E', respectively. The revised Planck scales are M_P' and R_P' if duality is violated but M_P and R_* if it is preserved [86]

In principle, such effects would permit the production of small black holes at the Large Hadron Collider (LHC), with their evaporation leaving a distinctive signature [79–81]. If the accessible energy is $E_{\max}$, then the extra dimensions can be probed for

$$
R_E > R_P \left(\frac{c^2 M_P}{E_{\max}} \right)^{(2+n)/n} \simeq 10^{-18+30/n} \left(\frac{E_{\max}}{10\,\mathrm{TeV}} \right)^{-(2+n)/n} \mathrm{cm}\,, \qquad (64)
$$

where $E_{\max}$ is normalised to 10 TeV, the energy associated with the LHC. Thus higher-dimensional black holes can be created at the LHC providing

$$
R_E > 10^{-18+30/n}\,\mathrm{cm} \simeq
\begin{cases}
10^{12}\,\mathrm{cm} & (n=1) \\
10^{-3}\,\mathrm{cm} & (n=2) \\
10^{-14}\,\mathrm{cm} & (n=7) \\
10^{-18}\,\mathrm{cm} & (n=\infty)\,.
\end{cases}
\qquad (65)
$$

Clearly, $n=1$ is excluded on empirical grounds but $n=2$ is possible. One expects $n=7$ in M-theory [82], so it is interesting that R_E must be of order a Fermi in this case. $R_E \to 10^{-18}$ cm as $n \to \infty$ since this is the smallest scale probed by the LHC.

The above analysis assumes that all the extra dimensions have the same size. One could also consider a hierarchy of compactification scales, $R_i = \alpha_i R_P$ with $\alpha_1 \geq \alpha_2 \geq \dots \geq \alpha_n \geq 1$, such that the dimensionality progressively increases as one goes to smaller distances [83]. In this case, the effective *average* length scale associated with the compact internal space is

$$\langle R_E \rangle = \left(\prod_{i=1}^{n} R_i \right)^{1/n} = R_P \left(\prod_{i=1}^{n} \alpha_i \right)^{1/n} \tag{66}$$

and the new effective Planck scales are

$$R_P' \simeq \left(R_P^2 \prod_{i=1}^{n} R_i \right)^{1/(2+n)} \simeq (R_P^2 \langle R_E \rangle^n)^{1/(2+n)} \tag{67}$$

$$M_P' \simeq \left(M_P^2 \prod_{i=1}^{n} M_i^n \right)^{1/(2+n)} \simeq (M_P^2 \langle M_E \rangle^n)^{1/(2+n)} , \tag{68}$$

where $M_i \equiv \hbar/(cR_i)$ and $\langle M_E \rangle \simeq \hbar/(c\langle R_E \rangle)$. For $R_{k+1} < R < R_k$, the effective Schwarzschild radius is then given by

$$R_S = R_{*(k)} \left(\frac{M}{M_P} \right)^{1/(1+k)} \quad \text{with} \quad R_{*(k)} = \left(R_P \prod_{i=1}^{k \leq n} R_i \right)^{1/(1+k)} . \tag{69}$$

This situation is represented in Fig. 10b. Clearly, for given n, the Planck scales are not changed as much as when all the extra dimensions have the same scale.

There is still no evidence for the extra dimensions [84], which suggests that either they do not exist or they have a compactification scale R_E which is so small that M_P' exceeds the energy attainable by the LHC. However, another possible reason is that the M dependence of R_C is affected by the extra dimensions. In addressing this question, one must appreciate that the Compton wavelength arises in several different contexts. The expression $R_C = h/(Mc)$ first appeared historically in the context of the Compton scattering of photons off electrons [85] and this is also relevant to processes (such as pair-production) which involve turning photon energy (hc/λ) into rest mass energy (Mc^2). On the other hand, the *reduced* Compton wavelength $\hbar/(Mc)$ appears naturally in the Klein-Gordon and Dirac equations. One can also associate the Compton wavelength with the *localisation* of a particle and this is most relevant in the present context. There are both non-relativistic and relativistic arguments for this notion, as discussed in detail in Ref. [86]. While all these arguments give roughly the same value in three dimensions, this may not apply in higher dimensions.

We now address the question of how the Compton wavelength of a fundamental particle—defined as the minimum possible positional uncertainty over measurements in *all* independent spatial directions—scales with mass if there exist n extra compact dimensions. Ref. [86] argues that the *effective* Compton wavelength depends on the form of the $(3 + n)$-dimensional wavefunction. The condition of spherical symmetry in the three large dimensions implies that the directly observable part of the wave-function is characterized by a single length scale, the 3-dimensional radius of the wave packet ΔR_{3D}. One can characterise the physical distribution of the wave packet by the $(1+n)$-dimensional volume

$$V_{(1+n)} \simeq \Delta R_{3D} \prod_{i=1}^{n} \Delta x_i \equiv (\Delta \mathscr{R})^{1+n}, \qquad (70)$$

where Δx_i characterizes the ith spatial dimension and $\Delta \mathscr{R}$ corresponds to the effective $(1+n)$-dimensional radius of the particle. In particular, for wave packets that are spherically symmetric in the large directions but irregular in the compact space, it is this length scale which controls pair-production rather than the geometric average over all $3+n$ dimensions. When extrapolating the usual arguments for the Compton wavelength to the case of compact extra dimensions, we therefore argue that $\Delta \mathscr{R}$ is relevant but identify the spread in the large dimensions of momentum space with the rest mass:

$$R_C \simeq \Delta \mathscr{R}, \quad \Delta P_{3D} \simeq Mc. \qquad (71)$$

This is also consistent with the relativistic interpretation of the Compton wavelength as the minimum localization scale for the wave packet below which pair-production occurs. If the wave-function is spherically symmetric in *all* the dimensions, then one has $R_C \propto M^{-1}$ (as usually assumed). However, if the wave function is pancaked in the extra dimensions and maximally asymmetric, then Eqs. (70) and (71) imply $R_C \propto M^{-1/(1+n)}$.

This has important implications for the duality between the Compton wavelength and the Schwarzschild radius. In $(3+1)$-dimensional spacetime, the first scales as $R_C \sim M^{-1}$, whereas the second scales as $R_S \sim M$, so the two are related via $R_S \sim R_P^2/R_C$. In higher-dimensional spacetimes with n compact extra dimensions, $R_S \sim M^{1/(1+n)}$ on scales smaller than the compactification radius R_E, which breaks the symmetry between particles and black holes if the Compton scale is unchanged. However, we have argued that the effective Compton scale depends on the form of the wavefunction in the higher-dimensional space. If this is maximally asymmetric (i.e. pancaked) in the full $3+n$ spatial dimensions, then we have argued that the effective radius scales as $R_C \sim M^{-1/(1+n)}$ rather than M^{-1} on scales less than R_E, which preserves the duality.

On the other hand, as indicated in Fig. 11, Eq. (63) no longer applies. Instead, the higher-dimensional Planck mass is unchanged but the Planck length is increased to the value R_* given by Eq. (62), which is even larger than before. So in this scenario, quantum gravity and microscopic black hole production are associated with the standard Planck energy, as in the 3-dimensional scenario. On the other hand, while there is no TeV quantum gravity in this scenario, the preservation of duality has interesting physical implications. In particular, the onset of pair-production is 'lifted' relative to the 3-dimensional case in the range $R_P < R < R_E$. This means that extra-dimensional effects may become visible via enhanced pair-production rates for

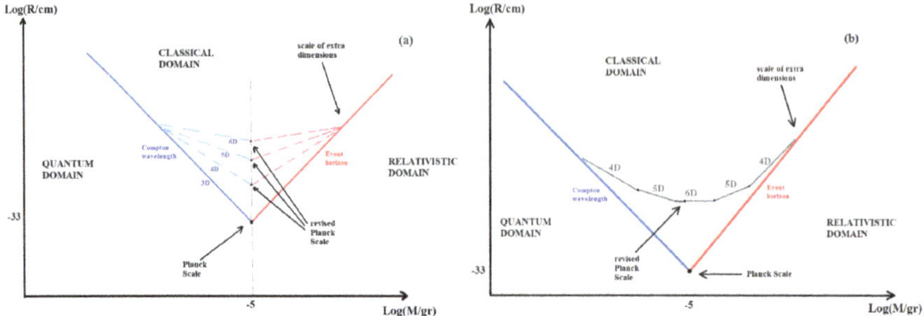

Fig. 12 Modifications of Fig. 2 for extra dimensions compactified on a single length scale for (a) a hierarchy of length scales or (b) a quasi-spherically symmetric higher-dimensional wave packet, preserving the duality between the Compton and Schwarzschild expressions [86]

particles with energies $E > M_E c^2 = \hbar c/R_E$. This prediction is consistent with minimum length uncertainty relations obtained from D-particle scattering amplitudes in string theory.

To summarize, for n extra dimensions compactified on a length scale R_E, we have

$$
R_C \simeq \begin{cases} R_P \frac{M_P}{M} & (R_C > R_E) \\ R_* \left(\frac{M_P}{M}\right)^{1/(1+n)} & (R_C < R_E) \end{cases} \tag{72}
$$

$$
R_S \simeq \begin{cases} R_P \frac{M}{M_P} & (R_S > R_E) \\ R_* \left(\frac{M}{M_P}\right)^{1/(1+n)} & (R_S < R_E) \end{cases} \tag{73}
$$

if duality is preserved and these lines intersect at (R_*, M_P). The effective Planck length is increased but this does not allow the production of higher-dimensional black holes at accelerators. Thus, the constraint on R_E given by Eq. (65) no longer applies. For a hierarchy of scales, the expressions for $R_C < R_E$ become

$$
R_C \simeq R_{*(k)} \left(\frac{M_P}{M}\right)^{1/(1+k)} , \quad R_S \simeq R_{*(k)} \left(\frac{M}{M_P}\right)^{1/(1+k)} \quad (R_{k+1} < R_C < R_k) , \tag{74}
$$

where $R_{*(k)}$ is defined in Eq. (69). This scenario is illustrated in Fig. 12a for extra dimensions compactified on a single length scale R_E and in Fig. 12b for a hierarchy of length scales.

We now consider the temperature of a higher-dimensional black hole. If all the extra dimensions have the same compactification scale R_E and the standard expression for the Compton wavelength applies, then the temperature is modified to [87, 88]

$$
T_H \simeq T_P' \left(\frac{M_P'}{M}\right)^{1/(1+n)} \simeq T_* \left(\frac{M_P}{M}\right)^{1/(1+n)} . \tag{75}
$$

Here M'_P is given by Eq. (63) and we have used the definitions

$$T'_\mathrm{P} \equiv M'_\mathrm{P} c^2 / k_B, \quad T_* \equiv (T_\mathrm{P} T_E^n)^{1/(1+n)}, \quad T_E \equiv M_E c^2 / k_B \,. \tag{76}$$

The M-dependence in Eq. (75) can be derived heuristically from the relations

$$\Delta R_\mathrm{T} \simeq R_\mathrm{S} \propto M^{1/(1+n)}, \quad \Delta P_\mathrm{T} \propto 1/(\Delta R_\mathrm{T}), \quad T_H \propto \Delta P_\mathrm{T} \propto M^{-1/(1+n)} \,, \tag{77}$$

where ΔP_T and ΔR_T represent the uncertainties in the full space, in which the wave function is assumed to be spherically symmetric. The temperature can also be obtained from the surface gravity, which gives the same result:

$$T_\mathrm{H} \propto \kappa \propto M / R_S^{2+n} \propto M^{-1/(1+n)} \,. \tag{78}$$

Note that Eq. (75) extends all way down to the reduced Planck scale M'_P, where the temperature has the maximum possible value ($T'_\mathrm{P} = M'_\mathrm{P}$).

If the form of the Compton wavelength is modified to preserve duality, there are two ways to generalize the above result. The first way assumes that ΔR_T is replaced by $\Delta \mathscr{R}$ in Eq. (77) but that ΔP_{3D} is still the momentum associated with the temperature. One then has

$$\Delta \mathscr{R} \simeq R_S \propto M^{1/(1+n)}, \quad \Delta P_{3D} \propto 1/(\Delta \mathscr{R})^{1+n}, \quad T_\mathrm{H} \propto \Delta P_{3D} \propto M^{-1} \,, \tag{79}$$

where the second relation comes from Eq. (70). In this case, the black hole temperature reverts to the standard Hawking expression, without any dependence on n, and the largest black hole temperature is just T_P. The second way corresponds to replacing the last two conditions in Eq. (79) with

$$\Delta P_{3D} \propto 1/(\Delta R_{3D}) \propto 1/M, \quad T_H \propto 1/\Delta \mathscr{R} \propto M^{-1/(1+n)} \,, \tag{80}$$

and is equivalent to the surface gravity argument:

$$\kappa \propto M / R_S^{2+n} \propto \Delta \mathscr{R}^{1+n} / \Delta \mathscr{R}^{2+n} \propto 1/\Delta \mathscr{R} \,. \tag{81}$$

Therefore the black hole temperature is still given by Eq. (75) and has a maximum value of T_*. Since both the above arguments are heuristic, we cannot be sure which one is correct. The issue is whether one associates the black hole temperature with a length or momentum scale in higher dimensions, these being inequivalent if the black hole is spherically symmetric but the particle wave-function is not.

We now consider the consequences of these results for PBH evaporation. In the 3-dimensional model ($n = 0$), PBHs complete their evaporation at the present epoch if they have an initial $M_0 \simeq 10^{15}$g and an initial radius $R_0 \simeq 10^{-13}$ cm, comparable to the size of a proton [89]. For most of their lifetime these PBHs are producing photons with energy $E_0 \simeq 100$ MeV, so the extragalactic γ-ray background at this energy places strong constraints on their number density and current explosion rate [90]. In

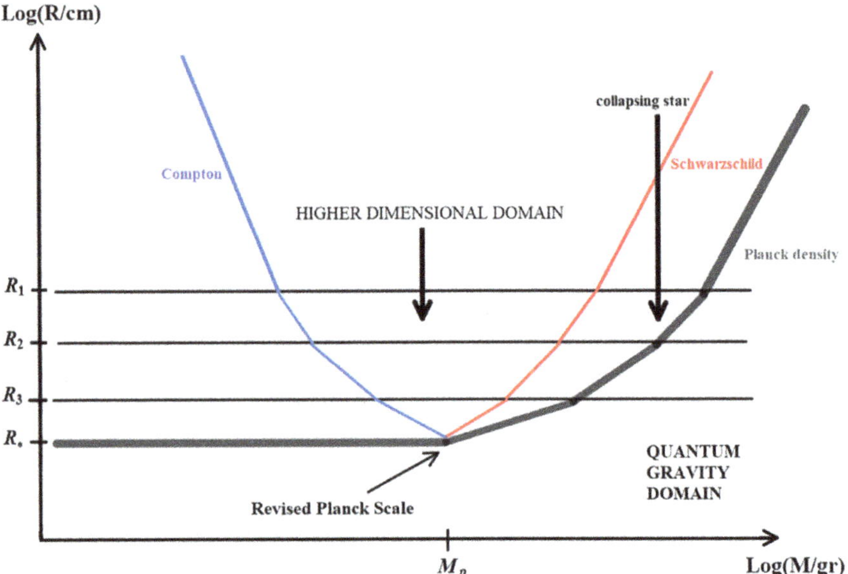

Fig. 13 Showing the form of the Compton and Schwarzschild scales for a hierarchical model with three compactified dimensions in which the Compton-Schwarzschild duality is preserved. In this case, the Planck length but not the Planck mass is modified and the collapsing matter may enter the quantum gravity regime at the modified Planck density [86]

principle, these PBHs could also contribute to cosmic-ray positrons and antiprotons, although there are other possible sources of these particles. However, the black holes evaporating at the present epoch are necessarily higher dimensional if $R_E > 10^{-13}$ cm. In the TeV quantum gravity scenario, for example, Eq. (65) implies that this condition is always satisfied for $n < 7$ and this is expected in M-theory because the maximum number of compactified dimensions is 7. Figure 13 shows the (M, R) diagram for a hierarchical scenario with three extra dimensions, compactified on scales R_1, R_2 and R_3. We therefore need to recalculate the critical mass and temperatures of PBHs evaporating at the present epoch, distinguishing between the standard case in which duality is broken and the alternative case in which it is preserved. However, as indicated above, there is some ambiguity here.

Our discussion of higher-dimensional black holes has assumed that the simple power-law forms for R_S and R_C apply all the way to their intersect at the (modified) Planck scale. However, the BHUP correspondence suggests that they should be unified in some way, which would smooth the minima in Figs. 10, 12 and 13. The simplest interpretation of the BHUP correspondence suggests the the modified higher-dimensional Compton wavelength should become

$$R'_C = \frac{\hbar}{Mc}\left[1 + \left(\frac{M}{M'_P}\right)^{(n+2)/(n+1)}\right] \quad (R < R_E) \tag{82}$$

if the Compton wavelength preserves its 3-dimensional form or

$$R'_{\mathrm{C}} = R_* \left(\frac{M_{\mathrm{P}}}{M}\right)^{1/(1+n)} \left[1 + \left(\frac{M}{M'_{\mathrm{P}}}\right)^{2/(n+1)}\right] \quad (R < R_E) \tag{83}$$

if duality between R_{S} and R_{C} is preserved. However, the literature on this is indecisive and there are subtleties of interpretation.

Finally, if we interpret the Compton wavelength as marking the boundary on the (M, R) diagram below which pair-production rates becomes significant, we expect the presence of compact extra dimensions to affect pair-production rates at high energies. Specifically, we expect pair-production rates at energies above the mass scale associated with the compact space, $M_E \equiv \hbar/(cR_E)$, to be enhanced relative to the 3-dimensional case. There is tentative theoretical evidence that enhanced pair-production may be a generic feature of higher-dimensional theories in which some directions are compactified [91, 92].

7 Brane Cosmology and Higher Dimensional Embeddings

In the type II Randall-Sundrum picture the flat spacetime of special relativity is regarded as a 4D 'brane' embedded in a 5D 'bulk' with an extended extra dimension. In the cosmological version of this picture (brane cosmology) the brane is curved and its curvature and the cosmic expansion are generated by the brane's motion through the 5th dimension. More precisely, the metric of the bulk corresponds to the 5D Schwarzschild—Anti de Sitter (AdS) solution, which can be written in the static form [93–96]:

$$ds_5^2 = -F(R)dT^2 + F(R)^{-1}dR^2 + R^2 \left[\frac{dr^2}{1 - Kr^2} + r^2 d\Omega^2\right] \tag{84}$$

where
$$F(R) = K - m/R^2 + (R/L)^2 \tag{85}$$

and we choose units with $G = c = 1$. Here (r, θ, ϕ, T) are the coordinates within the brane, T being cosmic time, and the R coordinate corresponds to the fifth dimension. The brane itself is located at $R = a(T)$, where $a(T)$ is the cosmic scale factor, so this describes its trajectory through the bulk. Each value of T define a hypersurface of homogeneity (as in the FRW case) and the constant K, which is 0 or ± 1, describes the curvature of these hypersurfaces. This solution only applies for $R < a(T)$ because one must impose Z_2 symmetry across the brane.

The constant L in Eq. (85) is the length-scale $\sqrt{3/|\Lambda|}$ associated with the 5D cosmological constant Λ. This is negative, corresponding to the AdS case, in order that gravity is confined to the brane and this is also relevant to the AdS/CFT correspondence [97]. The solution with $\Lambda = 0$ (or infinite L) corresponds to the Dvali-Gabadadze-Porrati (DGP) solution [98], in which a 4D Minkowski space is embedded in a 5D Minkowski space. This model was proposed to reproduce the cosmic acceleration without the need for a non-zero vacuum energy density but it may no longer be compatible with observations [99].

One can also choose coordinates in which the brane is fixed but the bulk metric is no longer static. This is related to the way in which the de Sitter solution—which can be represented as a 5D hypersphere—may itself be regarded as static or dynamic, depending on how one slices the hypersphere. In this case, the 5D metric can be written as

$$ds_5^2 = -N^2(t, y)dt^2 + dy^2 + A^2(t, y)\left[\frac{dr^2}{1 - Kr^2} + r^2 d\Omega^2\right], \tag{86}$$

where y is the 5th coordinate and t is another time coordinate. The brane is at $y = 0$, so that $N(t, 0) = 1$, and one can show that

$$A = a(t)\exp(-|y|/L), \tag{87}$$

where $a(t) = A(t, 0)$ is the FRW scale factor. So the metric is locally 'warped', with the matter being confined to a hypersurface at $y = 0$ of thickness L.

The constant m in Eq. (85) is the mass of the 5D black hole. The Randall-Sundrum solution has $m = 0$ and constant a, so the brane is static. If $K = 1$ and m is non-zero, the metric corresponds to a 5D Schwarzschild black hole in a de Sitter background, with the 5th coordinate playing the role of the 'radial' distance. For non-zero Λ, one still has the AdS solution asymptotically but there is now a range of R for which $F \approx 1$ providing $L \gg \sqrt{m}$, so the 5D metric in this region takes the form

$$ds_5^2 \approx -dT^2 + dR^2 + R^2[(1 - r^2)^{-1}dr^2 + r^2 d\Omega^2]. \tag{88}$$

(This situation would not pertain for $K = 0$ or $K = -1$ or for $K = +1$ and $L \ll \sqrt{m}$.) Apart from the dR^2 term, this corresponds to the Friedmann solution. However, the full $K = +1$ solution has an event horizon where $F = 0$, corresponding to

$$R_{\mathrm{H}} = \sqrt{[(L^4 + 4mL^2)^{1/2} - L^2]/2} \approx \begin{cases} m^{1/2} & (m \ll L^2) \\ m^{1/4}L^{1/2} & (m \gg L^2) . \end{cases} \tag{89}$$

The first expression corresponds to the standard radius of a 5D event horizon, so the Universe emerges out of a 5D white hole and in the time-reversed solution collapses into a 5D black hole. The second expression cannot apply if one requires a range of R for which $F \approx 1$. More precisely, the black hole part of the 5D metric

$$- F(R)dT^2 + F(R)^{-1}dR^2 + R^2 r^2 d\Omega^2 \tag{90}$$

turns into the FRW part of the 4D metric

$$-dT^2 + R^2[(1 - r^2)^{-1}dr^2 + r^2 d\Omega^2] \,, \tag{91}$$

in that $F \approx 1$ for $L \gg R \gg R_{\rm H}$.

One reason for discussing this solution is that it explains the inclusion of the "Universal" black hole in Fig. 1. It also justifies the link of the top of the Uroborus with higher-dimensional black holes on both the left and right. Since this is where the micro and macro domains meet, it is perhaps unsurprising that they are both associated with 5D black hole solutions but it is interesting to see the solutions explicitly. However, the considerations of Sect. 6 suggest an extension of this scenario. If there are a hierarchy of compactified dimensions on the left, this must also apply on the right. So the creation of the Universe could also be hierarchical, with a sequence of emergences from higher dimensional black holes (i.e. the 4D universe emerges from a 5D black hole, the 5D universe emerges from a 6D black hole etc.) This suggests that quantum gravity itself may involve higher-dimensional black holes, which is part of the motivation for the title of this chapter.

Another reason for discussing brane cosmology is that it throws light on the concept of an embedding space in general relativity (GR). One way to envisage the curvature of spacetime is to regard it as a curved hypersurface in a higher dimensional embedding space (often assumed to be flat). One can even classify solutions in GR according to the dimensionality of the embedding space. For example, the FRW and dS models are embeddable in 5D flat space and the Schwarzschild solution is embeddable in 6D flat space. Indeed, all GR solutions in n-dimensions are *locally* embeddable in a flat space of $n(n + 1)/2$ dimensions, this being the number of independent components of the n-dimensional metric. For $n = 4$, this gives a 10D embedding space, which raises the question of whether this relates to the 10D internal space of superstring theory. However, many more dimensions may be needed for a *global* embedding [100]. For values values of n between 5 and 11, the dimension of the embedding space is 15, 21, 28, 36, 45, 55 and 66.

The higher-dimensional embedding space is usually regarded as a mathematical tool, introduced merely for ease of visualization, but one can also regard it as "real" in some circumstances. For example, this is clearly the case for brane cosmology. In the standard "balloon" analogy, one can regard the 3D spatial hypersurface of the FRW solution as expanding into an extra dimension but one does not attribute physical reality to this because the curvature of the hypersurface is defined intrinsically. However, in brane cosmology the embedding direction is physical and one must regard the cosmological spacetime as a 4D hypersurface within the 5D space.

Another interesting feature of brane cosmology is that the extra dimension appears to be *locally* compactified but it is *globally* extended and this allows it to play a role as an embedding dimension. In the context of M-theory, it is the 11th dimension which is extended in brane cosmology and the other dimensions remain compactified, so one still has an internal space at each point of the brane. However, this raises the question of whether the other dimensions might also be extended. Indeed, one could one envisage a scenario in which *every* extra dimension is extended but locally warped and this would offer the prospect of using the extra dimensions to provide an embedding space.

One of the motivations for this approach is that one can regard the higher-dimensional space as empty, with the stress-energy tensor of the 4D matter resulting from the projection of the higher-dimensional field equations onto a 4D hypersurface. This is sometimes called the "induced matter" approach. For example, the 5th dimension is associated with mass in one approach and many cosmological solutions can be interpreted in this way [101–106]. This is in the spirit of Einstein's dream of having a purely geometrical description of nature. This approach does not necessarily require that the embedding space be flat; it might also be an Einstein space (satisfying $R_{\mu\nu} = 0$). Some interesting theorems have been established in this context [107–110], many of them dating back to Campbell [111]. The relationship between Induced Matter Theory and Membrane Theory has been described by Wesson [112] and this leads to the natural emergence of the Klein-Gordon and Dirac equations. These considerations suggest a curious link between GR and particle physics, although we do not explore this further here.

8 Conclusion: Linking Quantum Theory and General Relativity

This chapter has reviewed a range of problems involving black holes and higher dimensions. The selection of problems reflects the topics I have worked on over the last few decades but the common theme is that they all involve some sort of link between quantum theory and general relativity. The first link is between black holes and the Uncertainty Principle or (more precisely) the duality between the Compton and Schwarzschild scales, which we have seen has particularly interesting implications in the higher-dimensional context. The second link is between black holes and elementary particles, which is reminscent of the strong-gravity models of the 1970s but now invokes sub-Planckian black holes and higher dimensions. The third link is between the local embedding space of general relativity and the internal space of superstring theory.

Each of these links diminishes the traditional dichotomy between quantum theory and general relativity. Indeed, the existence of higher dimensions may amalgamate these theories in a rather surprising way. In some sense, the higher dimensional extension of general relativity permits a classical-type interpretation of quantum theory. For example, quantum non-locality may be a consequence of higher-dimensional connectedness and Everett's "many worlds" may be viewed as residing in the higher-dimensional space. It is even possible that this may elucidate the old philosophical problem of the passage of time. This is still not understood within either relativity theory or quantum theory but I have argued elsewhere that the extended 5th dimension of brane cosmology can describe this [113]. In any case, this is such a profound problem that it is not unreasonable to anticipate that some final theory may elucidate it in some way.

Acknowledgements I thank my collaborators in the work reported here: Matthew Lake, Heather Mentzner, Leonardo Modesto, Jonas Mureika, Piero Nicolini and Isabeau Prémont-Schwarz. Although I did not know Antonio Aurilia well, he pioneered some of the ideas described here and I am very grateful that contributing to this volume has allowed me to connect with him.

References

1. B. J. Carr, in *2nd Karl Schwarzschild Meeting on Gravitational Physics*, ed. by P. Nicolini, M. Kaminski, J. Mureika, M. Bleicher. Springer Proceedings in Physics, vol. 208 (Springer International Publishing, Switzerland, 2018), pp. 85–94
2. J. Mureika, in *Touring the Planck Scale: Antonio Aurilia Memorial Volume*, ed. by P. Nicolini. Fundamental Theories of Physics vol. 219 (Springer-Nature, 2025)
3. J. A. Wheeler, Phys. Rev. **97**, 511–536 (1955)
4. C. Rovelli, F. Vidotto, Int. J. Mod. Phys D **23**, 1442026 (2014)
5. A. Barrau, B. Bolliet, F. Vidotto, C. Weimar, JCAP **02**, 022 (2016)
6. B. Zwiebach, *A First Course in String Theory* (Cambridge University Press, Cambridge, UK, 2009)
7. B. Carr, J. Mureika, P. Nicolini, JHEP **07**, 052, (2016)
8. B. Carr, in *1st Karl Schwarzschild Meeting on Gravitational Physics*, ed. by P. Nicolini, M. Kaminski, J. Mureika, M. Bleicher. Springer Proceedings in Physics, vol. 170 (Springer International Publishing, 2016), pp. 23—31
9. B. Carr, in *1st Karl Schwarzschild Meeting on Gravitational Physics*, ed. by P. Nicolini, M. Kaminski, J. Mureika, M. Bleicher. Springer Proceedings in Physics, vol. 170 (Springer International Publishing, 2016), pp. 159–167
10. A. Aurilia and E. Spalluci, Planck's uncertainty principle and the saturation of Lorentz boosts by Planckian black holes (2013). arXiv:1309.7186 [gr-qc]
11. A. Aurilia, E. Spalluci, Adv. High Energy Phys. **2013**, 531696 (2013)
12. G. Dvali, S. Folkerts, C. Germani, Phys. Rev. D **84**, 024039 (2011)
13. G. Dvali, C. Gomez, Phys Lett B **719**, 419 (2913)
14. G. Dvali D. Flassig, C. Gomez, A. Pritzel, N. Wintergerst, Phys Rev D **88**, 124041 (2913)
15. G. Dvali, C. Gomez, D. Lüst, Phys. Lett. B **753**, 173–177 (2016)
16. G. Dvali, C. Gomez, Astrophys. J. **781**, 2, 112 (2014)
17. G. Dvali, G.F. Giudice, C. Gomez, A. Kehagias, JHEP **1108**, 108 (2011)
18. A. M. Frassino, S. Koppel, P. Nicolini, Entropy **18**, 181 (2016)
19. R. J. Adler, D.I. Santiago, Mod. Phys. Lett. A**14**, 1371 (1999)
20. R. J. Adler, P. Chen, D.I. Santiago, Gen. Rel. Grav. **33**, 2101 (2001)
21. P. Chen, R.J. Adler, Nucl. Phys. Proc. Suppl. **124** 103 (2003)
22. R.J. Adler, Am. J. Phys. **78**, 925 (2010)
23. M. Maggiore, Phys. Lett. B **304**, 65 (1993)
24. M. Maggiore, Phys. Lett. B **319**, 83 (1993)
25. M. Maggiore, Phys. Rev. D. **49**, 5182 (1994)
26. A. Ashtekar, S. Fiarhurst, J. L. Willis, Class. Quant. Grav. **20**, 1031 (2003)
27. G. M. Hossain, V. Husain, S. S. Seahra, Class. Quant. Grav. **27**, 165013 (2010)
28. B. S. Kay, Class. Quant. Grav. **15**, L89–L98 (1998)
29. B. S. Kay and V. Abyaneh, Expectation values, experimental predictions, events and entropy in quantum gravitationally decohered quantum mechanics (2007). arXiv:0710.0992 [quant-ph]
30. G. Veneziano, Europhys. Lett. **2**, 199 (1986)
31. E. Witten, Phys. Today April **49** 4, 24–30 (1996)
32. F. Scardigli, Phys. Lett. B **452**, 39–44 (1999)
33. D. J. Gross, P.F. Mende, Nuc. Phys. B **303**, 407–454 (1988)

34. D. Amati, M. Ciafaloni, G. Veneziano, Phys. Lett. B **216**, 41 (1989)
35. T. Yoneya, Mod. Phys. Lett. A **4**, 1587 (1989)
36. K. Konishi, G. Paffuti, P. Proverpo, Phys. Lett. B **234**, 276–284 (1990)
37. M. J. Lake and B. Carr, JHEP **1511**, 105 (2015)
38. S.W. Hawking, Nature **248**, 30–31 (1974)
39. B. Carr, L. Modesto, I. Premont-Schwarz, Generalized Uncertainty Principle and Selfdual Black Holes (2011). arXiv: 1107.0708 [gr-qc]
40. C. Rovelli, *Quantum Gravity* (Cambridge University Press, Cambridge, UK, 2004)
41. A. Ashtekar, Class. Quant. Grav. **21**, R53 (2004)
42. T. Thiemann, Lect. Notes Phys. **721**, 185-263 (2007)
43. T. Thiemann, Introduction to Modern Canonical Quantum General Relativity (2001). arXiv:gr-qc/0110034 [gr-qc]
44. T. Thiemann, Lect. Notes Phys. **631**, 41–135 (2003)
45. A. Ashtekar, Phys. Rev. Lett. **57**, 2244–2247 (1986)
46. M. Bojowald, Living Rev. Rel. **8**, 11 (2005)
47. A. Ashtekar, M. Bojowald, J. Lewandowski, Adv. Theor. Math. Phys. **7**, 233–268 (2003)
48. M. Bojowald, Phys. Rev. Lett. **86**, 5227–5230 (2001)
49. L. Modesto, Int. J. Theor. Phys. **49**, 1649–1683 (2010)
50. L. Modesto, I. Premont-Schwarz, Phys. Rev. D **80**, 064041 (2009)
51. L. Modesto, Adv. High Energy Phys. **2008**, 459290 (2008)
52. L. Modesto, Class. Quant. Grav. **23**, 5587–5602 (2006)
53. A. Ashtekar, M. Bojowald, Class. Quant. Grav. **23**, 391–411 (2006)
54. L. Modesto, Phys. Rev. D **70**, 124009 (2004)
55. L. Modesto, Int. J. Theor. Phys. **45**, 2235–2246 (2006)
56. F. Caravelli, L. Modesto, Class. Quant. Grav. **27**, 245022 (2010)
57. S. Hossenfelder, L. Modesto, I. Premont-Schwarz, Phys. Rev. D **81**, 044036 (2010)
58. J. Magueijo, L. Smolin, Class. Quant. Grav. **21**, 1725–1736 (2004)
59. , 319–325 (2003)
60. G. 't Hooft, Conf. Proc. C **930308**, 284–296 (1993)
61. J. R. Mureika, P. Nicolini, Phys. Rev. D **84**, 044020 (2011)
62. A. Tzikas, P. Nicolini, J. Mureika, B. Carr, JCAP **1812**, 033 (2018)
63. J. R. Mureika, R. B. Mann, Mod. Phys. Lett. A **26**, 171–181 (2011)
64. B. Carr, H. Mentzer, J. Mureika, P. Nicolini, Eur. Phys. J. C **80**, 12, 1166 (2020)
65. P. Nicolini, A. Smailagic, E. Spallucci, Phys. Lett. B **632**, 547 (2006)
66. P. Nicolini, Int. J. Mod. Phys. A **24**, 1229 (2009)
67. P. Nicolini, E. Spallucci, M. F. Wondrak, Phys. Lett. B **797**, 134888 (2019)
68. S. A. Hayward, Phys. Rev. Lett. **96**, 031103 (2006)
69. P. Nicolini, E. Spallucci, Adv. High Energy Phys. **2014**, 805684 (2014)
70. M. Isi, J. Mureika, P. Nicolini, JHEP **1311**, 139 (2013)
71. M. Knipfer, S. KÂ¨oppel, J. Mureika, P. Nicolini, JCAP **1908**, 008 (2019)
72. P. Kanti, Int. J. Mod. Phys. A **19**, 4899-4951 (2004)
73. L. Randall, R. Sundrum, Phys. Rev. Lett **83**, 3370 (1999)
74. L. Randall, R. Sundrum, Phys. Rev. Lett **83**, 4690 (1999)
75. N. Arkani-Hamed, S. Dimopoulos, G. Dvali, Phys. Lett. B. **428**, 263 (1998)
76. R. Maartens, K. Koyama, Living Rev. Rel. **13**, 5 (2010)
77. R. Maartens, Rept. Prog. Phys. **67**, 2183–2232 (2004)
78. S. Koppel, M. Knipfer, M. Isi, J. Mureika, P. Nicolini in *2nd Karl Schwarzschild Meeting on Gravitational Physics*, ed. by P. Nicolini, M. Kaminski, J. Mureika, M. Bleicher. Springer Proceedings in Physics, vol. 208 (Springer International Publishing, Switzerland, 2018), pp. 141–147
79. S. Dimopoulos, G. Landsberg, Phys. Rev. Lett. **87**, 161602 (2001)
80. S. Giddings, S. Thomas, Phys. Rev. D. **65**, 056010 (2002)
81. L. A. Anchordoqui, J. L. Feng, H. Goldberg, A. D. Shapere, Phys. Rev. D **65**, 124027 (2002)

82. K. Becker, M. Becker, J. H. Schwarz, *String theory and M-theory*: A modern introduction (Cambridge University Press, Cambridge, UK, 2006)
83. B. Carr, Mod. Phys. Lett. A **28**, 1340011 (2013)
84. ATLAS Collaboration, Eur. Phys. J. C **76**, 541 (2016)
85. A. I. M. Rae, *Quantum Mechanics, Fourth Edition* (Institute of Physics Publishing, Bristol and Philadelphia, USA, 2002)
86. M. J. Lake, B. J. Carr, Int. J. Mod. Phys. D **28**, 1930001 (2019)
87. M. Cavaglia, S. Das, R. Maartens, Class. Quant. Grav. **20**, L205 (2003)
88. M. Cavaglia, S. Das, Class. Quant. Grav. **21**, 4511–4522 (2004)
89. B. Carr, K. Kohri, Y. Sendouda, J. Yokoyama, Phys. Rev. D. **81**, 104019 (2010)
90. D. Cline, S. Otwinowsk, Evidence for Primordial Black Hole Final Evaporation: Swift, BATSE and KONUS and Comparisons of VSGRBs and Observations of VSB That Have PBH Time Signatures (2009). arXiv:0908.1352 [astro-ph.CO]
91. X. G. He, Phys. Rev. D **60**, 115017 (1999)
92. O. J. P. Eboli, T. Han, M. B. Magro, P. G. Mercadante, Phys. Rev. D **61**, 094007 (2000)
93. R. Maartens, K. Koyama, Living Rev. Rel. **13**, 5 (2010)
94. P. Bowcock, C. Charmousis, R. Gregory, Class. Quant. Grav. **17**, 4745–4764 (2000)
95. S. Mukhoyama, T. Shiromizu, K. Maeda, Phys Rev D **62**, 024028 (2000)
96. J. Garriga, M. Sasaki, Phys. Rev. D **62**, 043523 (2000)
97. S. Kachru, R. Kallosh, A. Linde, S. R. Trivedi, Phys. Rev. D **68**, 046005 (2003)
98. G. Dvali, G. Gabadadze, M. Porrati, Phys. Lett. B **485**, 208–214 (2000)
99. L. Lombriser, W. Hu, W. Fang, U. Seljak, Phys. Rev. D **80**, 063536 (2009)
100. C. Clarke, in *Proc. Roy. Soc. London Ser. A 314* (London, UK, 1970), pp. 417–428
101. P. Wesson, Astrophys. J. **394**, 19–24 (1994)
102. P. Wesson, Class. Quant. Grav. **19**, 2825–2834 (2002)
103. P. Wesson, Phys. Lett. B **538**, 151–163 (2002)
104. P. Wesson, J. Ponce de Leon, J. Math. Phys. **33**, 3883–3887 (1992)
105. P. Wesson, Mod. Phys. Lett. A **19**, 1995–2000 (2004)
106. B. Mashhon, P. Wesson, Gen. Rel. Grav. **39**, 1403-1412 (2007)
107. C. Romero, R. K. Tavakol, R. Zalaletdinov, Gen. Rel. Grav. **28**, 365–376 (1996)
108. S. Rippl, C. Romero, R. K. Tavakol, Class. Quant. Grav. **12**, 2411-2422 (1995)
109. J. E. Lidsey, C. Romero, R. K. Tavakol, S. Rippl, Class. Quant. Grav. **14**, 865 (1997)
110. J. E. Lidsey, R. K. Tavakol, C. Romero, Mod. Phys. Lett. A **12**, 2319–2323 (1997)
111. E. Campbell, *A Course of Differential Geometry* (Clarendon Press, Oxford, UK, 1926)
112. P. Wesson, *Five-dimensonal physics: Classical and Quantum Consequences of Kaluza-Klein Cosmology*, (World Scientific, Singapore, 2006)
113. B. J. Carr, in *The Philosophy of Cosmology*, ed. by K. Chamcham, J. Silk, J. Barrow and S. Saunders (Cambridge University Press, UK, 2017), pp. 40-65

Occam's Razor for Black Holes

Gia Dvali

Abstract Black hole quantum N-portrait is a microscopic theory that realizes the Occam's razon approach to black hole physics. Instead of modifying gravity, it describes a black hole as saturated state of soft gravitons within ordinary effective theory. This saturation is the key both for understanding the well-established black hole properties, as well as, for predicting new phenomena, such as a drastic departure from Hawking evaporation after half decay and the concept of internal entanglement. Using this theory, we explain our view of what it means from particle physicist's perspective to understand black holes and why having a microscopic theory based on universal mechanisms is crucial for this. This unifying thinking allows to nail why all saturated systems, whether black holes in gravity or baryons in QCD, store and reveal the quantum information in similar ways.

1 Introduction

Black holes are known to be the most efficient devices for storing quantum information in nature. It suffices to say that a black hole with the memory capacity of a human brain, would have a size of 10^{-27} centimeters [1, 2]. This note is about understanding of how a black hole manages to achieve such an incredible capacity of information storage and whether the effect is limited to them. For this purpose, we wish to put a certain framework, developed over series of papers starting from [3], in an unified perspective. All the formal results will therefore be borrowed from the original work to which the reader shall be referred for more technical details.

As a model-builder particle physicist, I am following the well-approved road for understanding physical phenomena. This consists of the following main ingredients: (1) Identifying the framework, usually represented by an effective quantum field

G. Dvali (✉)
Arnold Sommerfeld Center, Ludwig-Maximilians-Universität, Theresienstraße 37, 80333 Munich, Germany
e-mail: Georgi.Dvali@physik.uni-muenchen.de

Max-Planck-Institut für Physik, Föhringer Ring 6, 80805 Munich, Germany

P. Nicolini (ed.), *Touring the Planck Scale*, Fundamental Theories of Physics 219, https://doi.org/10.1007/978-3-031-76066-2_5

theory (EQFT); (2) Identifying the relevant degrees of freedom; and (3) Understanding the generic dynamical mechanisms that operate within the theory. Putting all these together, we built a microscopic theory. We then try to predict new phenomena without further alterations of the model.

This is the way the Standard Model works. First, it incorporates several universal mechanisms, such as, the Higgs mechanism or the asymptotic freedom with dimensional transmutation. Although crucial, these mechanisms are not bound to the Standard Model and are generic in the sense that they can successfully be implemented in variety of gauge theories. So, what defines the standard model is the combination of all three ingredients, with very clearly identified set of degrees of freedom, in form of quarks, leptons and gauge and Higgs bosons. Of course, their choice is subject to standard consistency requirements of QFT, such as, the positivity of norm and anomaly cancellations. Once we have the theory, we no longer touch the structure, but instead, try to calculate and derive predictions. As it is well known, for the Standard Model and QCD this turned out to be an extraordinarily successful program.

2 Black Hole Quantum N-Portrait

In [3] we have taken the same approach to black holes, but with one important simplification. In search of the theory, we short-cut a selection of the QFT degrees of freedom. This is possible thanks to the guiding principle of *Occam's razor*. Namely, instead of trying to modify gravity and/or abandon some basic concepts of QFT, we describe a black hole as a state within the ordinary EFT of Einstein gravity. Then, since black holes are states in theory of pure gravity, not much dispute about the identification of the right degrees of freedom is required. The relevant degree of freedom is of course *graviton*, a massless particle of spin-2, which is an unique propagating QFT degree of freedom in Einstein gravity.

So, the key idea of black hole quantum N-portrait [3] is that a black hole is a bound-state of soft gravitons. By *soft* we mean the wavelength of size of a black hole, R. The rest then follows. For example, the occupation number N is uniquely determined both by matching the energy of a black hole, as well as, by the self-sustainability condition. Indeed, each graviton contributes energy $\epsilon_0 \sim 1/R$, so that the black hole mass is $M_{BH} \sim N/R$. This determines $N \sim M_{BH}R$.

Independently, the same number is obtained from the self-sustainability condition of the bound-state which demands that the kinetic energy of each graviton is balanced by the collective attraction from the rest. This fixes

$$N = \alpha^{-1} = (M_P R)^2 = M_{BH} R \,, \tag{1}$$

where α is the quantum coupling of gravitons of wavelength R. Notice, this coupling is always equal to the *area of a black hole* in Planck units. The above relation tells us that the black hole is a *saturated* state, i.e., state in which the particles are *maximally packed*. That is, although the quantum coupling among the constituents is minuscule, the *collective* coupling, $\lambda_c \equiv N\alpha = 1$, is critical. That is, viewed as a bound-state of gravitons, the black hole is a system at a quantum critical point [4].

It is this criticality that is responsible for black hole's well known properties. For example, the Hawking radiation [5] is a result of the depletion of the graviton condensate due to their quantum re-scattering. As a consequence, at initial times, the bound-state looses on average one graviton of momentum $1/R$ per time R. The emission of harder modes of energy $E \gg 1/R$ is exponentially suppressed because it takes re-scattering of many soft quanta for producing a hard one. This suppression imitates the thermal Boltzmann factor, $\exp(-ER)$.

For an external observer, this creates a effect of a thermal spectrum with temperature $T = 1/R$. However, the microscopic theory reveals that the thermality is only a leading order effect in $1/N$. This allows us to move beyond the Hawking's original computation which corresponds to $N = \infty$ limit of our picture. Therefore, in microscopic theory we can compute corrections and make predictions that are invisible in a semi-classical theory. One of the key predictions is a drastic deviation from the thermal evaporation latest by the time the black hole looses half of its mass [6, 7].

From (1), we see that with no additional assumptions, the saturated value of N came out equal to black hole's Bekenstein-Hawking entropy [8]. This is another indication that we are on a right track. However, we need to understand the underlying reason for this coincidence. That is, we wish to explain the mechanism by which the black hole reaches its extraordinary capacity of information storage. It is as important for the black hole portrait as the Higgs mechanism is for the Standard Model.

The above mechanism was introduced in [4]. For simplicity, we shall reduce it to bare essentials, as in [1]. We shall refer to the effect as the *assisted gaplessness* [9]. This mechanisms is universal in the sense that it is operative in wide range of saturated systems. Again, this property is similar to the Higgs mechanism that can be implemented in various gauge theories for generating a mass gap. The difference is that the assisted gaplessness generates *gapless* modes. These modes then act as qubits (or qudits) and store quantum information at a very low energy cost, thereby, promoting the system into an enhanced capacitor of memory storage.

The essence of the mechanism is that occupation of a soft graviton to a critical level N, renders N other modes of graviton gapless. In other words, system delivers N different species of gapless quantum oscillators. In the absence of a highly populated soft mode, these gapless modes would carry the Planck scale frequencies. It therefore would cost Planck energy to excite any of them around the vacuum. An observer around the Minkowski vacuum would never dare to store any quantum information in such modes, due to enormity of their energy cost. However, the soft mode suppresses their gaps. That is, the soft mode acts as the *master* mode: Being in a highly occupied state, it assists many different Planckian modes to become gapless. The black hole literally trades excitement for diversity.

The bare essentials of the phenomenon of assisted gaplessness can be explained by the following simple Hamiltonian (see, [1])

$$\hat{H} = \epsilon_0 \hat{n}_0 + \left(1 - \frac{\hat{n}_0}{N}\right) \sum_{k=1}^{N} \epsilon_k \hat{n}_k + \cdots , \tag{2}$$

where $\hat{n}_k = \hat{a}_k^\dagger \hat{a}_k$, $k = 0, 1, \ldots, N$ are number operators for the oscillator modes that satisfy the usual commutation relations, $[\hat{a}_j, \hat{a}_k^\dagger] = \delta_{jk}$, $[\hat{a}_j, \hat{a}_k] = 0$. The inessential interaction terms are not shown explicitly. Also, the particle number (non)conservation is not important for the present discussion. The Hamiltonian (2) is rather general and captures the essence of many different systems exposed to assisted gaplessness. The mode $\hat{a}_0$ corresponds to a soft mode that is occupied to near-saturation.

When applied to a black hole, this mode impersonates the constituent graviton of frequency $\epsilon_0 = 1/R$. We shall refer to it as the *master mode*. On the other hand, the set of modes $\hat{n}_{k \neq 0}$ impersonate the graviton modes of Planck momenta. Of course, the term *Planck momenta* must be understood as the modes around a maximal frequency describable within the EFT for a system. So, we shall take $\epsilon_{k \neq 0} \sim M_P$. We shall refer to $k \neq 0$ oscillators as the *memory modes* [1]. As it is clear from (2), the energy gaps of the later modes around the vacuum, $n_0 \ll N$, are very large, $\sim M_P$. Using the modes $\hat{n}_k$, one can form the number eigenstates

$$|n_0, n_1, \ldots, n_N\rangle \equiv |n_0\rangle \otimes |n_1\rangle \otimes, \ldots, \otimes |n_N\rangle , \tag{3}$$

where $n_0, \ldots n_k$ are occupation numbers that can take different values. In them, we can store various quantum memory patterns in form of the number sequences. We shall therefore refer to the space spanned over such states as *memory space*.

Clearly, for $n_0 = 0$ such states cost enormous energy. Therefore, they are useless for storing quantum information. In particular, the energy of the information pattern $|0, 1, \ldots, 1\rangle$ would be $E_{\text{pattern}} \sim M_P N$, which is too high. For example, such a device with the memory capacity of a human brain would weight $\sim 10^4$ tons.

However, increasing the population of the soft master mode changes the picture dramatically. Indeed, for $\hat{n}_0 \neq 0$, the effective gaps $\mathscr{E}_{k \neq 0}$ of the $\hat{n}_{k \neq 0}$-modes are lowered:

$$\mathscr{E}_k = \left(1 - \frac{n_0}{N}\right) \epsilon_k. \tag{4}$$

The black hole state corresponds to a saturated state on which the occupation number of the master mode reaches the critical value $n_0 = N$. As it is clear from (2) and (4), at this point the effective energy gaps of memory modes collapse to zero. Thus, the information pattern that near the vacuum costed energy $M_P N$, in the critical state costs only $E_{\text{pattern}} \sim \epsilon_0 N$. This is a gain by a factor of $\sqrt{N}$.

The same happens when we apply the effect to a black hole. Thanks to the assisted gaplessness, all the degenerate micro-states cost only the energy $\epsilon_0 N$ which is exactly the black hole mass. This explains why black holes are very efficient storers of information. Indeed, an information pattern that on the vacuum would cost energy $\sim NM_P$, in the black hole is stored for the price that is $M_P R \sim \sqrt{N}$ times less. This is an extraordinary profit. For example, for an earth mass black hole the gain is by a factor of $\sim 10^{33}$.

Obviously, the number of degenerate information patterns (3) is exponential in N. Correspondingly, the micro-state entropy scales as $S \sim N$. This is the qualitative origin of black hole entropy in N-portrait.

3 Quantum Break-Time, Inner Entanglement and Information

The effect of saturation and assisted gaplessness is the key to understanding the black hole properties. For example, let us ask: Why a black hole, at the initial stages of its evolution, emits energy but very little information? This behaviour is *universal* for all objects that achieve the enhanced capacity of information storage by the assisted gaplessness [10]. At the same time, such systems are fundamentally different from the ordinary thermal objects such as, for example, a burning piece of wood in the oven. The reason is simple: The piece of wood is very far from being a saturated state and is not exposed to the phenomenon of the assisted gaplessness. The wood therefore starts releasing information at order-one rate from the onset of burning.

Equally importantly, the microscopic theory reveals that certain black hole properties that were (and still are) taken for granted, in reality are not there. For example, the naive assumptions that the black hole evaporation is self-similar turns out to be unjustified [7, 11]. An old black hole by no means is a copy of an younger one with an equal mass. Instead, the black hole *ages* due to an internal quantum clock. This clock forces a full departure from the standard semi-classical picture latest by half decay. This marks the *quantum break-time* of a black hole.

This is a fundamentally new phenomenon brought in by the N-portrait. The effect is due to the quantum back-reaction. At the beginning of black hole decay, this effect is of order $1/N$ per each emission. Therefore, in order to grasp this effect, we need to have a theory that can resolve $1/N$-corrections. This is impossible to achieve in a standard picture in which one things in terms of small fluctuations on top of a classical geometry. Such a picture is fully equivalent to the limit $N = \infty$. From this point of view, it in not surprising that the standard picture is blind to $1/N$-corrections. Correspondingly, it misses effects that are the keys for understanding of how black hole stores and emits information.

In order to illustrate this, we shall consider examples of two important effects revealed by the microscopic theory. The first example is entanglement. The N-portrait brings in a new concept of *inner entanglement* which is different from the concept of *external entanglement* usually considered in the semi-classical picture. Indeed, already in the latter picture, one can guess that the internal quantum state of a black hole (call is $|BH\rangle$) becomes entangled with the emitted radiation [12]. For example, consider a non-rotating black hole that is emitting a particle with two possible polarization states $|+\rangle$ and $|-\rangle$. One may expects that after the emission, the black hole state evolves into something like

$$|BH\rangle \rightarrow |BH\rangle_+ \otimes |-\rangle + |BH\rangle_- \otimes |+\rangle .\tag{5}$$

In this state, the remaining black hole, described by $|BH\rangle_\pm$, becomes entangled with the emitted particle. However, without a microscopic theory, one would never know what the states $|BH\rangle$, $|BH\rangle_+$, $|BH\rangle_-$ stand for. Therefore, one would miss an important part of the story.

The story told by the N-portrait goes as follows. First, the states $|BH\rangle$ and $|BH\rangle_\pm$ are multi-particle states. They of course differ by the occupation number of the soft master mode. On average, the states $|BH\rangle_\pm$ contain $N-1$ soft gravitons. The crucial thing however is that, not only the states $|BH\rangle_\pm$ are entangled with $|\mp\rangle$, but, in addition, each of them is *internally-entangled*. That is, the modes $\hat{n}_k$ participating in each state $|BH\rangle_\pm$ are entangled among each other. This means that these states represent the entangled superpositions of several basic memory states (3).

The concept of *internal entanglement* in only possible to define for multi-particle states. This concept, therefore, is missed by the standard semi-classical picture which is not capable of resolving the black hole state as a multi-graviton state.

Next important thing is that the inner entanglement of the black hole state grows in time. It reaches the maximum by the time the black hole looses half of its constituents. It is clear that the phenomenon of inner entanglement is different from the standard notion of entanglement between the remaining black hole state and an outgoing radiation. It is this inner entanglement that leads to a complete breakdown of semi-classical treatment after a half-decay.

Another new phenomenon operating in quantum N-portrait is an effect of *memory burden* [10, 11]. This phenomenon is universal for the saturated systems of enhanced capacity of information storage. The essence of it is that the memory modes $\hat{n}_{k\neq0}$, that are "comfortably-gapless" at the saturation point ($n_0 = N$), resist to departures from this point. As it is clear from (4), any excursion away from $n_0 = N$, generates the energy gaps for the memory modes. This makes the information pattern (3) carried by these modes costly in energy. This effectively results into a back-reacting force of the *memory burden*,

$$\mu = \frac{1}{N} \sum_{k=1}^{N} \epsilon_k n_k.\tag{6}$$

This force tries to confine the system to the saturation point. It also explains why a black hole cannot reveal information at the initial stage of its decay. Since the memory modes are initially gapless, their mixing with the free quanta of the same momenta is suppressed by $1/N$. Correspondingly, only a negligible fraction of information is encoded in the outgoing radiation. As a consistency check, notice that this fraction matches the model-independent lower bound on non-thermality of Hawking radiation of a black hole of mass M given by the quantity [13],

$$\frac{\dot{T}}{T^2} \sim \frac{M_P^2}{M^2}, \tag{7}$$

where dot is the time-derivative. This parameter encodes no assumptions about the microscopic theory and describes an intrinsic fuzziness of any thermal bath due to time variation of the temperature. Yet, it exactly matches the $1/N$ corrections to black hole thermality predicted by the N-portrait.

Now, since the memory force resists to a decrease of the occupation number n_0, it therefore resists to black hole evaporation. The tendency increases with the decrease of n_0. The study of the prototype systems [11] reveals that the memory burden becomes unbearable latest after a half decay. What happens beyond this point is currently under investigation. However, the numerical studies of the prototype models indicate that system tends to be stabilized. This opens up an interesting possibility [11] of stabilizing unusually light primordial black holes [14–16] and promoting them into viable dark matter candidates.

Naturally, having a microscopic theory with universal mechanisms is essential for understanding black holes. Such mechanisms provide the descriptions in terms of general phenomena that operate in variety of systems. This is similar to the way the Higgs or Goldstone mechanisms explain behaviours of very different systems in universal languages. Indeed, irrespectively of the symmetry group and the particle content, the Higgs mechanism always generates a mass gap and the Goldstone mechanism always delivers some massless fields.

Likewise, the phenomenon of saturation and assisted gaplessness always enhances the information storage capacity, independently of particularities of the system. The recent studies [17–19] indicate that this capacity is maximal for systems that saturate unitarity. Namely, for any object in QFT with coupling α, the maximal entropy compatible with unitarity of scattering amplitudes is

$$S_{max} = \frac{1}{\alpha}. \tag{8}$$

Here α has to be understood as the running coupling evaluated at the scale given by the size of the object. For gravity this gives a black hole entropy. This universality is the key reason for why all the saturated systems share black hole properties. Remarkably, it turns out [17–19] that all known QFT objects, such as solitons, baryons, instantons, at the saturation point of the entropy bound (8) behave in the same way as black holes. First, the saturation of (8) implies the saturation of Bekenstein entropy bound [20]. Moreover, the entropy is given by the area in units of the corresponding scale. At $N = \infty$ the objects possess the information horizons, whereas at finite N they reveal information very slowly.

As evidence for the universality [17–19], let us consider an example of a baryon in QCD with N-colors and N_F quark flavors, both numbers being large. We shall keep in mind the 't Hooft's planar limit [21] in which the gauge coupling α is infinitesimal but $N\alpha$ is finite. As it is well-understood [22], a baryon is an N-quark bound-state of the size R given by the QCD scale $R = 1/\Lambda_{QCD}$ and the mass $M_{\text{baryon}} = N/R$. Alternatively, the baryon can be described as skyrmion [23], a soliton that can be viewed as a bound-state of N soft pions [24]. Already this fact points [3] towards the analogy between a large-N baryon and a black hole of N-portrait. In this analogy the role of gravitons is played by pions. Their decay constant, $f_\pi = \sqrt{N}/R$, is the analog of the Planck mass in gravity. However, the connection goes deeper. At the point of saturating the entropy bound (8), the baryon properties become identical to a black hole. For a detailed analysis we refer the reader to [17–19] and shall only sketch some key points here.

The idea is to think of a baryon as a device that stores quantum information very efficiently. This is the case, because a baryon transforms as an exponentially large representation of the flavor symmetry group $SU(N_F)$. Thus, baryon stores a large amount of quantum information in its flavor content. The capacity of information-storage is measured by the log of dimensionality of this flavor representation. This defines the micro-state entropy of the baryon. The micro-states are different flavor states of the flavor multiplet.

Now, the first remarkable thing is that the baryon entropy becomes maximal when theory saturates unitarity. This happens for $N \sim N_F$ and the resulting baryon entropy is $S_{\text{baryon}} \sim N$. At this point, the baryon in QCD behaves like a black hole in gravity. For example, its entropy is equal to its surface area measured in units of the pion decay constant f_π,

$$S_{\text{baryon}} \sim N \sim (f_\pi R)^2 \tag{9}$$

This is similar to Bekenstein-Hawking entropy of a black hole, $S \sim N \sim (M_P R)^2$. As already noted, the scale f_π plays exactly the same role for the theory of pions, as the Planck mass plays for the theory of gravitons.

Now, how fast the quantum information stored in baryon flavor can be retrieved? It is obvious that any measurement process must be suppressed by the quantum coupling. Thus, it can only take place at order $1/N$ or higher. So, the minimal time-scale of the start of the information-retrieval is

$$t > NR. \tag{10}$$

Note, the information becomes unreadable in strict 't Hooft limit $N = \infty$, as it should. We thus observe that the property of very slow release of information is not unique to black holes. It is equally hard to extract information stored within the baryon.

4 Final Remarks

This way of looking at things makes one tempted to say that black hole story is almost trivial. Indeed, we understand very well why a readout of baryon flavor content is suppressed by $1/N$: This is the strength of the quantum coupling α in 't Hooft's limit. It is the basics of QFT that a read-out of quantum information comes with the price of suppression by a small quantum coupling. Then, it must be obvious that a readout of black hole quantum information is suppressed by $1/N$, given that this is the strength of the gravitational quantum coupling at scale R. This is also the essence of black hole's $1/N$ hair [6]. This hair vanishes in the semi-classical limit, $N = \infty$, recovering the standard no-hair theorems [25–31]. In this limit, the black hole becomes eternal and there is no conflict with information storage.

For a finite N, the minimal readout time of information is (10). In this expression one can recognize the well known Page's time [12]. The presented picture reveals its microscopic origin: The equation (10) is simply a minimal time of any quantum measurement process suppressed by the coupling $\alpha = 1/N$.

Acknowledgements It is a pleasure to thank Goran Senjanović for discussions and comments. This work was supported in part by the Humboldt Foundation under Humboldt Professorship Award, by the Deutsche Forschungsgemeinschaft (DFG, German Research Foundation) under Germany's Excellence Strategy-EXC-2111-390814868, and Germany's Excellence Strategy under Excellence Cluster Origins.

Note Added: Since the submission of this chapter a substantial progress was made in the discussed topics and in particular in implications of the memory burden effect and saturation for physics of primordial black holes. The corresponding later references are therefore missing from the text.

References

1. G. Dvali, Fortschr. Phys. **66**, no. 4, 1800007 (2018). [arXiv:1801.03918 [hep-th]]
2. G. Dvali, Critically excited states with enhancedmemory and pattern recognition capacities in quantum brain networks: Lesson from black holes (2017). [arXiv:1711.09079 [quant-ph]]
3. G. Dvali, C. Gomez, Fortschr. Phys. **61**, 742 (2013). [arXiv:1112.3359 [hep-th]]
4. G. Dvali, C. Gomez, Eur. Phys. J. C **74**, 2752 (2014). [arXiv:1207.4059 [hep-th]]
5. S. Hawking, Commun. Math. Phys. **43**, 199 (1975)
6. G. Dvali, C. Gomez, Phys. Lett. B **719**, 419 (2013). [arXiv:1203.6575 [hep-th]]
7. G. Dvali, C. Gomez, JCAP **01**, 023 (2014). [arXiv:1312.4795 [hep-th]]
8. J. D. Bekenstein, Phys. Rev. D **7**, no. 8, 23336 (1973)
9. G. Dvali, M. Michel, S. Zell, Eur. Phys. J. Quant. Technol. **6**, no. 1 (2019). [arXiv:1805.10292 [quant-ph]]
10. G. Dvali, A Microscopic Model of Holography: Survival by the Burden of Memory (2018). [arXiv:1810.02336 [hep-th]]
11. G. Dvali, L. Eisemann, M. Michel, S. Zell, Phys. Rev. D **102**, no.10, 103523 (2020). [arXiv:2006.00011 [hep-th]]
12. D. N. Page, Phys. Rev. Lett. **71**, 3743 (1993) [arXiv:9306083 [hep-th]]

13. G. Dvali, Fortsch. Phys. **64**, 106 (2016). [arXiv:1509.04645 [hep-th]]
14. S. Hawking, Mon. Not. Roy. Astron. Soc. **152**, 75 (1971)
15. B. J. Carr, S. Hawking, Mon. Not. Roy. Astron. Soc. **168**, 399 (1974)
16. B. Carr, F. Kuhnel, M. Sandstad, Phys. Rev. D **94**, no. 8, 083504 (2016). [arXiv:1607.06077 [astro-ph.CO]]
17. G. Dvali, Fortsch. Phys. **69**, no. 1, 2000090 (2021). [arXiv:1906.03530 [hep-th]]
18. G. Dvali, Fortsch. Phys. **69**, no. 1, 2000091 (2021). [arXiv:1907.07332 [hep-th]]
19. G. Dvali, JHEP **03**, 126 (2021). [arXiv:2003.05546 [hep-th]]
20. J. D. Bekenstein, Phys. Rev. D **23**, 287 (1981)
21. G. 't Hooft, Nucl. Phys. B **72**, 461 (1974)
22. E. Witten, Nucl. Phys. B **160**, 57-115 (1979)
23. T. H. R. Skyrme, Nucl. Phys. **31**, 556-569 (1962)
24. E. Witten, Nucl. Phys. B **223**, 433–444 (1983)
25. R. Ruffini, J.A. Wheeler, Phys. Today **24**, no. 1, 30 (1971)
26. J. Hartle, Phys. Rev. D **3**, 2938 (1971)
27. J. D. Bekenstein, Phys. Rev. Lett. **28**, 452 (1972)
28. J. D. Bekenstein, Phys. Rev. D **5**, 1239 (1972)
29. J. D. Bekenstein, Phys. Rev. D **5**, 2403 (1972)
30. C. Teitelboim, Lett. Nuovo. Cim. **3**, 326 (1972)
31. C. Teitelboim, Lett. Nuovo. Cim. **3**, 397 (1972)

Quantum Black Holes and the Higgs Mechanism at the Planck Scale

Euro Spallucci and Anais Smailagic

Abstract In this paper we present a suitably adjusted Higgs-like mechanism producing black holes at, and beyond, the Planck energy. Planckian objects are difficult to classify either as "particles", or as "black holes", since the Compton wavelength and the Schwarzschild radius are comparable. Due to this unavoidable ambiguity, we consider more appropriate a quantum field theoretical (QFT) approach rather than General Relativity (GR), which is known to break down as the Planck scale is approached. *A posteriori*, a connection between the two description can be established for masses large enough with respect to the Planck mass, though always describing black holes at the microscopic level.
We adopt a QFT inspired by the Higgs mechanism in the sense that a massive scalar field develops a non-trivial vacuum for $m > \mu_{Pl}$. Exictations around this vacuum are Planckian objects we name *"black particles"* to remark the ambiguous identity of these objects, as it has been mentioned above. A black particle eventually turns into a " quantum black hole " when the Schwarzschild radius becomes *larger* than its Compton wavelength. However, for $m = \mu_{Pl}$ the scalar field is massless at the tree-level, but develops a non-trivial vacuum at one-loop through a Coleman-Weinberg mechanism. In this case, excitations describe Planck mass black particles.

1 Introduction

Physics at the Planck scale is still an open challenge: the semi-classical approach, where matter is quantized but gravity is not, cannot be applied anymore. Even String Theory, which is up to now the only self-consistent way to quantize gravity, does not provide fully satisfactory answers to questions about the physical nature of black hole degrees of freedom.

E. Spallucci (✉)
Dipartimento di Fisica Teorica, Università di Trieste and INFN, Sezione di Trieste, Italy
e-mail: Euro.Spallucci@ts.infn.it

A. Smailagic
INFN, Sezione di Trieste, Italy
e-mail: anais@ts.infn.it

P. Nicolini (ed.), *Touring the Planck Scale*, Fundamental Theories of Physics 219,
https://doi.org/10.1007/978-3-031-76066-2_6

The problems are not only technical, e.g. non-renormalizability of gravity, but also conceptual in nature. A remarkable example is the ambiguity in the very distinction between " elementary particles " and quantum black holes, whatever is meant by this term, when the Compton wave length and Schwarzschild radius become comparable. This energy regime corresponds to a " *strong-coupling* " phase of the gravitational field where the *effective coupling* is of order one, i.e. $M_{BH}^2 G_N \simeq 1$. This regime is analogous to QCD *confining* phase, in both cases perturbative techniques fail and the physical nature of dynamical degrees of freedom is substantially different[1]. By analogy with gluons, one considers "*gravitons*" as physical excitations in the weak-coupling phase. It is tempting to argue that in the strong-coupling phase the role of hadrons is played by objects similar to black-holes. In a series of recent papers, black holes has been described in terms of "graviton condensates" as a possible realization of the idea of composite gravitational objects [1–16]. Before considering any specific model of "quantum gravity", we think a basic question should be answered: what do we physically mean by " quantum black hole " ?

Our answer is that the fundamental description of such an object should be in terms of an "*uncertain*" event horizon subject to quantum fluctuations.

This, apparently, " obvious " consideration has an immediate and substantial consequence: any geometric description of a quantum black hole, assigning a *definite* position to the event horizon, is inadequate. Quantum oscillations completely delocalize the horizon in the vicinity of the Planck mass. The horizon " freezes " at the classical position when the mass becomes large enough with respect to the Planck mass, thus making the geometrical description feasible again. A quantum mechanical formulation of the fluctuating horizon has been recently considered in [17–26].

To avoid possible misunderstanding, we stress that we are referring to Planckian black holes which are not the result of a gravitational collapse of an astrophysical object, but are produced by a genuine quantum process.

In this paper we want to make a step forward from quantum mechanical towards a quantum field theory description.

The Higgs mechanism is a cornerstone of the Standard Model of Elementary particles. Its role is instrumental in providing masses without spoiling renormalizability of the theory. On the other hand, mass/energy is the source of gravity and it is intriguing to investigate the possibility of a gravitational role of the Higgs field itself. Many papers have studied the Higgs field during the inflationary phase of the early universe [27–37]. Here we speculate on the Higgs field as a source of Planck scale black holes.

In Sect. (2) we introduce an adapted, two-phase, scalar field theory; below the Planck mass the field remains massless, while above this threshold the field develops a non-trivial vacuum expectation value and becomes massive. Then, we define an *effective geometry* induced by this massive object, without resorting to classical Einstein equation. Instead, we build the line element starting from the very concept

---

[1] QCD dynamics can be perturbatively formulated only at high energy, where quarks and gluons are weakly coupled. At low energy confinement switches-on and dynamical degrees of freedom are composite hadrons.

of "gravitational radius". For the physical mass of the Higgs field few times larger than the Planck mass, the geometry is well approximated by the standard Schwartzschild metric. An interesting result of this approach is that, even in the classical limit, the horizon entropy keeps memory of the quantum origin of this object in the form of a logarithmic correction to the area law.

In Sect. (3) we extend the tree-level analysis of Sect. (2) to include the one-loop level contributions. It turns out that these quantum corrections play the dominant role when the classical mass is zero. In this case, the Coleman-Weinberg mechanism [38–57] provides a non-vanishing vacuum expectation value. The one-loop induced mass can be identified with the Planck mass itself. In this case, one finds that the gravitational radius cannot be shorter than the Planck length.

In Sect. (4) we give a brief summary and discussion of the results.

2 Tree-Level Higgs Mechanism

One of the heuristic pictures of black hole formation at the Planck scale is through the gravitational collapse of vacuum energy fluctuations. The problem with this view is that it is described as a *classical* gravitational collapse within a purely quantum framework. To our knowledge, there is no available description of the vacuum energy fluctuation gravitational collapse.

An alternative process for micro black hole creation has been formulated in the framework of "large extra-dimension quantum gravity" where the unification energy can be lowered down to the TeV scale [58–60]. In this case, a "true" quantum collapse of a pair of particles can be represented as an hadronic collision where the impact parameter b is shorter than the effective Schwartzschild radius of the colliding pair, i.e. $b \leq 2G_N \sqrt{-s}$, where s is the Mandelstam invariant mass of the system. Under this condition, the unelastic production channel: hadron + hadron $\longrightarrow$ black hole can open [61–66]. Unfortunately, up to now no signal of this process has been observed at LHC.

Here, we present a different possibility for black hole creation through a Higgs mechanism, rather than an hadronic collision. Being the way in which elementary particles get their masses, one can wonder if the same mechanism can also generate the mass of Planckian black holes. So far, there is no phenomenological evidence of micro black holes with a mass up to $E \simeq 10\,TeV$. Thus, the eventual dynamical mechanism producing these structures should activate only above some threshold energy E^*. For the sake of simplicity, we opt for the (very) conservative choice $E^* = \mu_{Pl} \simeq 10^{19}\,GeV$, and consider the simplest case of a single scalar field with quartic self-interaction and a non-conventional quadratic coupling. The kinetic term has the standard form and will be understood in what follows.

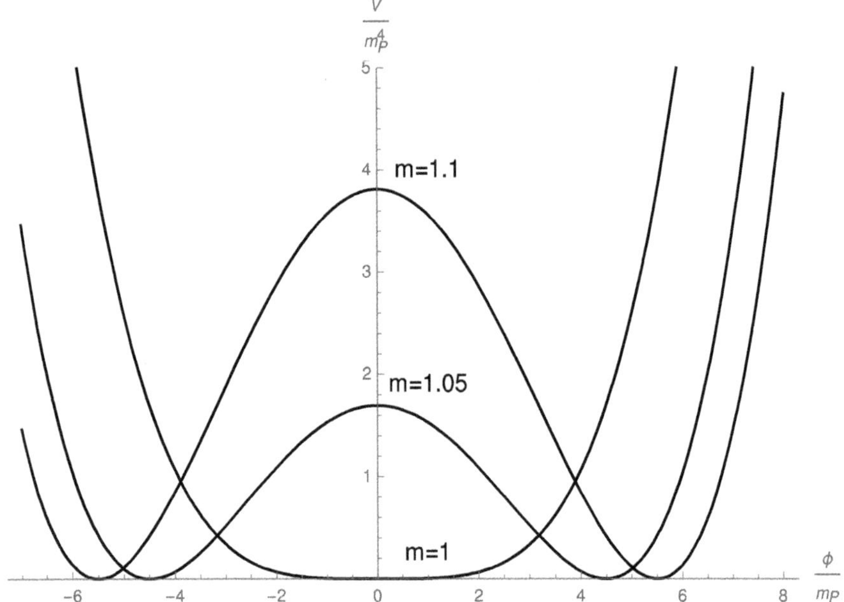

Fig. 1 Plot of the classical potential (1) in Planck units, with $\lambda = 0.1$. Curves correspond to $m = \mu_{Pl}$; $m = 1.05\mu_{Pl}$; $m = 1.1\mu_{Pl}$

Let us consider the potential Fig. 1.

$$V_{cl}(\phi) = -\frac{1}{2}m\sqrt{m^2 - \mu_{Pl}^2}\,\phi^2 + \frac{\lambda}{4!}\phi^4 + V_0\,, \quad 0 < \lambda << 1 \tag{1}$$

where ϕ is a scalar field with canonical mass dimensions; μ_{Pl} is the Planck mass defined, in natural units $h = 1$, $c = 1$, as $2G_N = \mu_{Pl}^{-2} = l_{Pl}^2$, and V_0 is a normalization constant in order to guarantee that the vacuum energy density at the minimum $\phi = \phi_0$ is zero, i.e. $V(\phi_0) = 0$.

The non-canonical quadratic coupling is real only for $m \geq \mu_{Pl}$. When the parameter m equals μ_{Pl} the quadratic term vanishes, the origin $\phi = 0$ is the stable minimum and the field excitations around it are massless objects. For $m > \mu_{Pl}$ the minimum shifts to a new position $\phi_0 \neq 0$

$$\frac{dV_{cl}}{d\phi} = 0 \longrightarrow \phi\left[-m\left(m^2 - \mu_{Pl}^2\right)^{1/2} + \frac{\lambda}{6}\phi^2\right] = 0 \tag{2}$$

$$\phi_0 = 0 \longleftrightarrow \quad m \leq \mu_{Pl} \tag{3}$$

$$\phi_0^2 = \frac{6m}{\lambda}\sqrt{m^2 - \mu_{Pl}^2} \longleftrightarrow \quad m > \mu_{Pl} \tag{4}$$

In this case, the mass of the field excitations around the minimum is given by

$$m^2_{phys} = \left[\frac{d^2 V_{cl}}{d\phi^2} \right]_{\phi=\phi_0} = 2\,m\sqrt{m^2 - \mu^2_{Pl}} \tag{5}$$

The vacuum energy density of the true vacuum is normalized to zero by choosing

$$V_0 = \frac{3\,m^2}{2\lambda} \left(m^2 - \mu^2_{Pl} \right) \tag{6}$$

2.1 *Effective Geometry*

Up to this point we proposed an adapted Higgs mechanism with the only novelty that a non-vanishing vacuum expectation value appears only above some energy scale that we pushed to the Planck mass. We did not mention gravity, space-time curvature, black holes or General Relativity. Indeed, the physical mass (5) can be very small and gravitational effects physically negligible. In order to have a relevant back-reaction on the surrounding space-time geometry we expect $m_{phys} \geq \mu_{Pl}$. Let us consider this point a little more in detail.

We start from the basic idea that to each particle of mass m_{phys} one can associate two different length scales:

(i) the Compton wave length $\lambda_C \equiv 1/m_{phys}$;

(ii) the "gravitational radius" $r_h \equiv 2m_{phys} G_N$.

For $\lambda_C > r_h$ we call the object an *"elementary particle"*; for $\lambda_C < r_h$ we call it a *"black hole"*. On the border line $\lambda_C \simeq r_h$, well ... nobody knows! Some people call these objects " Maximon(s) "[67, 68], others call them "precursors"[69, 70], "Planckion(s)"[71], etc. As an additional contribution to this dictionary of exotic names, we add the colloquial term *"black-particle "*.

In spite of its particle-like character, we can still try to associate a " metric " to a black-particle without referring to the Einstein equations. It is worth reminding that the idea of gravitational radius predates General Relativity [72, 73], and can be used as a starting point to recover an *effective* geometry. This description has a physical meaning only for distance larger than l_{Pl}.

To the physical mass (5) we associate a gravitational radius r_h as:

$$\frac{r^2_h}{4G^2_N} = \left[\frac{d^2 V_{cl}}{d\phi^2} \right]_{\phi=\phi_0} = 2\,m\sqrt{m^2 - \mu^2_{Pl}} \tag{7}$$

Let us notice that r_h will *eventually* be identified with the radius of the horizon in the limit $m^2 \geq 0.5 \left(1 + \sqrt{2} \right) \mu^2_{Pl} \equiv m^2_*$, where the gravitational length scale is larger than the Compton wavelength.

Solving (7) for m in terms of r_h we find

$$m = \frac{\mu_{Pl}}{\sqrt{2}}\left(1 + \sqrt{1 + \frac{r_h^4}{l_{Pl}^4}}\right)^{1/2} \tag{8}$$

From (8) we construct the effective metric as

$$ds^2 = -f(r)dt^2 + f(r)^{-1}dr^2 + r^2 d\Omega^2 , \tag{9}$$

$$f(r) = 1 - \frac{\sqrt{2}m}{\mu_{Pl}}\left(1 + \sqrt{1 + \frac{r^4}{l_{Pl}^4}}\right)^{-1/2} , \quad m > m_* , r >> l_{Pl} \tag{10}$$

At large distance, $r >> l_{Pl}$, the metric coincides with the Scwarzschild one:

$$f(r) \approx 1 - \frac{\sqrt{2}m}{\mu_{Pl}}\frac{l_{Pl}}{r} = 1 - 2G_N\frac{\sqrt{2}m}{r} \tag{11}$$

It is customary to describe a black hole in thermodynamical terms. The first quantity to be introduced is the Hawking temperature. In our case we find

$$T_H = \frac{1}{4\pi l_{Pl}^4}\frac{1}{1 + \sqrt{1 + r_h^4/l_{Pl}^4}}\frac{r_h^3}{\sqrt{1 + r_h^4/l_{Pl}^4}} \tag{12}$$

T_H approaches the standard form of the Hawking temperature for m large with respect μ_{Pl}. In this case $r_h >> l_{Pl}$ and

$$T_H \longrightarrow \frac{1}{4\pi r_h} \tag{13}$$

as it is expected. Furthermore, as m decreases, T_H reaches a maximum value $T_H \approx 0.03 T_{Pl}$ at $r_H = 1.59\, l_{Pl}$. For $r_h \to l_{Pl}$, the temperature decreases to $T_H \approx 0.02 T_{Pl}$. This is how far we can push T_H to have physically meaningful results. If one *formally* considers the limit $r_h \to 0$ ($m \to \mu_{Pl}$) one gets

$$T_H \approx \frac{r_h^3}{8\pi l_{Pl}^4} \longrightarrow 0 \tag{14}$$

This result has to be taken with care: physically it means that we are approaching the region where particles and black holes are indistinguishable and the thermodynamical description loses its meaning.

Another interesting thermodynamical quantity is the entropy, given by the First Law as:

$$dS = \frac{2\pi}{l_{Pl}} \left(1 + \sqrt{1 + r_h^4/l_{Pl}^4} \right)^{1/2} dr_h \tag{15}$$

Integration of (15) is non-trivial, but can be carried out in the limit $r_h > l_{Pl}$, where we find

$$S \simeq \pi \left[r_h \sqrt{r_h^2 + l_{Pl}^2} + l_{Pl}^2 \ln \left(\frac{r_h + \sqrt{r_h^2 + l_{Pl}^2}}{l_{Pl}} \right) \right] \tag{16}$$

The result (16) shows the appearance of a logarithmic correction with respect the area contribution. Furthermore, even in the " classical limit " $m >> \mu_{Pl}$, this correction survives giving:

$$S \longrightarrow \pi r_h^2 + \frac{\pi}{2} l_{Pl}^2 \ln \left(\frac{4r_h^2}{l_{Pl}^2} \right) \tag{17}$$

The first term is the celebrated Area Law, $S = A_H/4$, while the logarithmic correction can be traced to the quantum nature of black particles.

So far, we have considered only the tree-level Higgs potential (1). It is possible to further include one-loop corrections. These corrections are physically negligible in the regime $m > \mu_{Pl}$, but become dominant for $m = \mu_{Pl}$ due to the absence of the quadratic term in the tree-level potential.

3   One-Loop Effects

The Higgs mechanism works at the tree-level and quantum corrections do not alter in a significant way the classical results. However, starting from a quartic potential, no "wrong sign" quadratic term, a non-vanishing vacuum expectation value and mass can be dynamically generated through the Coleman-Weinberg effect [38]. In our case, this would correspond to start with $m = \mu_{Pl}$. In order to include this case, we shall consider, in this section, one-loop corrections to the tree-level potential. This is a standard calculation which can be found in quantum field theory textbooks and will not be repeated here. The general form of the one-loop effective potential is given by

$$V_1 (\phi) = V_{cl} + \frac{V_{cl}''^2}{64\pi^2} \left(\ln \frac{V_{cl}''}{\mu^2} - \frac{1}{2} \right) \tag{18}$$

where, μ is the renormalization scale. The order $\hbar$ corrected potential reads in our case Fig. (2):

$$V_1(\phi) = V_0 - \frac{1}{2}m\sqrt{m^2 - \mu_{Pl}^2}\,\phi^2 + \frac{\lambda}{4!}\phi^4$$

$$+\frac{1}{64\pi^2}\left(m\sqrt{m^2 - \mu_{Pl}^2} + \frac{\lambda}{2}\phi^2\right)^2$$

$$\left[\ln\left(-\frac{m}{\mu^2}\sqrt{m^2 - \mu_{Pl}^2} + \frac{\lambda\phi^2}{2\mu^2}\right) - \frac{1}{2}\right] \tag{19}$$

The quartic coupling constant is generally assumed to be small, i.e. $\lambda << 1$. Thus, order λ^2 one-loop corrections are smaller than the tree-level term. In this regime, the previous analysis is essentially not significantly modified by quantum effects. This is true for $m > \mu_{Pl}$, but for $m = \mu_{Pl}$ the minimum of the quantum corrected potential is no more $\phi = 0$.

Let us look for the extremal points of (19) by equating to zero the first derivative of $V_1(\phi)$:

$$\left[\frac{dV_1}{d\phi}\right]_{\phi=\phi_0} = -m\sqrt{m^2 - \mu_{Pl}^2}\,\phi_0 + \frac{\lambda}{6}\phi_0^3$$

$$+\frac{\lambda\phi_0}{32\pi^2}\left(-m\sqrt{m^2 - \mu_{Pl}^2} + \frac{\lambda}{2}\phi_0^2\right)$$

$$\ln\left(-\frac{m}{\mu^2}\sqrt{m^2 - \mu_{Pl}^2} + \frac{\lambda\phi_0^2}{2\mu^2}\right)$$

$$= 0 \tag{20}$$

The physical mass is defined as the second derivative of V_1 at the eventual minimum:

$$\left[\frac{d^2V_1}{d\phi^2}\right]_{\phi=\phi_0} = -m\sqrt{m^2 - \mu_{Pl}^2}$$

$$+\frac{\lambda}{2}\phi_0^2 + \frac{\lambda M^2}{32\pi^2}\ln\frac{M^2}{\mu^2} + \frac{\lambda^2\phi_0^2}{32\pi^2}\ln\frac{M^2}{\mu^2} + \frac{\lambda^2\phi_0^2}{32\pi^2} \tag{21}$$

where M^2 is a shorthand for the second derivative of the tree-level potential in ϕ_0, i.e. $M^2 \equiv -m\sqrt{m^2 - \mu_{Pl}^2} + \lambda\phi_0^2/2$.

We can use Eq. (20) to get rid of the arbitrary mass scale μ as follows:

$$\frac{\lambda}{32\pi^2}\ln\left(-\frac{m}{\mu^2}\sqrt{m^2 - \mu_{Pl}^2} + \frac{\lambda\phi_0^2}{2\mu^2}\right) = \frac{-m\sqrt{m^2 - \mu_{Pl}^2} - \frac{\lambda}{6}\phi_0^2}{M^2(\phi_0)} \tag{22}$$

Now we can replace the logarithmic term in (21):

$$
\left[\frac{d^2 V_1}{d\phi^2}\right]_{\phi=\phi_0} = -m\sqrt{m^2 - \mu_{Pl}^2} + \frac{\lambda}{2}\phi_0^2 + \frac{\lambda^2}{32\pi^2}\phi_0^2 - \frac{\lambda}{6}\phi_0^2 + m\sqrt{m^2 - \mu_{Pl}^2}
$$

$$
+ \frac{\lambda\phi_0^2}{M^2(\phi_0)}\left(m\sqrt{m^2 - \mu_{Pl}^2} - \frac{\lambda}{6}\phi_0^2\right)
$$

$$
= \frac{\lambda}{3}\phi_0^2 + + \frac{\lambda^2}{32\pi^2}\phi_0^2 + \frac{\lambda\phi_0^2}{M^2(\phi_0)}\left(m\sqrt{m^2 - \mu_{Pl}^2} - \frac{\lambda}{6}\phi_0^2\right) \quad (23)
$$

By defining the one-loop physical mass as:

$$
\left[\frac{d^2 V_1}{d\phi^2}\right]_{\phi=\phi_0} = m_{phys}^2 \equiv 2\,m\sqrt{m^2 - \mu_{Pl}^2} + \mu_{Pl}^2 \quad (24)
$$

we see that:

- for $m^2 - \mu_{Pl}^2 >> \mu_{Pl}^2$ (24) reproduces Eq. (5);
- for $m = \mu_{Pl}$ it gives raise to Planck mass black-particle (Fig. 2).

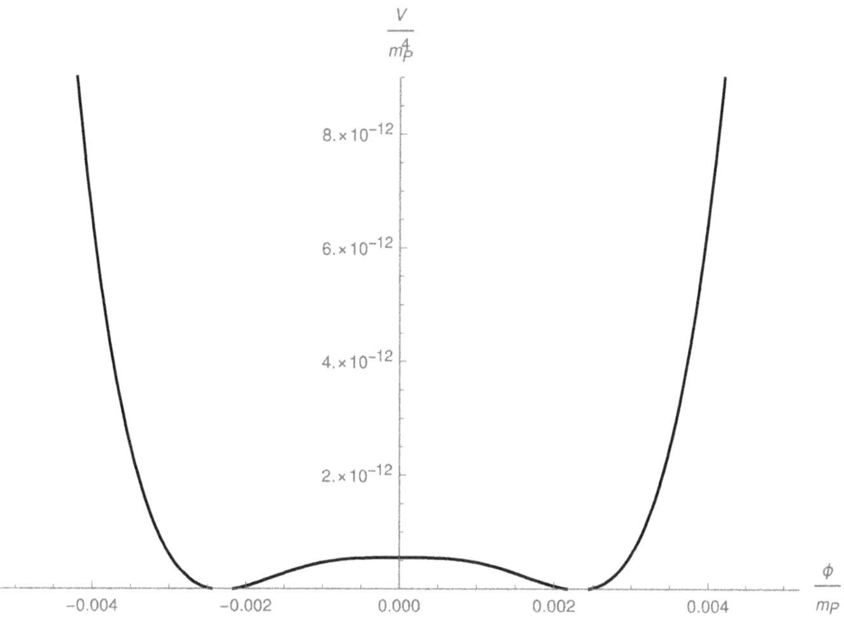

Fig. 2 Plot of the Coleman-Weinberg potential in Planck units, with $\lambda = 0.1$ and $m = \mu_{Pl}$

By inserting (24) into (23) we obtain an algebraic equation for the minimum :

$$\frac{\lambda^3}{64\pi^2}\phi_0^4 + \frac{2\lambda}{3}m\phi_0^2\sqrt{m^2 - \mu_{Pl}^2} - \frac{\lambda}{2}m_{phys}^2\phi_0^2$$

$$-\frac{\lambda^2}{32\pi^2}m\phi_0^2\sqrt{m^2 - \mu_{Pl}^2} + m_{phys}^2 m\sqrt{m^2 - \mu_{Pl}^2} = 0 \tag{25}$$

We stress that the quartic term, originating from the one-loop contribution, is relevant *only* very close to μ_{Pl}. The limiting case $m = \mu_{Pl}$ gives

$$\frac{\lambda^3}{64\pi^2}\phi_0^4 - \frac{\lambda}{2}\mu_{Pl}^2\phi_0^2 = 0 \Rightarrow \phi_0^2 = \frac{32\pi^2}{\lambda^2}\mu_{Pl}^2 \tag{26}$$

A non-zero vacuum expectation value ϕ_0 is generated by the Coleman-Weinberg mechanism.

As m increases the quartic term in (25) becomes negligible and the equation reduces to a quadratic one

$$\frac{2\lambda}{3}m\sqrt{m^2 - \mu_{Pl}^2}\phi_0^2 - \lambda m\sqrt{m^2 - \mu_{Pl}^2}\phi_0^2 + 2m^2\left(m^2 - \mu_{Pl}^2\right) = 0 \tag{27}$$

The final result is the tree-level minimum (4)

$$\phi_0^2 = \frac{6m^2}{\lambda}\sqrt{m^2 - \mu_{Pl}^2} \tag{28}$$

As m approaches μ_{Pl} form above, i.e. $m \to \mu_{Pl}$, one has to keep also the order λ^0 correction. The result is

$$\phi_0^2 \simeq \frac{6}{\lambda}m\sqrt{m^2 - \mu_{Pl}^2}\left(1 - \frac{3\lambda}{32\pi^2}\right) \tag{29}$$

At this point, it is tempting to see as the one-loop correction modify the effective metric (9), (10). The gravitational length Eq. (7) is now

$$\frac{r_h^2}{4G_N^2} = 2m\sqrt{m^2 - \mu_{Pl}^2} + \mu_{Pl}^2 \tag{30}$$

Equation (30) shows that the minimal value of r_h is no more zero, but equal to l_{Pl} (Fig. 3):

$$(r_h)_{\min} = l_{Pl} \tag{31}$$

Therefore, any black hole has to be larger than l_{Pl}.

The corresponding metric function is given by

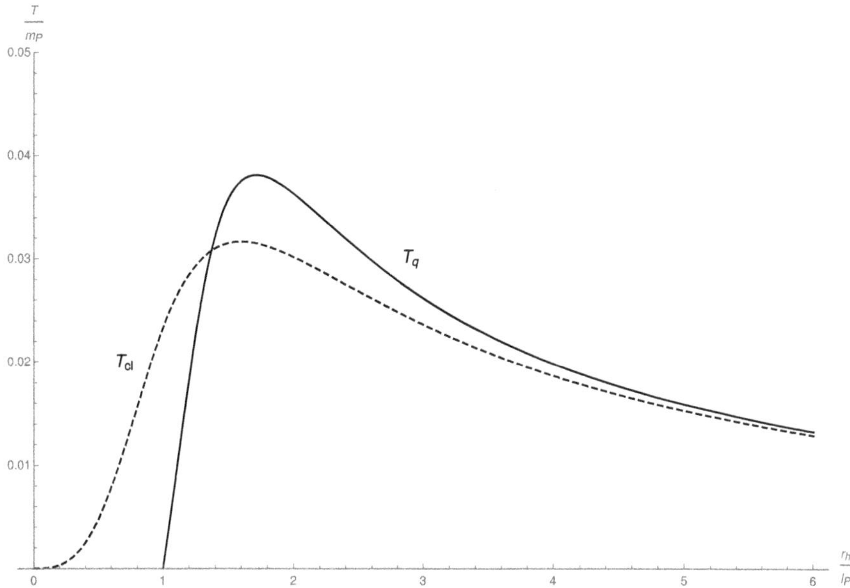

Fig. 3 Plot of the classical temperature (12), dashed line, and the one-loop temperature (33), continuous line

$$
f(r) = 1 - \frac{\sqrt{2}m}{\mu_{Pl}} \left(1 + \sqrt{1 + \left(\frac{r^2}{l_{Pl}^2} - 1 \right)^2} \right)^{-1/2}
\tag{32}
$$

Once again, we recall that this semi-classical description has the same physical limitation, $r > l_{Pl}$, as in the tree-level case. Nevertheless, one finds a significant difference is considering the Hawking temperature in the latter case Fig. (3).

$$
T_H = \frac{1}{4\pi l_{Pl}^4} \frac{r_h \left(r_h^2 - l_{Pl}^2 \right)}{1 + \sqrt{1 + \left[(r_h/l_{Pl})^2 - 1 \right]^2}} \frac{1}{\sqrt{1 + \left[(r_h/l_{Pl})^2 - 1 \right]^2}}
\tag{33}
$$

The expression (33) vanishes for $r_h \to l_{Pl}$, instead of $r_h \to 0$ as in the tree-level case. This is the main effect of the one-loop quantum corrections. This behavior confirms that near Planck scale the distinction between particles and black holes fades away and the thermodynamical description is no more an adequate one.

4 Summary and Discussion

In this paper we have described a possible formation of microscopic black holes through an Higgs-like mechanism operating at the Planck scale. The model we discussed contains only one scalar field. It is clear that further extensions, involving scalar multiplets, gauge fields, etc. can be considered. However, our intention was to test the feasibility of this alternative approach in the simplest possible framework, before attempting to account for phenomenological implications. From this perspective, the choice of the Planck energy as the lower bound for black hole production is very traditional. Nothing prevents that in more elaborate models one can lower, or raise , this threshold energy.

The preliminary results obtained in this simple model are encouraging to proceed towards more involved realizations of this type of Higgs mechanism.

We have also shown that, near the threshold energy, one-loop effects play an important role and modify the behavior of the Hawking temperature which vanishes as $r_h \to l_{Pl}$. It is important to notice that the very concept of temperature has to be taken with caution as in this critical region there is no clear distinction between particles and black holes. This hybrid object is what we named " black particle ", a quantum lump of energy with a Compton wavelength and a gravitational radius which are of the same order of magnitude.

Keeping in mind all the limitations of the geometrical description, we found that the horizon entropy (16), even in the classical limit, "recalls" quantum effects in the form of a logarithmic correction to the Area Law.

References

1. G. Dvali, C. Gomez, Self-Completeness of Einstein Gravity (2010). arXiv:1005.3497 [hep-th]
2. G. Dvali, G.F. Giudice, C. Gomez, A. Kehagias, JHEP **1108**, 108 (2011)
3. G. Dvali, C. Gomez, S. Mukhanov, Black Hole Masses are Quantized (2011). arXiv:1106.5894 [hep-ph]
4. G. Dvali, C. Gomez, A. Kehagias, JHEP **1111**, 070 (2011)
5. G. Dvali, C. Gomez, JCAP **1207**, 015 (2012)
6. G. Dvali, C. Gomez, R.S. Isermann, D. Lust, S. Stieberger, Nucl. Phys. B **893**, 187-235 (2015)
7. G. Dvali, C. Gomez, Fortsch. Phys. **61**, 742 (2013)
8. G. Dvali, C. Gomez, Phys. Lett. B **716**, 240 (2012)
9. G. Dvali, C. Gomez, Phys. Lett. B **719**, 419 (2013)
10. G. Dvali, C. Gomez, Eur. Phys. J. C **74**, 2752 (2014)
11. G. Dvali, D. Flassig, C. Gomez, A. Pritzel, N. Wintergerst, Phys. Rev. D **88**, no. 12, 124041 (2013)
12. P. Nicolini, Phys. Lett. B **778**, 88 (2018)
13. R. Casadio, F. Scardigli, Eur. Phys. J. C **74**, no. 1, 2685 (2014)
14. R. Casadio, Eur. Phys. J. C **75**, no. 4, 160 (2015)
15. R. Casadio, O. Micu, D. Stojkovic, JHEP **1505**, 096 (2015)
16. R. Casadio, O. Micu, D. Stojkovic, Phys. Lett. B **747**, 68 (2015)
17. R. Casadio, A. Giugno, O. Micu, A. Orlandi, Phys. Rev. D **90**, no. 8, 084040 (2014)
18. R. Casadio, A. Giugno, O. Micu, A. Orlandi, Entropy **17**, 6893 (2015)

19. R. Casadio, A. Orlandi, JHEP **1308**, 025 (2013)
20. R. Casadio, O. Micu, F. Scardigli, Phys. Lett. B **732**, 105 (2014)
21. E. Spallucci, A. Smailagic, in *Advances in Black Hole Research*, ed. by A. Barton (Nova Science Pub, Inc., New York, USA, 2015), pp. 1–26
22. E. Spallucci, A. Smailagic, Phys. Lett. B **743**, 472 (2015)
23. E. Spallucci, A. Smailagic, in *Quantum Gravity: Theory and Research*, ed. by B. Mitchell (Nova Science Pub. Inc. ,New York, USA, 2017)
24. E. Spallucci, A. Smailagic, Int. J. Mod. Phys. D **26**, no. 07, 1730013 (2017)
25. R. Casadio, A. Giugno, A. Giusti, Phys. Lett. B **763**, 337 (2016)
26. R. Casadio, A. Giugno, A. Giusti, O. Micu, Eur. Phys. J. C **77**, no. 5, 322 (2017)
27. B. J. W. van Tent, J. Smit, A. Tranberg, JCAP **0407**, 003 (2004)
28. N. Kaloper, L. Sorbo, J. Yokoyama, Phys. Rev. D **78**, 043527 (2008)
29. F. L. Bezrukov, A. Magnin, M. Shaposhnikov, Phys. Lett. B **675**, 88 (2009)
30. J. L. F. Barbon, J. R. Espinosa, Phys. Rev. D **79**, 081302 (2009)
31. A. O. Barvinsky, A. Y. Kamenshchik, C. Kiefer, C. F. Steinwachs, Phys. Rev. D **81**, 043530 (2010)
32. R. N. Lerner, J. McDonald, JCAP **1004**, 015 (2010)
33. C. Germani, A. Kehagias, JCAP **1005**, 019 (2010)
34. R. N. Lerner, J. McDonald, Phys. Rev. D **82**, 103525 (2010)
35. K. Nakayama, F. Takahashi, JCAP **1102**, 010 (2011)
36. F. Bezrukov, A. Magnin, M. Shaposhnikov, S. Sibiryakov, JHEP **1101**, 016 (2011)
37. G. F. Giudice, H. M. Lee, Phys. Lett. B **694**, 294 (2011)
38. S. R. Coleman, E. J. Weinberg, Phys. Rev. D **7**, 1888 (1973)
39. J. S. Kang, Phys. Rev. D **10**, 3455 (1974)
40. A. J. Paterson, Nucl. Phys. B **190**, 188 (1981)
41. M. A. Sher, Nucl. Phys. B **183**, 77 (1981)
42. L. F. Abbott, Nucl. Phys. B **185**, 233 (1981)
43. A. Billoire, K. Tamvakis, Nucl. Phys. B **200**, 329 (1982)
44. M. D. Pollock, M. Calvani, Phys. Lett. B **117**, 392 (1982)
45. A. D. Linde, Phys. Lett. B **114**, 431 (1982)
46. A. D. Linde, Phys. Lett. B **116**, 340 (1982)
47. H. Kleinert, Phys. Lett. B **128**, 69 (1983)
48. H. Aoyama, Phys. Rev. D **29**, 1763 (1984)
49. J. Okada, Phys. Lett. B **144**, 183 (1984)
50. P. F. Gonzalez-Diaz, Phys. Lett. B **141**, 314 (1984)
51. T. Futamase, Phys. Rev. D **29**, 2783 (1984)
52. G. Denardo, E. Spallucci, Nuovo Cim. A **83**, 35 (1984)
53. Y. Hosotani, Phys. Rev. D **30**, 1238 (1984)
54. F. Cooper, R. W. Haymaker, T. Matsuki, S.W. Wang, Phys. Rev. D **32**, 2049 (1985)
55. L. P. Chimento, A. S. Jakubi, J. Pullin, Class. Quant. Grav. **6**, L45 (1989)
56. R. Floreanini, R. Percacci, E. Spallucci, Class. Quant. Grav. **8**, L193 (1991)
57. W. H. Huang, Class. Quant. Grav. **8**, 83 (1991)
58. N. Arkani-Hamed, S. Dimopoulos, G. R. Dvali, Phys. Rev. D **59**, 086004 (1999)
59. T. G. Rizzo, JHEP **0609**, 021 (2006)
60. P. Nicolini, Int. J. Mod. Phys. A **24**, 1229 (2009)
61. R. Casadio, P. Nicolini, JHEP **0811**, 072 (2008)
62. M. Bleicher, P. Nicolini, in *Proceedings of the 14th International Symposium Symmetries in Science*, Bregenz, Austria, July 19-24, 2009, ed. by D. Schuch, M. Ramek. J. Phys. Conf. Ser. **237**, 012008 (2010)
63. J. Mureika, P. Nicolini, E. Spallucci, Phys. Rev. D **85**, 106007 (2012)
64. E. Spallucci, A. Smailagic, Phys. Lett. B **709**, 266 (2012)
65. P. Nicolini, J. Mureika, E. Spallucci, E. Winstanley, M. Bleicher, in *Proceedings of the 13th Marcel Grossmann Meeting on Recent Developments in Theoretical and Experimental General Relativity, Astrophysics, and Relativistic Field Theories* (MG13), ed. by R. Ruffini, R. Jantzen, K. Rosquist (Stockholm, Sweden, 2012). arXiv:1302.2640 [hep-th]

66. R. Casadio, O. Micu, P. Nicolini, Fundam. Theor. Phys. **178**, 293 (2015)
67. M. A. Markov, Sov. Phys. JETP **24**, no. 3, 584 (1967)
68. M. A. Markov, V.P. Frolov, Teor. Mat. Fiz. **13**, 41 (1972)
69. X. Calmet, Mod. Phys. Lett. A **29**, no. 38, 1450204 (2014)
70. X. Calmet, R. Casadio, Eur. Phys. J. C **75**, no. 9, 445 (2015)
71. A. Aurilia, S. Ansoldi, E. Spallucci, Class. Quant. Grav. **19**, 3207 (2002)
72. P. S. Laplace, *Exposition du système du monde* (Cambridge University Press, UK, 1827)
73. J. Michell, Phil. Trans. Roy. Sov. **74**, 35–57 (1784)

Membranes and Gauged Supergravity

Paul K. Townsend

Abstract In 1980, Antonio Aurilia, Hermann Nicolai and I constructed an $N = 8$ supergravity with a positive exponential potential by adapting the dimensional reduction of 11D supergravity to allow for a non-zero 4-form field-strength. This was the first example of what came to be known as "massive" supergravity theory, but it has since been interpreted as a special case of gauged maximal supergravity; it had little influence at the time because there is no maximally-symmetric vacuum. However, as shown here, there is a domain-wall solution, which lifts to the M2-brane solution of 11D supergravity. A similar construction for other M-branes is also explored.

1 Introduction

Early in 1980, Antonio Aurilia came to my office at CERN to tell me about his version, with Christodoulou and Legovini, of the "bag model" of hadrons; their bag was the space enclosed by a membrane coupled to a 3-form gauge potential such that the 4-form field strength was zero outside the bag but a non-zero constant inside it [1, 2].

That was my introduction to the relativistic membrane, but my interest at the time was in the antisymmetric tensor fields that arise in supergravity theories, and I was intrigued by the idea that a 3-form potential could be physically relevant even though it has no propagating modes. While listening to Antonio, it occurred to me that a constant 4-form field strength would be a cosmological constant in a gravitational context. We soon worked out the details, in which the cosmological constant emerges

P. K. Townsend (✉)
Department of Applied Mathematics and Theoretical Physics, Centre for Mathematical Sciences, University of Cambridge, Wilberforce Road, Cambridge CB3 0WA, UK
e-mail: pkt10@cam.ac.uk

P. Nicolini (ed.), *Touring the Planck Scale*, Fundamental Theories of Physics 219,
https://doi.org/10.1007/978-3-031-76066-2_7

as an integration constant in the solution of the field equation for the 3-form potential,[1] and we then had the idea of applying the result to 11D supergravity [5]; the 3-form gauge potential of that theory implies, in the context of dimensional reduction, a 3-form gauge potential for 4D $N = 8$ supergravity. Cremmer and Julia had recently found the full $N = 8$ supergravity action in this way [6, 7], but they had set to zero the 4D 4-form field strength. Working with Herman Nicolai, we found a more general $N = 8$ supergravity action with an exponential scalar potential for one of the 70 scalar fields [8]; this became known as a "massive" N=8 supergravity following the 1986 construction by Romans [10] of a "massive N=2a" 10D supergravity with a similar scalar potential.[2] The cosmological constant had effectively been traded for the expectation value of a scalar field; this was interesting but the absence of any maximally symmetric 4D vacuum was disappointing.[3]

The idea that a 3-form potential could replace a cosmological constant soon attracted attention. An example that is noteworthy here, because it has elements in common with the Aurilia-Chistodoulou-Legovini bag model, is the 1987 work of Brown and Teitelboim [11, 12] in which it is shown that the cosmological constant, interpreted as the magnitude F of a 4-form field strength, could be dynamically reduced by nucleation of membrane bubbles if F is lower inside the bubble than outside. In contrast, our new $N = 8$ supergravity had very little impact, presumably because it was not related to other ideas of the time, and thus did not appear to be part of some bigger picture.

The bigger picture began to emerge a few months later, when Freund and Rubin showed that 11D supergravity has an $AdS_4 \times S^7$ solution with the 4-form field strength proportional to the volume form of AdS_4 [13]. It was soon conjectured, and later proved, that the associated 4D theory is a gauged maximal supergravity with an $SO(8)$ gauge group, which was constructed by de Wit and Nicolai in 1982 [14, 15]. These developments were reviewed in 1986 by Duff, Nilsson and Pope [16]; by then it was understood that the de Wit-Nicolai theory is but one of many gauged maximal supergravity theories. A classification was achieved in relatively recent times [17]; this was reviewed in 2008 by Samtleben [18], who points out that the modified $N = 8$ supergravity of [8] is included, which means that the $N = 8$ 'massive' supergravity found in 1980 by Antonio, Hermann and myself was actually the first gauged maximal supergravity theory.

The aim of this article is to show how this ANT supergravity theory (as I shall call it here, after its authors) fits into the bigger picture of M-theory. Although it has no Minkowski vacuum, it does have a membrane/domain-wall 'vacuum' solution that preserves half the supersymmetry. Considering that the starting point of my collaboration with Antonio was his idea that a relativistic membrane is a source for a

---

[1] As was independently discovered around the same time by Duff and van Nieuwenhuizen [3], and (in the context of a superspace formulation of $N = 1$ supergravity) by Ogievetsky and Sokatchev [4].

[2] This use of "massive" should not be confused with its more recent use in which it indicates that the graviton has a non-zero mass.

[3] We addressed this issue briefly in a conference report [9], where it was observed that a positive potential is what one might expect from a spontaneous partial breaking of the local supersymmetry.

3-form gauge potential,[4] it now seems surprising that we did not immediately look for a domain-wall solution of our new $N = 8$ supergravity theory. Of course, there was then no understanding of how a membrane coupling to the 3-form of 11D supergravity could be compatible with local supersymmetry; the 11D supermembrane lay seven years in the future [21, 22]. However, if we had looked for, and found, the half-supersymmetric 4D domain wall solution, it would surely have been obvious that this must lift to a membrane solution of 11D supergravity. In fact, as will be shown here, it lifts to the the membrane solution found by Duff and Stelle in 1990 [23].

I will begin with an exposition of the construction of [8] but, in contrast to the detailed exposition there, no attempt will be made here to include all supergravity fields, including fermions. Instead, I will start with a simplified model of gravity in a spacetime of general dimension $D = d + n$, coupled to a d-form field strength F_d. This suffices for an explanation of the basic idea, which yields (in this simplified but also generalized context) a d-dimensional dilaton-gravity model; for $(d, n) = (4, 7)$ it is a consistent truncation of the modified $N = 8$ supergravity theory found in [8]. For certain other values of (d, n) a similar construction may apply for other M-theory branes, as will be discussed.

2 The ANT Construction

Our starting point will be the following Lagrangian density

$$\mathcal{L}_D = \sqrt{-g^{(D)}} \left\{ R^{(D)} - \frac{1}{2} |F_d|^2 \right\}, \tag{1}$$

where $|F_d|^2$ is defined such that, in local coordinates for which the D-metric is g_{MN} and F_d has components $F_{M_1 \cdots M_d}$,

$$|F_d|^2 = \frac{1}{d!} g^{M_1 N_1} \cdots g^{M_d N_d} F_{M_1 \ldots M_d} F_{N_1 \ldots N_d}. \tag{2}$$

We will be interested in an n-dimensional reduction to a d-dimensional spacetime $\mathcal{M}_d$ (so $D = d + n$) with a reduction/truncation ansatz for which

$$ds_D^2 = e^{-2\alpha\phi} ds^2(\mathcal{M}_d) + e^{2[(d-2)/n]\alpha\phi} ds^2(\mathbb{E}^n)$$
$$F_d = \mathrm{vol}\,(\mathcal{M}_d)\, f, \tag{3}$$

where $\mathrm{vol}(\mathcal{M}_d)$ is the volume d-form on $\mathcal{M}_d$, and α is an arbitrary constant. Although both ϕ and f are scalar fields on $\mathcal{M}_d$ that are constant on the Euclidean n-space $\mathbb{E}^n$, the scalar f is constrained to be the Hodge dual of a d-form field strength on $\mathcal{M}_d$.

---

[4] To my knowledge, this idea originated in Antonio's 1977 paper with Legovini [19] as a generalization of the Kalb-Ramond coupling of strings to a 2-form potential [20].

The metric ansatz has been chosen such that $ds^2(\mathcal{M}_d)$ is the Einstein conformal frame metric. A choice of the constant α is equivalent to a normalization for ϕ, and if we choose

$$\alpha = \sqrt{\frac{n}{2(d-2)(D-2)}},\tag{4}$$

then the Lagrangian density for the d-dimensional theory is

$$\mathcal{L}_d = \sqrt{-g}\left\{R - \frac{1}{2}(\partial\phi)^2 + \frac{1}{2}e^{2(d-1)\alpha\phi}f^2\right\}.\tag{5}$$

For $(d,n) = (4,7)$ this is a consistent truncation of the full $N=8$ supergravity found by dimensional reduction of 11D supergravity. If the scalar f were unconstrained, its equation of motion would be $f=0$; we could then back substitute to trivially eliminate f from the action, which would then be, for $(d,n) = (4,7)$, a consistent truncation of the standard Cremmer-Julia $N=8$ supergravity.

However, f is not unconstrained; the unconstrained variable is the $(d-1)$-form potential for (the Hodge dual of) f and *its* field equation has the general solution

$$f = \mu\, e^{-2(d-1)\alpha\phi},\tag{6}$$

for arbitrary mass parameter μ. For $\mu \neq 0$ it is not legitimate to substitute for f into the Lagrangian density of (5) because the variation of the action is not zero for this solution of the field equations. This can be remedied by first adding to the Lagrangian density a μf term, which is a total derivative because of the constraint on f (which implies that this addition preserves the local supersymmetry of the full supergravity theory). It is now legitimate to substitute for f, and this yields the new 'dual' Lagrangian density

$$\tilde{\mathcal{L}}_d = \sqrt{-g}\left\{R - \frac{1}{2}(\partial\phi)^2 - 2\Lambda e^{-a\phi}\right\},\tag{7}$$

where

$$a = 2(d-1)\alpha, \qquad \Lambda = (\mu/2)^2.\tag{8}$$

Notice that the sign of the potential term is *opposite* to what one would find by the *illegitimate* substitution for f in (5). This sign change may be checked by verifying that the field equations of (5) are equivalent, for f given by (6), to the field equations of (7).

Our conventions are now those of [24] except that Λ here is $-\Lambda$ there; defined here, Λ becomes the usual cosmological constant (negative for anti-de Sitter) when $a = 0$. Let us record here that, as a consequence of (4),

$$a^2 = \frac{2n(d-1)^2}{(d-2)(D-2)}.\tag{9}$$

3 Domain Walls and M-Branes

The domain-wall solutions of the field equations that follow from the Lagrangian density of (7) take the form [24, 25]

$$ds_d^2 = H^{\frac{4}{(d-2)\Delta}} ds^2(\text{Mink}_{d-1}) + H^{\frac{4(d-1)}{(d-2)\Delta}} dy^2$$
$$e^\phi = H^{\frac{2a}{\Delta}}, \tag{10}$$

where H is linear in y with $dH = \mp\sqrt{\Lambda\Delta}\, dy$, and

$$\Delta \equiv a^2 - \frac{2(d-1)}{(d-2)}. \tag{11}$$

From (9) we see that, for us,

$$\Delta = \frac{2(n-1)(d-1)}{d+n-2}. \tag{12}$$

Our aim now is to lift these domain wall solutions to solutions of the D-dimensional theory from which we started, using the ansatz (3). The spacetime $\mathcal{M}_d$ appearing in this ansatz is now the above domain-wall spacetime, and

$$\text{vol}\,(\mathcal{M}_d) = H^{\frac{4(d-1)}{(d-2)\Delta}} \text{vol}(\text{Mink}_{d-1}) \wedge dy. \tag{13}$$

Let us also record here that for the domain-wall solution, the function $H(y)$ is such that

$$dH^{-1} = \pm\sqrt{\Lambda\Delta}\, H^{-2} dy. \tag{14}$$

On substituting the domain-wall metric and dilaton configurations of (10) into the ansatz (3), one finds that

$$ds_D^2 = H^{-\frac{2}{d-1}} ds^2(\text{Mink}_{d-1}) + H^{\frac{2}{n-1}} ds^2(\mathbb{E}^{n+1})$$
$$F_d = \pm\frac{2}{\sqrt{\Delta}}\, \text{vol}(\text{Mink}_{d-1}) \wedge dH^{-1}. \tag{15}$$

We have been assuming that H is a function only of y, but this is now a special case of a more general $(d-1)$-brane solution of the the D-dimensional theory from which we started; in general H is a harmonic function on the $(n+1)$-dimensional transverse Euclidean space. A simple choice, with $SO(n+1)$ symmetry, is

$$H = 1 + (r_0/r)^{n-1}, \tag{16}$$

for constant r_0. Near the singularity of this function at $r = 0$, the D-metric takes the asymptotic form

Table 1 Supergravity solutions with an 'AdS × S' near-horizon geometry

Solution type	D	d	n	Δ
M2	11	4	7	4
M5	11	7	4	4
D3	10	5	5	4
Self-dual black string	6	3	3	2
Magnetic black string	5	3	2	4/3
Tangerlini black hole	5	2	3	4/3
Reissner-Nordström black hole	4	2	2	1

$$H \sim e^{-(n-1)\rho/r_0} ds^2 (\mathrm{Mink}_{d-1}) + d\rho^2 + r_0^2 d\Omega_n \qquad [\rho = r_0 \ln(r/r_0)], \qquad (17)$$

where $d\Omega_n$ is the $SO(n+1)$-invariant metric on S^n. This is an $\mathrm{AdS}_d \times S^n$ metric, which implies that $r = 0$ is a Killing horizon of the full D-metric, near which the solution asymptotes to $\mathrm{AdS}_d \times S^n$.

The configuration (15) is a solution of a D-dimensional supergravity in those cases of Table 1, which is adapted from a similar table in [26]. The first column gives the type of solution. The second column gives the spacetime dimension D and the next two columns gives the values of d and n. The last column gives the value of Δ according to the formula (12).

Notice that entries with $d \neq n$ occur in 'dual' pairs for which the values of d and n are interchanged; these have the same value of Δ, as is manifest from the formula (12).

For $(d, n) = (4, 7)$, in which case $\Delta = 4$, we recover the half-supersymmetric Duff-Stelle membrane solution of 11D supergravity [23] as a lift to $D = 11$ of a domain-wall solution of the modified $N = 8$ supergravity theory of [8].

3.1 Other Cases

What about the other cases of the table? For $(d, n) = (7, 4)$, we recover the half-supersymmetric fivebrane solution of 11D supergravity, but in terms of a 7-form field strength for the 6-form 'dual' potential that is defined only on solutions of the 11D supergravity equations. This is probably only a technical difficulty since the ANT construction could be recast as a purely on-shell construction, but this will not be attempted here. The remaining $\Delta = 4$ case involves the self-dual 5-form field strength of IIB 10D supergravity, and we now face the complication that a non-zero

Table 2 Intersecting M-branes with 'AdS $\times S \times \mathbb{E}$' near-horizon geometries

$M2 \perp M5$	$AdS_3 \times S^3 \times \mathbb{E}^5$	1/4
$M2 \perp M2 \perp M2$	$AdS_2 \times S^3 \times \mathbb{E}^6$	1/8
$M5 \perp M5 \perp M5$	$AdS_3 \times S^2 \times \mathbb{E}^6$	1/8
$M2 \perp M2 \perp M5 \perp M5$	$AdS_2 \times S^2 \times \mathbb{E}^7$	1/8

5-form on the 5-dimensional spacetime of the maximal 5D supergravity obtained by dimensional reduction must be accompanied by a non-zero 5-form field strength on the Euclidean 5-space on which we reduce; again, some modification of the ANT construction may allow for this but this will not be investigated here either.

What about the $D < 10$ cases with $\Delta < 4$? The are various intersecting M-brane solutions of 11D supergravity for which the asymptotic geometry near the intersection is of 'AdS $\times S \times \mathbb{E}$' form. The four geometries of this type obtainable in this way, with a particular realization in terms of intersecting M2-branes and M5-branes are given in Table 2, which is adapted from a similar table in [27]. The symbol $\perp$ indicates either (i) an orthogonal intersection in which p-branes self-intersect on $(p - 2)$-planes, which may then also intersect on $(p - 4)$-planes, or (ii) the orthogonal intersection of an M2 with an M5 on a line; these intersections are among those that preserve some fraction of supersymmetry and the fraction preserved is given in the last column of the table.

Consider the 6D self-dual black string solution of the minimal $(1, 0)$ 6D supergravity, which is a consistent truncation of the maximal 6D supergravity obtained by dimensional reduction of 11D supergravity. An application of the ANT construction would, if successful, lead to a $d = 3$ $N = 4$ supergravity with a domain-wall vacuum that lifts to the 6D self-dual string. However, the 3-form field strength that supports this solution is self-dual, and this presents the same difficulty that we found for the D3 case. Next is the magnetic black string: a half-supersymmetric solution of minimal 5D supergravity, which is a consistent truncation of the maximal 5D supergravity obtained by dimensional reduction of 11D supergravity. Here we have the same issue that arose for the M5 case, the 3-form field strength that we need for an application of the ANT construction is defined only on-shell. These difficulties (which may be purely technical) do not arise for the remaining, extreme black hole, cases but for these the ANT construction would require us to dimensionally reduce to 2D, which is the dimension for which there is no Einstein-frame metric.

So, the ANT construction of [8] leading to a modified $N = 8$ supergravity with a domain-wall 'vacuum' that lifts to the M2-brane of 11D supergravity does not immediately generalise to the other M-theory possibilities, although a modified construction may work in some of these other cases.

4 Comments

My interaction with Antonio Aurilia was brief but intense. Apart from our joint work with Hermann Nicolai, we also co-authored a paper with Antonio's long-term (and long distance) collaborator Yasushi Takahashi (of Ward-Takahashi fame) on another application of a non-propagating 3-form gauge potential [28]. I eventually met my co-author Takahashi at the Edmonton Summer Institute of 1987, although he was speaking at a parallel event on superconductivity; my talk at the other event was on membranes, as was that of Kellogg Stelle, with whom I co-wrote a conference proceedings with the title "Are 2 branes better than 1?" [29]. I think Antonio and I met only once again after our 1980 collaboration at CERN, but his influence on me was very significant. I paid attention to his ideas about membranes at the same time that I was working with him on 11D supergravity, and I returned to wondering how to connect these topics after the Green-Schwarz superstring revolution of 1984. Eventually, in Trieste in January 1987, six months before the conference in Edmonton, Eric Bergshoeff, Ergin Sezgin and I were able to solve the problem of how to couple 11D supergravity to a membrane [21]. This set the course for much of my subsequent research, although branes only became mainstream in 1996. Before then, Antonio was a fellow heretic; see, for example, his 1993 paper with Spallucci [30]. He would have been thrilled to know that our joint work on 11D supergravity would also turn out to be related to the topic so close to his heart—the relativistic membrane.

Acknowledgements Partial support from the UK Science and Technology Facilities Council (consolidated Grant ST/P000681/1) is gratefully acknowledged.

References

1. A. Aurilia, D. Christodoulou, F. Legovini, Phys. Lett. B **73**, 429 (1978)
2. A. Aurilia, D. Christodoulou, Phys. Lett. B **78**, 589 (1978)
3. M. J. Duff, P. van Nieuwenhuizen, Phys. Lett. B **94**, 179 (1980)
4. V. Ogievetsky, E. Sokatchev, Sov. J. Nucl. Phys. **32**, 589 (1980)
5. E. Cremmer, Phys. Lett. B **76**, 409 (1978)
6. E. Cremmer, B. Julia, Phys. Lett. B **80**, 48 (1978)
7. E. Cremmer, B. Julia, Nucl. Phys. B **159**, 141 (1979)
8. A. Aurilia, H. Nicolai, P.K. Townsend, Nucl. Phys. B **176**, 509 (1980)
9. A. Aurilia, H. Nicolai, P. K. Townsend, in *Nuffield Workshop on Superspace and Supergravity*, ed. by S. W. Hawking and M. Roček (Cambridge University Press, Cambridge, UK, 1981), pp. 403–412
10. L. J. Romans, Phys. Lett. B **169**, 374 (1986)
11. J. D. Brown, C. Teitelboim, Phys. Lett. B **195**, 177 (1987)
12. J. D. Brown, C. Teitelboim, Nucl. Phys. B **297**, 787 (1988)
13. P. G. O. Freund, M. A. Rubin, Phys. Lett. B **97**, 233 (1980)
14. B. de Wit, H. Nicolai, Phys. Lett. B **108**, 285 (1982)
15. B. de Wit, H. Nicolai, Nucl. Phys. B **208**, 323 (1982)
16. M. J. Duff, B. E. W. Nilsson, C. N. Pope, Phys. Rept. **130**, 1 (1986)

17. B. de Wit, H. Samtleben, M. Trigiante, JHEP **06**, 049 (2007)
18. H. Samtleben, Class. Quant. Grav. **25**, 214002 (2008)
19. A. Aurilia, F. Legovini, Phys. Lett. B **67**, 299 (1977)
20. M. Kalb, P. Ramond, Phys. Rev. D **9**, 2273 (1974)
21. E. Bergshoeff, E. Sezgin, P. K. Townsend, Phys. Lett. B **189**, 75 (1987)
22. E. Bergshoeff, E. Sezgin, P. K. Townsend, Annals Phys. **185**, 330 (1988)
23. M. J. Duff, K. S. Stelle, Phys. Lett. B **253**, 113 (1991)
24. H. Lü, C.N. Pope, E. Sezgin, K.S. Stelle, Nucl. Phys. B **456**, 669 (1995)
25. H. Lü, C. N. Pope, P. K. Townsend, Phys. Lett. B **391**, 39 (1997)
26. P. Claus, R. Kallosh, J. Kumar, P. K. Townsend, A. Van Proeyen, JHEP **9806**, 004 (1998)
27. H. J. Boonstra, B. Peeters, K. Skenderis, Nucl. Phys. B **533**, 127 (1998)
28. A. Aurilia, Y. Takahashi, P. K. Townsend, Phys. Lett. B **95**, 265 (1980)
29. K. S. Stelle, P. K. Townsend, in *Proceedings of the 1987 Edmonton Summer Institute in Theoretical Physics*, ed. by F.C. Khanna, H. Umezawa, G. Kunstatter, H.C. Lee (World Scientific Publishing, Singapore, 1988), pp. 321–345
30. A. Aurilia, E. Spallucci, Class. Quant. Grav. **10**, 1217 (1993)

Exotic Quantum Gravity Effects

Some Considerations on BEC Analogue Black Holes

Roberto Balbinot and Alessandro Fabbri

Abstract In this paper, dedicated to the memory of A. Aurilia, we will review some basic features of Hawking's black hole radiation and compare them with the corresponding ones present in Bose-Einstein condensate analogue black holes.

1 Preface

This paper is dedicated to the memory of Antonio Aurilia, a theoretical physicist with a wide range of interests. He could move easily with competence from Quantum Gravity to QCD. One of us (R.B.) had the pleasure to work with him many years ago on a project dedicated to classical and quantum aspects of shell dynamics. Here we give a personal account on a field, analogue gravity, which has tremendously grown in recent years and has allowed to verify experimentally, although in an indirect way, one of the most spectacular prediction of modern theoretical physics, obtained by an encounter of General Relativity and Quantum Mechanics, namely Hawking's Black Hole (BH) emission.

R. Balbinot
Dipartimento di Fisica dell'Università di Bologna and INFN sezione di Bologna,
Via Irnerio 46, 40126 Bologna, Italy
e-mail: Roberto.Balbinot@bo.infn.it

A. Fabbri (✉)
Departament de Física Teórica and IFIC, Universidad de Valencia-CSIC,
C. Dr. Moliner 50, 46100 Burjassot, Spain
e-mail: afabbri@ific.uv.es

P. Nicolini (ed.), *Touring the Planck Scale*, Fundamental Theories of Physics 219,
https://doi.org/10.1007/978-3-031-76066-2_8

2 Hawking Radiation and Analogue Models

One of the most striking and unexpected features of BHs is that, despite their name, they emit radiation. This is the astonishing discovery of S. Hawking in 1974 [1, 2]. The number of particles emitted follows a thermal law

$$N_\omega = \frac{1}{e^{\frac{\hbar\omega}{k_B T_H}} - 1} \, ,\tag{1}$$

where T_H is the emission temperature of the hole, k_B the Boltzmann's constant and $\hbar$ the reduced Planck's constant. T_H is the Hawking temperature which is proportional to the surface gravity κ of the BH horizon

$$T_H = \frac{\hbar\kappa}{2\pi k_B} \, .\tag{2}$$

This discovery has allowed a beautiful synthesis between gravity and thermodynamics mediated by Quantum Mechanics, as can be appreciated by Eq. (2), and led to the identification of the laws of BH mechanics [3] with the fundamental principles of thermodynamics. One can consider this synthesis as the basic support for the existence of Hawking BH radiation since a direct experimental verification seems till now quite impossible. Indeed, the emission temperature T_H scales with the inverse of the BH mass M, roughly

$$T_H \sim 10^{-6} \left(\frac{M_s}{M} \right)\tag{3}$$

where M_s is the solar mass. Since BHs formed by stellar gravitational collapse have masses bigger than M_s, T_H appears by far much lower than the cosmic microwave background radiation ($\simeq 2.7K$) to allow a direct observation. Furthermore, so far, there is no evidence of an emission of this type related to a primordial population of BHs whose mass can be much lower then M_s [4].

Beside this, also from the theoretical point of view Hawking's derivation of his famous result is not free of criticisms. The theoretical scheme assumed by Hawking in his seminal work is that of Quantum Field Theory (QFT) in curved space-time. In this hybrid framework matter is quantized according to QFT while gravity is treated classically according to General Relativity. The limit of validity of this approximation of a yet unknown quantum gravity theory is expected to be set by the Planck scale

$$l_P = \left(\frac{G\hbar}{c^3} \right)^{1/2} \sim 10^{-33} \mathrm{cm}\tag{4}$$

or in terms of energy $E_P \sim 10^{19}$ GeV.

Hawking radiation can be viewed as a conversion of vacuum fluctuations near the horizon into pairs of real on shell particles, of which one escapes to infinity

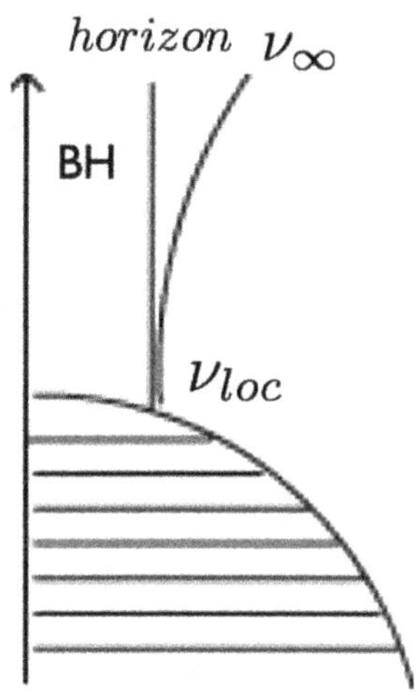

Fig. 1 For the Hawking particle, the local frequency close to the horizon is exponentially amplified with respect to its frequency measured at infinity and the corresponding energy becomes transplanckian

and constitutes Hawking radiation, the other, called "partner", is swallowed by the BH. Now, suppose one observes at infinity a Hawking particle of frequency ν_∞. Tracing back in time the particle's trajectory till the horizon region where it originates (see Fig. 1) one finds that there the local frequency of the particle is exponentially amplified and rapidly the corresponding energy becomes much bigger than E_P. This regime is outside the range of validity of the framework used (QFT in curved space-time). So one might wonder if Hawking's BH radiation is a real phenomenon or just an artifact of an approximation pushed far beyond its limits. This is the so called "transplanckian problem" [5].

Partially motivated by this incongruence, in recent years a new branch of research has grown called "analogue gravity models" [6] which aims to reproduce, in condensed matter systems (Bose-Einstein condensates, BEC, are the most investigated) manageable in laboratory, some features of gravitational systems, most notably Hawking's BH thermal radiation. The analogy relies on an exact mathematical correspondence between the propagation of sound waves in the condensed matter system under the hydrodynamical approximation and the propagation of light in a gravitational field.

As shown by Unruh [7], the propagation of sound in an inhomogeneous fluid is governed by an effective curved space-time metric (the acoustic metric) which, if the flow turns from subsonic to supersonic, has exactly a BH form. The transition occurs at the corresponding horizon called "sonic horizon". Basically, sound waves are trapped and dragged by the supersonic flow as light inside a BH. So, once the sound waves are quantized one can envisage the occurrence of Hawking like radiation of phonons in a fluid which turns supersonic.

A problem similar to the transplanckian one seems to appear also here because of the very short wavelengths of the emitted phonons near the sonic horizon breaking down the hydrodynamical (large wavelength) approximation. However for the analogue models, unlike in the gravitational case, we know exactly the short distance quantum description of the system and so we can perform an ab initio analysis based on this fundamental description without any use of the gravitational analogy. One can then, at the end, compare the results with the predictions based on the analogy to test how this is reliable. One finds that there is a regime in which the gravitational analogy based on QFT in curved space-time does indeed predict quite well the exact result, while in other regimes the analogy fails dramatically.

3 BEC and Gravitational Black Holes

Let us analyse in some detail how a supersonically flowing BEC can mimic some of the characteristic features of a BH.

The fundamental bosonic field operator $\hat{\Psi}(t, \mathbf{x})$ in a BEC can be splitted as a sum of a mean-field classical field Ψ_0 describing the condensate and a quantum operator describing quantum fluctuations above the condensate [8]

$$\hat{\Psi}(t, \mathbf{x}) = \Psi_0(t, \hat{x}) \left[1 + \hat{\phi}(t, \hat{x}) \right] . \tag{5}$$

Ψ_0 satisfies the Gross-Pitaevskii equation

$$i\hbar \frac{\partial \Psi_0}{\partial t} = \left(-\frac{\hbar^2}{2m} \nabla^2 + V_{ext} + g|\Psi_0|^2 \right) \Psi_0 , \tag{6}$$

where V_{ext} is the external potential, m the mass of the atoms and g the effective coupling constant. The fluctuations quantum operator $\hat{\phi}$ satisfies the Bogoliubov-de Gennes (BdG) equation

$$i\hbar \frac{\partial \hat{\phi}}{\partial t} = -\left(\frac{\hbar^2}{2m} \nabla^2 + \frac{\hbar^2}{m} \frac{\nabla \Psi_0}{\Psi_0} \nabla \right) \hat{\phi} + gn(\hat{\phi} + \hat{\phi}^\dagger) \tag{7}$$

where $n = |\Psi_0|^2$ is the condensate density.

Let us assume that the condensate is flowing from right to left making a transition from a subsonic flow to a supersonic one (Fig. 2).

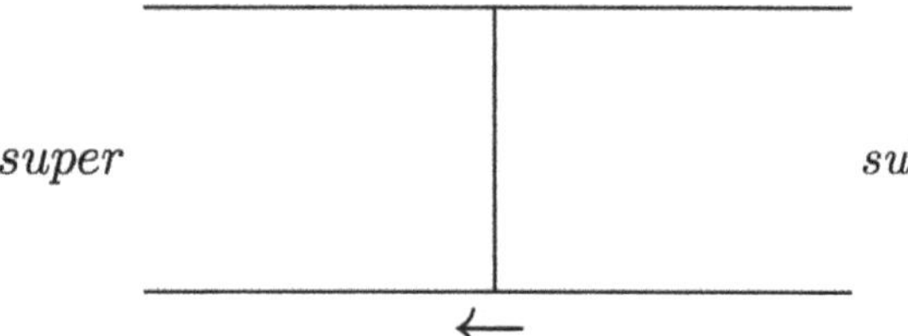

Fig. 2 Schematic representation of an acoustic black hole

Fig. 3 Black hole formed by gravitational collapse

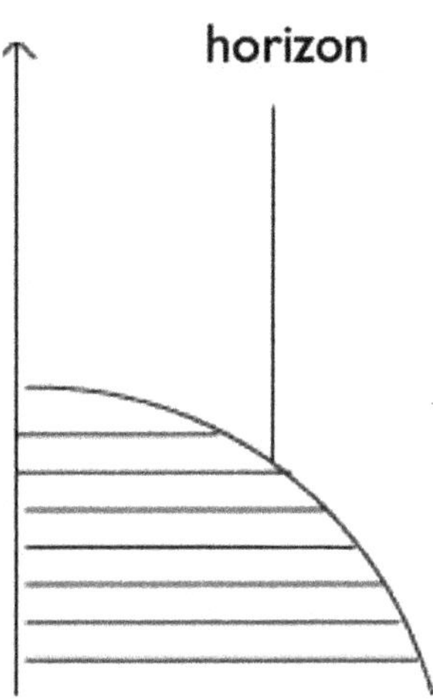

Consider now a BH formed by the gravitational collapse of a star (Fig. 3). The resulting gravitational field is described by the metric tensor $g_{\mu\nu}$ solution of the Einstein equations

$$R_{\mu\nu} - \frac{R}{2}g_{\mu\nu} = 8\pi G T_{\mu\nu} \tag{8}$$

where $R_{\mu\nu}$ is the Ricci tensor, R is the scalar curvature and $T_{\mu\nu}$ the energy momentum tensor of the matter distribution. In the vacuum space outside the star $T_{\mu\nu} = 0$ and Eq. (8) reduce to $R_{\mu\nu} = 0$. In this part of spacetime the solution, describing the BH, can be written in Painlevé-Güllstrand form

$$ds^2 = -(c^2 - V^2(x))dt^2 - 2V_i dx^i dt + dx_i dx^i \tag{9}$$

where $V(x)$ is some function; for example for a Schwarzschild BH one has $V = -\sqrt{\frac{2GM}{c^2 r}}$. In this BH spacetime consider a massless scalar quantum field $\hat{\varphi}$ satisfying

$$\nabla_\mu \nabla^\mu \hat{\varphi} = 0 \,, \tag{10}$$

where the covariant derivative ∇_μ is calculated from the metric Eq. (9).

What do these two systems, one, the BEC described by Eqs. (6), (7), the other, the BH plus scalar field satisfying Eqs. (8), (10), have in common?

If we use a density-phase representation of the BEC bosonic field operator $\hat{\Psi}$ [6] (see also [9])

$$\hat{\Psi} = \sqrt{n + \hat{n}_1}\, e^{i(\theta + \hat{\theta}_1)} \simeq \Psi_0 \left(1 + \frac{\hat{n}_1}{2n} + i\hat{\theta}_1\right) \tag{11}$$

the BdG equation (7) reduces to a pair of equations of motion for the density fluctuation $\hat{n}_1$ and the phase fluctuation $\hat{\theta}_1$, namely

$$\hbar \partial_t \hat{\theta}_1 = -\hbar \mathbf{v} \, \nabla \hat{\theta}_1 - \frac{mc^2}{n} \hat{n}_1 + \frac{mc^2}{4n} \xi^2 \nabla \left[n \nabla (\frac{\hat{n}_1}{n}) \right] , \tag{12}$$

$$\partial_t \hat{n}_1 = -\nabla (\mathbf{v} \hat{n}_1 + \frac{\hbar n}{m} \nabla \hat{\theta}_1) . \tag{13}$$

Here a fundamental length enters, the "healing length" $\xi = \frac{\hbar}{mc}$ defined in terms of the local speed of sound $c_s = \sqrt{\frac{gn}{m}}$ and $\mathbf{v} = \frac{\hbar}{m} \nabla \theta$ is the local speed of the flow. If one is probing the system on scales $\gg \xi$ (hydrodynamical approximation) the last term in Eq. (12) can be neglected and we can write the density fluctuation $\hat{n}_1$ as

$$\hat{n}_1 = -\frac{\hbar n}{mc^2} [\mathbf{v} \nabla \hat{\theta}_1 + \partial_t \hat{\theta}_1] , \tag{14}$$

which, when inserted in (13), gives the equation of motion for the phase fluctuation

$$- (\partial_t + \nabla \mathbf{v}) \frac{n}{mc^2} (\partial_t + \mathbf{v} \, \nabla) \hat{\theta}_1 + \nabla (\frac{n}{m} \nabla \hat{\theta}_1) = 0 . \tag{15}$$

Remarkably, (15) can be rewritten as

$$\bar{\nabla}_\mu \bar{\nabla}^\mu \hat{\theta}_1 = 0 , \tag{16}$$

where the covariant derivative $\bar{\nabla}_\mu$ is calculated from the "acoustic metric"

$$d\bar{s}^2 = \frac{n}{mc} [-(c_s^2 - v^2) dt^2 - 2v^i dx_i dt + dx^i dx_i] \tag{17}$$

showing that the propagation of sound waves in a BEC is governed by an effective spacetime metric.

One sees at glance the amazing similitude between Eqs. (16), (17) for the BEC and Eqs. (9), (10) for the massless scalar field in the BH spacetime. This is the core of the analogy: in particular, both metrics exhibit a horizon, at $|V| = c$ for the BH and $|v| = c_s$ for the BEC (the sonic horizon). The corresponding surface gravity is

$$\kappa = \frac{1}{2c_s} \frac{d}{dn} (c_s^2 - v^2)|_{hor} , \tag{18}$$

where n is the normal to the horizon, and a similar expression for the BH (with $c_s \to c$, $v \to V$).

Equations (9, 10) are the starting point of Hawking derivation of his famous result which is completely kinematical. So, based on the analogy one expects the supersonic BEC of Fig. (2) to radiate thermal phonons in the subsonic region at a temperature given by Eq. (2), which would open the possibility to reveal the Hawking effect in BEC.

One should however proceed with care since Eq. (16) is just an approximation of the BdG Eqs. (12), (13), an approximation in the long wavelength regime $\lambda \gg \xi$. As announced we encounter here the same consistency problem of the transplanckian frequency in the BH case since the wave equations are the same (16) and (10) and both predict an unbounded growth of the frequency of the modes near the horizon. However, we have now the complete quantum microscopic theory, given by the BdG Eqs. (12), (13) and we can study, starting directly from these, the presence or not of Hawking like phonon radiation in the BEC system without making any use of the hydrodynamic approximation, i.e. of the gravitational analogy.

To simplify the exposition let us assume that the BEC flow is stationary and that asymptotically (both in the subsonic and in the supersonic regions) the condensate is homogeneous. In these regions the dispersion relation of the plane waves solutions of the BdG Eqs. (12), (13) take the form

$$\omega - vk = \pm c_s \sqrt{k^2 + \frac{\xi^2 k^4}{4}} , \tag{19}$$

where ω is the conserved frequency and k the corresponding wave vector. Neglecting the last term in Eq. (19), the relation becomes linear

$$\omega = k(v \pm c_s) , \tag{20}$$

which would correspond to Eq. (16) and similarly to Eq. (10): this is the "relativistic" dispersion relation. So the hydrodynamical approximation leading to Eq. (18) misses the dispersive character of wave propagation in a BEC.

Let us give a look at Figs. 4 and 5 representing, graphically, the dispersion relation (19) in, respectively, the subsonic and supersonic regions. This latter, as an equation in k, for fixed ω, has in general four roots. In the subsonic case, two of them are real and two complex conjugates. The real ones, $k_v^{(+)}$ and $k_u^{(+)}$, correspond one to a

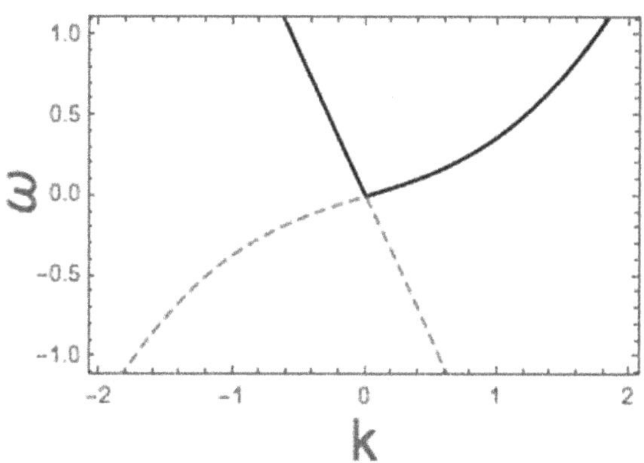

Fig. 4 BEC dispersion relation in the subsonic region

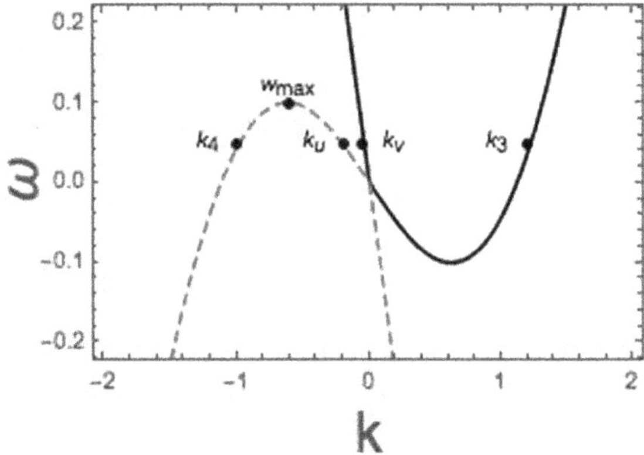

Fig. 5 BEC dispersion relation in the supersonic region

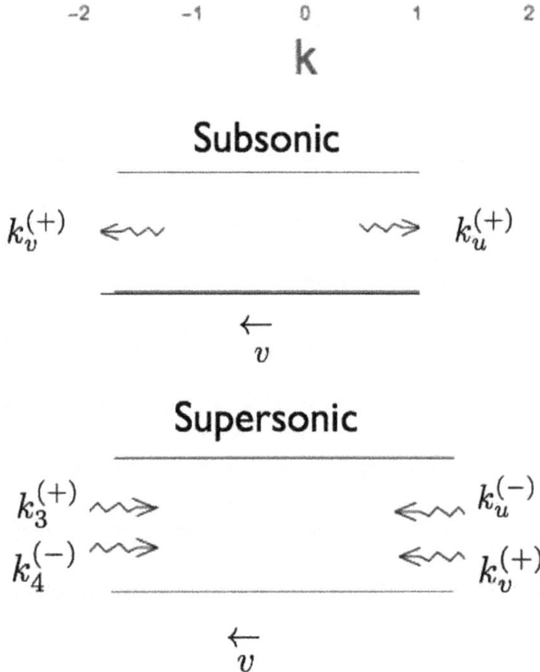

Fig. 6 Modes propagation in the subsonic (upper panel) and supersonic (lower panel) regions

wave propagating downstream along the flow ($k_v^{(+)}$), the other to a wave propagating upstream ($k_u^{(+)}$). See the upper panel of Fig. 6.

In the hydrodynamic approximation they reduce to the two solutions of the relativistic dispersion relation (20). The supersonic case is much more involved. There is a threshold frequency w_{max} which scales as ξ^{-1}. For $0 < \omega < \omega_{max}$ there are four real roots, of which $k_v^{(+)}$, $k_u^{(-)}$ are similar to the ones found in the subsonic case. The novelty is that $k_u^{(-)}$ is dragged by the flow and propagates downstream like $k_v^{(+)}$. These roots belong to the linear part of the dispersion relation. $k_3^{(+)}$ and $k_4^{(-)}$ propagate upstream, notwithstanding the supersonic character of the flow. See the lower panel of Fig. 6. So, while the propagation of the hydrodynamical modes $k_v^{(+)}$, $k_u^{(-)}$ can be well described by the wave equation (16) in the acoustic metric (17), the propagation

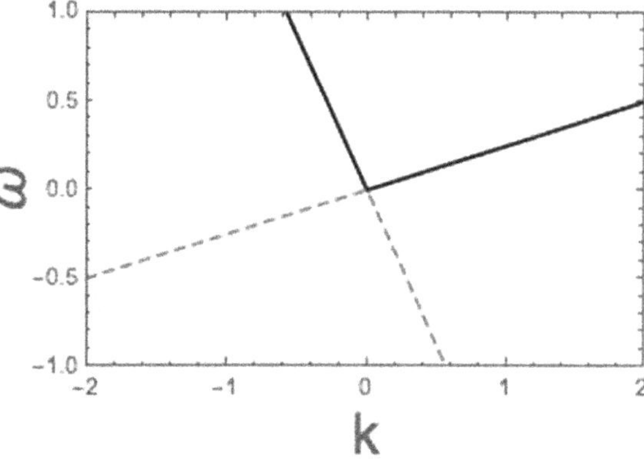

Fig. 7 Linear dispersion relation in the subsonic region

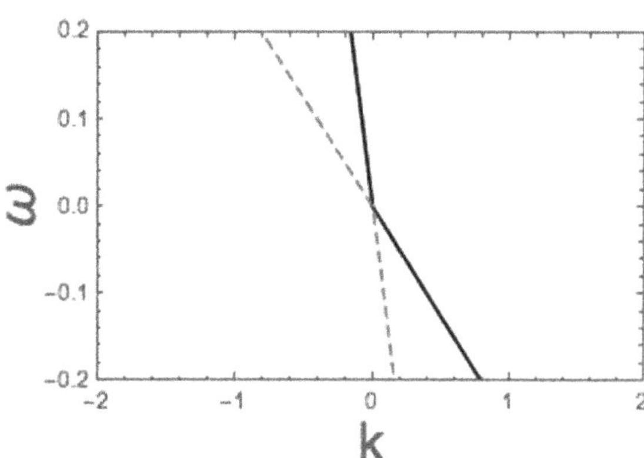

Fig. 8 Linear dispersion relation in the supersonic region

of $k_3^{(+)}$ and $k_4^{(-)}$ is completely dispersive. These modes do not feel at all the acoustic metric and the sonic horizon. Above ω_{max} only two solutions remain real.

For comparison, the relativistic dispersion relations are plotted in Figs. 7 and 8.

The dashed part in the plots of Figs. 4, 5, 7 and 8 represents modes with negative norm. One sees that—in the supersonic region—$k_u^{(-)}$ and $k_4^{(-)}$ have negative norm (this explains the superscript $^{(-)}$) and positive frequency ω or, equivalently, positive norm and negative frequency.

This is the key feature that a BH and a supersonically flowing BEC have in common: the presence in the spectrum of negative energy states. These lay inside the horizon for the BH (where the Killing vector associated to stationarity is spacelike) and in the supersonic region for the BEC. As a consequence, in both systems one can have production of particles out of the vacuum without violating energy conservation. Particles are produced in entangled pairs, one with positive energy and the other (the partner) with negative energy. This is how vacuum fluctuations are converted in real on shell particles (phonons in the BEC case). So without using the gravitational analogy the occurrence of Hawking like radiation in BEC can be firmly established starting from the Bogoliubov theory Eqs. (6), (7). One can then calculate, given

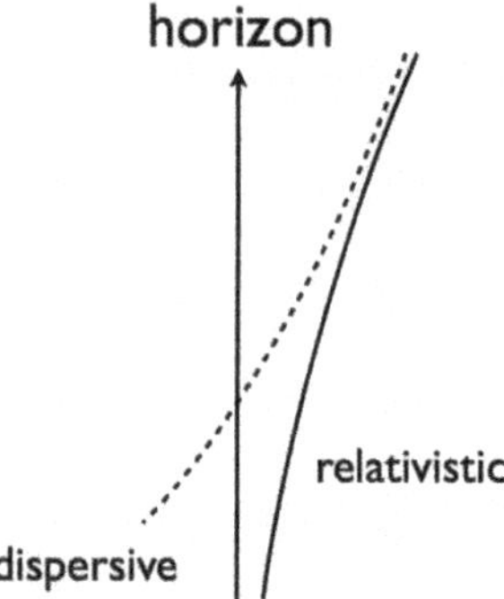

Fig. 9 Dispersion (dashed line) eliminates the piling up of the modes on the horizon (solid line)

an acoustic BH like profile for the flow, the spectrum of phonons emitted in the subsonic region of the BEC. In terms of S-matrix theory the process corresponds to the conversion of a negative energy $k_4^{(-)}$ incoming mode in the supersonic region to a positive energy $k_u^{(+)}$ outgoing mode in the subsonic one. There is no transplanckian problem since the dispersive character of Eq. (19) eliminates the piling up of the modes on the horizon characteristic of the relativistic dispersion relation (20) (see Fig. 9) responsible of the infinite blueshift of the frequency.

The obtained spectrum can then be compared to the one expected on the basis of the analogy, namely N_ω of Eq. (1) with a temperature given by Eq. (18) [10, 11].

First of all one should note that the emission is cut-off at $\omega_{\max}$. For $\omega > \omega_{\max}$ there are no more negative energy states (see Fig. 5). Without the presence of a partner a Hawking quantum can not materialize out of the vacuum. This is completely missed by the analogy. Given this, one finds that for $\xi \ll \frac{c_s}{\kappa}$ the emission is thermal at exactly the Hawking temperature (18). So in this regime the predictions of the analogy are reliable. On the other hand for $\xi \gg \frac{c_s}{\kappa}$ the emission at low ω is still thermal but the temperature differs from the one associated to the surface gravity. Furthermore, significant deviations from the prediction of the analogy emerge in limiting cases.

4 Limiting Cases

Consider the following step-like profile for the BEC flow (Fig. 10) for which the velocity v is constant and the sound velocity c_s equals $c_r > |v|$ in the subsonic region while equals $c_l < |v|$ in the supersonic region.

One finds at low ω an emission temperature proportional to ξ^{-1} [12, 13]

$$T_{step} = \frac{\hbar}{k_B} \frac{(c_r + v)(v^2 - c_r^2)}{(c_r - v)(c_r^2 - c_l^2)} \frac{2c_r}{c_l \xi} . \tag{21}$$

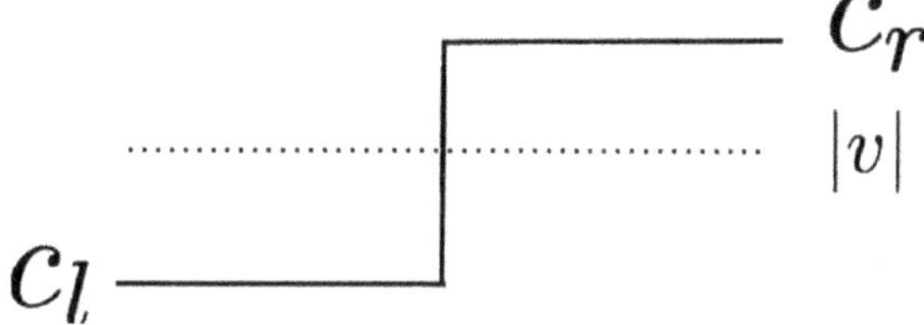

Fig. 10 Step-like profile for an acoustic black hole

T_{step} can be considered as the maximum temperature achievable for monotonic profiles with fixed asymptotic c_r and c_l. If the step like profile is obtained by a limiting procedure of smooth profiles, like for example $\tanh(x/L)$ with $L \to 0$, the analogy would predict a continuously increasing surface gravity and hence of T_H, diverging in the step limit. The healing length, which characterizes the dispersive regime of (19), provides therefore an upper cutoff for the emission temperature. So the analogy in this case is completely unreliable.

Consider now the opposite situation in which $\kappa = 0$. The usual example in the gravitational case would be the extreme Reissner-Nordström BH described, in Schwarzschild coordinates, by

$$ds^2 = -(1 - \frac{GM}{c^2 r})^2 dt^2 + \frac{dr^2}{(1 - \frac{GM}{c^2 r})^2} + r^2 d\Omega^2 \tag{22}$$

for which $\kappa = 0$ and $T_H = 0$, i.e. no Hawking emission. A corresponding BEC flow profile would be something like the one depicted in Fig. 11.

With this profile there is no phonons emission neither for the BEC. The reason is that there is no supersonic region and so there are no negative energy states. Similarly for the BH, the Killing horizon associated to stationarity is everywhere timelike (it becomes null just on the horizon $r = \frac{GM}{c^2}$).

More tricky is the following example. Consider a BEC with a profile like the one depicted in Fig. 12, for example $c(x) \propto \tanh(x^3)$.

A corresponding BH metric would be

$$ds^2 = -g_{00}dt^2 + g_{00}^{-1}dr^2 + r^2 d\Omega^2 \ , \tag{23}$$

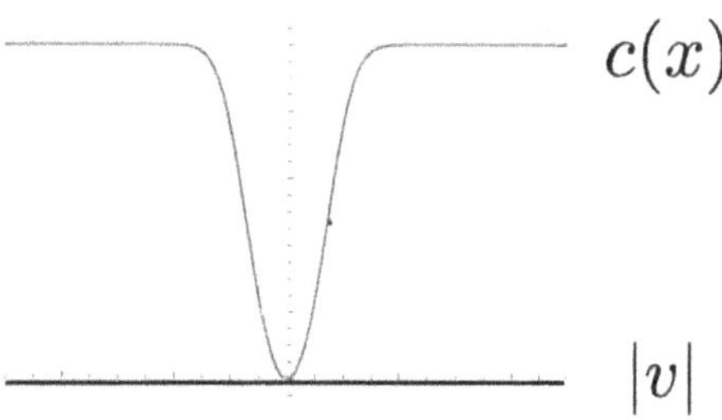

Fig. 11 Zero temperature flow profile with no supersonic region

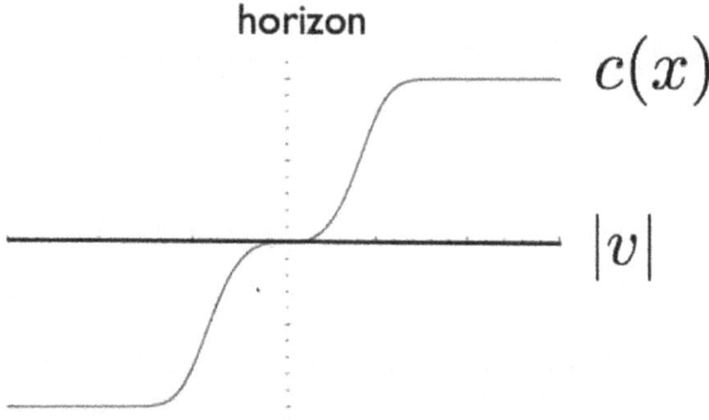

Fig. 12 Zero temperature flow profile with a supersonic region

where $g_{00} \sim (r - r_H)^3$ for r approaching the horizon r_H. Having $\kappa = 0$ this BH does not emit and the analogy would predict the same for the BEC. This is not the case since for the BEC, using the Bogoliubov theory, one finds an emission characterized by $N_\omega \sim const.$ at low ω [14]. The presence of a supersonic region with negative energy states available guarantees the emission in the BEC. The problem is to understand why the BH described by the metric (23) does not emit even having, unlike the extreme Reissner-Nordström BH of Eq. (22), an interior region where negative energy states exist (the Killing vector is spacelike there).

In general, for BHs the emission is triggered by the collapse of the star across the horizon. There is an initial transient emission which depends on the details of the collapsing star which rapidly decays in time leaving at late time a steady thermal flux independent on the collapse details being a function only of the surface gravity κ of the resulting BH horizon. This latter is Hawking radiation, described by Eqs. (1), (2). The presence of this steady thermal emission then requires a BH region and a nonvanishing κ. When $\kappa = 0$ there is just a transient emission which rapidly decreases to zero and no steady thermal flux. To get a steady emission in the BEC case it is not necessary to take into account the formation of the sonic horizon. Given the dispersive character of the dispersion relation Eq. (19) one has emission even in a stationary flow simply because of the existence of the $k_4^{(-)}$ mode capable of propagating upstream in the supersonic region as we have seen.

5 Low Energy-Emission and Gray-Body Factor

The thermality of Hawking radiation is not completely exact because of the backscattering of the modes caused by the curvature of the space-time, and the expression (1) of the emitted particles gets a correction, namely

$$N_\omega = \frac{\Gamma(\omega)}{e^{\frac{\hbar\omega}{k_B T_H}} - 1} , \tag{24}$$

where $\Gamma(\omega)$ is the so called "gray-body" factor which accounts for the backscattering (see for instance [15]). $\Gamma(\omega)$ is the probability for an outgoing mode created near the horizon to reach infinity. For asymptotically flat BHs [16]

$$\Gamma(\omega) \sim A_H \omega^2 , \tag{25}$$

where A_H is the area of the BH horizon. For $\omega \to 0$ outgoing modes are almost completely reflected back to the horizon. This correction makes N_ω of Eq. (24) infrared finite.

In the BEC case the small ω behaviour can be well analyzed by the hydrodynamical limit of the BdG equations namely Eqs. (16), (17) and one finds a quite different result: $\Gamma(\omega)$ goes to a constant at small ω [17, 18]

$$\Gamma(\omega) = \frac{c_r v}{(v + c_r)^2} . \tag{26}$$

The reason is that in the BEC the effective potential responsible for the backscattering of the modes allows for the existence of an everywhere bounded solution of the $\omega \to 0$ limit of the wave equation. In the BH case the corresponding limit gives a solution growing like r at infinity. So the graybody factor of Eq. (26) does not cancel the $\frac{1}{\omega}$ divergence coming from the Planckian distribution: the emission of the BEC black hole is dominated by soft phonons.

6 Correlations

Having established the solid foundations of the existence of Hawking like radiation in BEC one hopes this radiation to be revealable in laboratory experiments. The major problem one has to overcome is that in the most favourable case the emission temperature expected is at least one order of magnitude lower than the background temperature of the BEC ($10^{-9} K$). So thermal phonons can completely cover the Hawking signal one is searching for. However it was realized that, being Hawking radiation basically a pair production process, one can look for the correlation between the pair members, i.e. the Hawking particle and the partner inside the hole. The unambiguous signature of this correlation is a peak in the correlation functions, for example the density-density. It was predicted in [19], using the hydrodynamic approximation of the BEC theory, that thermal analog Hawking radiation gives rise to the (one-time) nonlocal correlator

$$\langle \hat{n}_1(t, x)\hat{n}_1(t, x) \rangle \sim -\frac{\kappa^2}{\cosh^2(\frac{\kappa}{2}(\frac{x}{c_r - |v|} + \frac{x'}{|v| - c_l}))} , \tag{27}$$

for points located on the both sides and sufficiently away from the horizon where the profile is to a good approximation homogeneous. We see that height and width of this correlator is fixed by the black hole surface gravity κ, and disappears when there is no emission. Such a profile has a stationary peak along the line

$$\frac{x}{c_r - |v|} + \frac{x'}{|v| - c_l} = 0, \tag{28}$$

and this in turns corresponds to a steady emission of phonons by the horizon ($x = 0$), one of which propagating upstream ($k_u^{(+)}$) in the region exterior to the horizon with velocity $c_r - |v|$ and the other (the partner, $k_u^{(-)}$) being dragged by the flow in the supersonic region and propagating with velocity $c_l - |v|$. It was then shown that this signature persists in the full BEC theory [20], and, importantly, that it is still there in the presence of a thermal bath at a temperature $T > T_H$. See Fig. 13. The peak at (28) is the smoking gun of the Hawking effect in BECs.

In a series of experiments, starting in 2016, Steinhauer [21] was indeed able to reveal this peak giving the first experimental evidence of Hawking like radiation in BECs (see also [22]). Finally, the quantum signature of Hawking radiation (i.e. the fact that the emitted pairs come from the vacuum, i.e. they are entangled, rather than the thermal background) is better characterized in terms of the Fourier transform of the density correlator [23, 24], or the momentum correlator [25, 26].

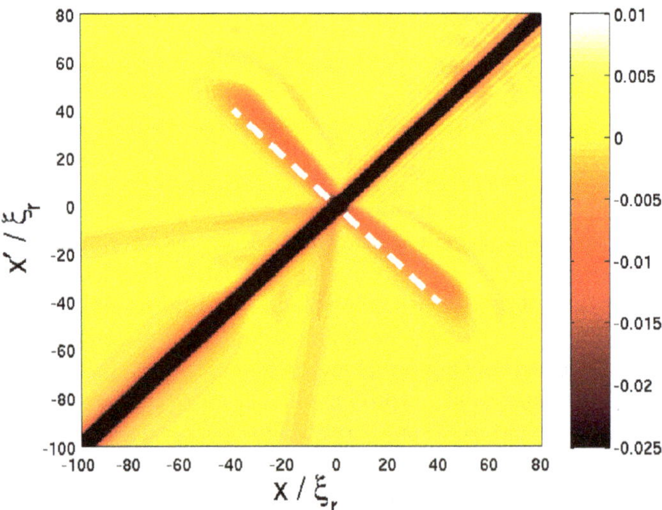

Fig. 13 Hawking quanta-partner peak in the density correlator (the dotted line is the hydrodynamical prediction). Reprinted from Balbinot, R., Carusotto, I., Fabbri, A., Mayoral, C., Recati, A. (2013). Understanding Hawking Radiation from Simple Models of Atomic Bose-Einstein Condensates. In: Faccio, D., Belgiorno, F., Cacciatori, S., Gorini, V., Liberati, S., Moschella, U. (eds) Analogue Gravity Phenomenology. Lecture Notes in Physics, vol 870. Springer, Cham. https://doi. org/10.1007/978-3-319-00266-8_9 with permission, ©Springer Nature BV. All rights reserved

Acknowledgements A.F. acknowledges partial financial support by the Spanish Grants PID2020-116567GB-C21, PID2023-149560NB-C21 funded by MCIN/AEI/10.13039/501100011033, and by the Severo Ochoa Excellence Grant CEX2023-001292-S.

References

1. S. W. Hawking, Nature **248**, 30 (1974)
2. S. W. Hawking, Commun. Math. Phys. **43**, 199 (1975)
3. J. M. Bardeen, B. Carter, S. W. Hawking, Commun. Math. Phys. **31**, 62 (1973)
4. B. J. Carr, Astrophys. J. **201**, 1 (1975)
5. T. Jacobson, Phys. Rev. D **44**, 1731 (1991)
6. C. Barceló, S. Liberati, M. Visser, Living Rev. Relativ. **14**, 3 (2011)
7. W. G. Unruh, Phys. Rev. Lett. **46**, 1351 (1981)
8. L. Pitaevski, S. Stringari, *Bose-Einstein Condensation* (Oxford University Press, Oxford, UK, 2003)
9. R. Balbinot, I. Carusotto, A. Fabbri, C. Mayoral, A. Recati, Lect. Notes. in Phys. **870**, 181 (2013)
10. J. Macher, R. Parentani, Phys. Rev. A **80**, 043061 (2009)
11. S. Finazzi, R. Parentani, Phys. Rev. D **83**, 084010 (2011)
12. A. Recati, N. Pavloff, I. Carusotto, Phys. Rev. A **80**, 043603 (2009)
13. P. E. Larré, A. Recati, I. Carusotto, N. Pavloff, Phys. Rev. A **85**, 013621 (2012)
14. S. Finazzi, private communication
15. B. de Witt, Physics Reports **19**, 295 (1975)
16. D. N. Page, Phys. Rev. D **13**, 198 (1976)
17. P. R. Anderson, R. Balbinot, A. Fabbri, R. Parentani, Phys. Rev. D **90**, 104044 (2014)
18. P. R. Anderson, A. Fabbri, R. Balbinot, Phys. Rev. D **91**, 064061 (2015)
19. R. Balbinot, A. Fabbri, S. Fagnocchi, A. Recati, I. Carusotto, Phys. Rev. A **78**, 021603 (2008)
20. I. Carusotto, S. Fagnocchi, A. Recati, R. Balbinot, A. Fabbri, New J. Phys. **10**, 103001 (2008)
21. J. Steinhauer, Nat. Phys. **12**, 959 (2016)
22. J. R. M. de Nova, K. Golubkov, V. I. Kolobov, J. Steinhauer, Nature **569**, 688–691, 7758(2019)
23. X. Busch, R. Parentani, Phys. Rev. D **89**, 105024 (2014)
24. J. Steinhauer, Phys. Rev. D **92**, 024043 (2015)
25. D. Boiron, A. Fabbri, P. E. Larré, C.I. Westbrook, P. Zin, Phys. Rev. Lett. **115**, 025301 (2015)
26. A. Fabbri, N. Pavloff, SciPost Phys. **4**, 019 (2018)

On a Gambini-Pullin-Inspired Maxwell Electrodynamics with Lorentz-Violating Dimension-5 Operators

Patricio Gaete and José A. Helayël-Neto

Abstract We explore the properties of a new Maxwell electrodynamics coupled to a Lorentz-violating background through the presence of higher-derivative terms. Physical implications of this alternative effective theory modified by Lorentz-violating operators of mass dimension-5 are considered, such as modified dispersion relations which exhibit the vacuum birefringence phenomenon. Subsequently, we analyze the energy-momentum tensor in the case we reproduce the Gambini-Pullin electrodynamics. We explicitly show that for a background time-like four-vector, n^{μ}, the equations of motion indeed reproduce those encountered by Gambini and Pullin in the framework of loop quantum gravity . Finally, extended expressions for Maxwell-type equations are written down in the case the external Lorentz-violating four-vector is space-like.

1 Introduction

As it is well known, the Standard Model (SM) describes the interactions among the fundamental constituents of the matter, that is, the quantum field theory of the smallest building blocks of nature interacting in a locally gauge-invariant way. Its success is based on predicting with an impressive accuracy every process in the accelerators to this day, over a wide range of energy. Despite this amazing success, the SM is far from perfect. The problems begin, as widely discussed, with the fact that the SM does not include gravity, an interaction for which we do not have to this date a quantum description in four space-time dimensions. In fact, this has led to

P. Gaete (✉)
Departamento de Física and Centro Científico-Tecnológico de Valparaíso, Universidad Técnica Federico Santa María, Avenida España, 1680 Valparaíso, Chile
e-mail: patricio.gaete@usm.cl

J. A. Helayël-Neto
Centro Brasileiro de Pesquisas Físicas, Rua Xavier Sigaud, 150, Urca, Rio de Janeiro 22290-180, Brazil
e-mail: helayel@cbpf.br

P. Nicolini (ed.), *Touring the Planck Scale*, Fundamental Theories of Physics 219,
https://doi.org/10.1007/978-3-031-76066-2_9

assume that these two theories will merge at the Planck scale, $m_P \cong 10^{19}$ GeV, in order to provide a consistent framework to unify all fundamental interactions [1].

Let us also mention here that, at these energies, Lorentz- and CPT-symmetry violation effects are expected to show up. Actually, the search for departures from Lorentz invariance is an active research area. The interest in studying these issues is mainly due to that in some candidate theories of a quantum description of gravity, such as string theory [2–5] and loop quantum gravity [6, 7], there are scenarios where the breakdown of Lorentz symmetry takes place. We mention, in passing, that the main idea of these developments is that there would be atoms of geometry of which fundamental building blocks of gravity arise (discreteness). However, independent from the quantum gravity theory, the focus of these studies lies on low-energies, where the effects of Lorentz-symmetry violation (LSV) can be unveiled. This discreteness would act as a dispersive medium for particles, such as photons, for example, propagating on this spacetime.

In this connection, it becomes of interest, to recall that low-energy effective field theories are a very suitable framework to probe effects of quantum gravity introduced as small (LSV) terms in the usual field theories. The Standard Model Extension (SME) is the key example, so that possible departures from exact Lorentz invariance can be implemented. In fact, this has provided a theoretical background from different points of view. For example, in connection with photon physics [8–25], also in effects of radiative corrections [26–36], systems of fermions [37–43], neutrino physics [44–52], topological deffects [53, 54], cosmic rays [55–58], and in magnetic monopoles studies [59]. As well as, Lorentz-violating theories with higher-dimensional operators [60–63]. Another possible scenario is a newly proposed theory of Lorentz-symmetry violation from first principles [64, 65]. In addition, in previous studies [66, 67], we have considered aspects of LSV physics in a supersymmetric scenario. On the one hand, we have studied the minimal supersymmetric extension of the Carroll-Field-Jackiw model for Electrodynamics with a topological Chern-Simons-like LSV term. In this case, we have shown that supersymmetry is naturally broken whenever a single scalar supermultiplet triggers Lorentz violation symmetry. On the other hand, we have discussed that if the scale of breaking of Lorentz covariance is above the supersymmetry breaking scale, we find that once Lorentz symmetry is violated, supersymmetry also takes place. In this case the photon and photino dispersion relations are split from one another with different profiles in terms of the supersymmetric background induced in the process of LSV.

As we have already expressed, a particularly interesting example that spacetime would be discrete, within loop quantum gravity, is the approximation developed in [6, 7] for the Einstein-Maxwell theory. In fact, these authors devised a method for obtaining effective Maxwell equations from the expectation values of the quantum Hamiltonian. As a consequence, they obtain modified dispersion relations which display the breaking of Lorentz symmetry. Interestingly, a similar result has been obtained in the context of an effective theory with Lorentz violating dimension-5 operators for the photon sector [61]. Inspired by the above observations, in this work we would like to recover the results obtained in [6, 7] within the familiar language of standard (effective) quantum field theory in four dimensions. This piece of information can be useful in view of establishing new connections among different models describing the same physical phenomena.

In Sect. 2, we propose and describe this modified Maxwell electrodynamics with Lorentz-violating dimension-5 operators and study aspects of birefringence and the energy-momentum tensor. In Sect. 3, we study Maxwell-type equations in the case of a space-like four-vector. Finally, our Concluding Remarks are cast in Sect. 4.

2 The Model Under Consideration

As already mentioned in the introduction, in this section we will propose and study the properties of this modified Maxwell electrodynamics in the language of usual field theory. With this purpose, let us consider first a new electromagnetic field strength tensor. This would not only provide the setup theoretical for our subsequent work, but also fix the notation.

2.1 Electromagnetic Field-Strength Tensor

We start off our analysis by considering a new four-dimensional spacetime electromagnetic field-strength tensor here extended as follows below

$$
\mathscr{F}_{\mu\nu} = F_{\mu\nu} + \frac{\alpha}{2} l_P n_\kappa \varepsilon^\kappa{}_{\mu\alpha\beta} \partial^\alpha F_\nu{}^\beta - \frac{\alpha}{2} l_P n_\kappa \varepsilon^\kappa{}_{\nu\alpha\beta} \partial^\alpha F_\mu{}^\beta + \beta l_P n_\kappa \varepsilon^\alpha{}_{\mu\nu\beta} \partial_\alpha F^{\kappa\beta}, \quad (1)
$$

where l_P is Planck's length, whereas α and β are dimensionless numerical coefficients. Here, $F_{\mu\nu} = \partial_\mu A_\nu - \partial_\nu A_\mu$, stands for the usual electromagnetic field-strength tensor, and n^μ is a dimensionless four-vector.

From this equation, it follows that the dual electromagnetic field-strength tensor takes the form

$$
\begin{aligned}
\tilde{\mathscr{F}}^{\mu\nu} &= \tilde{F}^{\mu\nu} - l_P \left(-\frac{\alpha}{2} + \beta \right) n_\kappa \partial^\mu F^{\kappa\nu} + l_P \left(-\frac{\alpha}{2} + \beta \right) n_\kappa \partial^\nu F^{\kappa\mu} \\
&\quad + l_P \frac{\alpha}{2} n^\nu \partial_\kappa F^{\kappa\mu} - l_P \frac{\alpha}{2} n^\mu \partial_\kappa F^{\kappa\nu},
\end{aligned} \quad (2)
$$

where $\tilde{F}^{\mu\nu} = \frac{1}{2} \varepsilon^{\mu\nu\rho\lambda} F_{\rho\lambda}$ is the usual dual electromagnetic field-strength tensor.

For the sake of completeness, these equations can be alternatively written in terms of electric (**E**) and magnetic (**B**) fields, as

$$
\mathscr{F}_{0i} = E_i - \left(-\frac{\alpha}{2} + \beta \right) l_P \varepsilon_{ijk} \partial_j E_k, \quad (3)
$$

$$\mathscr{F}_{ij} = -\varepsilon_{ijk}\left(B_k - \frac{\alpha}{2}l_P(\nabla \times \mathbf{B})_k - \beta\partial_t E_k\right), \tag{4}$$

$$\tilde{\mathscr{F}}_{0i} = B_i - \frac{\alpha}{2}l_P(\nabla \times \mathbf{B})_i - \beta l_P\partial_t E_i, \tag{5}$$

$$\tilde{\mathscr{F}}_{ij} = \varepsilon_{ijk}\left(E_k - \left(-\frac{\alpha}{2} + \beta\right)l_P(\nabla \times \mathbf{E})_k\right). \tag{6}$$

Having characterized the electromagnetic field-strength tensor, we can now consider a new effective Lagrangian density.

2.2 *Effective Lagrangian and Gambini-Pullin Equations*

In this case, the corresponding model is governed by the Lagrangian density:

$$\mathscr{L} = -\frac{1}{4}\mathscr{F}^2_{\mu\nu} + \frac{\xi}{2}\,l_P n_\kappa \varepsilon^\kappa{}_{\alpha\beta\gamma}\,n^\lambda \tilde{F}_\lambda{}^\alpha \partial^\beta n^\rho \tilde{F}_\rho{}^\gamma + a\,l_P n^\kappa n^\lambda n^\rho \partial_\kappa \tilde{F}_{\lambda\sigma}\partial_\rho A^\sigma, \tag{7}$$

where ξ and a are dimensionless numerical coefficients.

On simplification (after expansion in l_P) we find that Eq. (7) can be brought into the form

$$\begin{aligned}
\mathscr{L} &= -\frac{1}{4}F^2_{\mu\nu} - \frac{\alpha}{2}l_P\,F^{\mu\nu}n_\kappa \varepsilon^\kappa{}_{\mu\alpha\beta}\partial^\alpha F_\nu{}^\beta + \frac{\xi}{2}\,l_P\,n_\kappa \varepsilon^\kappa{}_{\alpha\beta\gamma}\,n^\lambda \tilde{F}_\lambda{}^\alpha \partial^\beta n^\rho \tilde{F}_\rho{}^\gamma \\
&\quad + a\,l_P n^\kappa n^\lambda n^\rho \partial_\kappa \tilde{F}_{\lambda\sigma}\partial_\rho A^\sigma.
\end{aligned} \tag{8}$$

Notice that to get this new equation, we have ignored quadratic terms in l_P, as it shall become clear below. In other words, this new effective theory provides us with a suitable starting point for our analysis.

The equations of motion following from the Lagrangian density (8) read

$$\partial_\mu F^{\mu\nu} - l_P\left[\left(\alpha - \xi n^2\right)\Delta + (\xi + 2a)\,(n \cdot \partial)^2\right]n_\kappa \tilde{F}^{\kappa\nu} = 0, \tag{9}$$

where $\Delta \equiv \partial_\mu \partial^\mu$.

From Eq. (9), it now follows, for a time-like four-vector $n_\mu = (1, \mathbf{0})$ and $\nu = 0$, that

$$\nabla \cdot \mathbf{E} = 0, \tag{10}$$

while for $\nu = i$, we find

$$(\nabla \times \mathbf{B}) + l_P(\nabla \times \nabla \times \mathbf{B}) = \frac{\partial}{\partial t}\mathbf{E}. \tag{11}$$

It should be further noticed that, to obtain the foregoing equations, we have adjusted the dimensionless numbers (α, a and ξ) so that they satisfy the algebraic equations $\alpha + 2a = 0$ and $\xi - \alpha = -1$.

Before going on, a remark is pertinent at this point. Observe that Eq. (2) yields

$$\partial_\mu \tilde{\mathscr{F}}^{\mu\nu} = \partial_\mu \tilde{F}^{\mu\nu} - l_P \left(-\frac{\alpha}{2} + \beta\right) n_\kappa \Delta F^{\kappa\nu} + l_P \left(-\frac{\alpha}{2} + \beta\right) n_\kappa \partial^\nu \partial_\mu F^{\kappa\mu}$$
$$+ l_P \frac{\alpha}{2} n^\mu \partial^\kappa \partial_\mu F_\kappa{}^\nu. \tag{12}$$

Here, we clearly see that if the right-hand side of Eq. (12) is zero, we re-obtain the Bianchi identities for the new field-strength ($\tilde{\mathscr{F}}^{\mu\nu}$). To examine this question let us recall that $\partial_\mu \tilde{F}^{\mu\nu} = 0$. Second, it should be noticed that, for theories described by $\mathscr{L} = -\frac{1}{4}\mathscr{F}_{\mu\nu}\mathscr{F}^{\mu\nu}$, the equations of motion read $\partial_\mu F^{\mu\nu} \sim \mathscr{O}(l_P)$. With this information, it is evident that the third and fourth terms on the right-hand side of Eq. (12) are terms $\mathscr{O}\left((l_P)^2\right)$. It is also of interest to remark that, from the Bianchi identity, $\partial_\mu F_{\nu\kappa} + \partial_\nu F_{\kappa\mu} + \partial_\kappa F_{\mu\nu} = 0$, it follows that $\Delta F_{\nu\kappa} \sim \mathscr{O}(l_P)$. Making use of the foregoing results, we can, therefore, write (to order l_P)

$$\partial_\mu \tilde{\mathscr{F}}^{\mu\nu} = 0. \tag{13}$$

Having made these observations, we can immediately write the field equations for a background time-like four-vector $n_\mu = (1, \mathbf{0})$. When $\nu = 0$, we obtain

$$\nabla \cdot \mathbf{B} = 0, \tag{14}$$

whereas for $\nu = i$,

$$(\nabla \times \mathbf{E}) + l_P (\nabla \times \nabla \times \mathbf{E}) = -\frac{\partial}{\partial t}\mathbf{B}, \tag{15}$$

to get this last line, we have used the algebraic equation $-\frac{\alpha}{2} + \beta = -1$.

In summary, then: with the aid of the Eqs. (11) and (15), we obtain the equations

$$\frac{\partial}{\partial t}\mathbf{E} = (\nabla \times \mathbf{B}) - l_P \nabla^2 \mathbf{B}, \tag{16}$$

and

$$\frac{\partial}{\partial t}\mathbf{B} = -(\nabla \times \mathbf{E}) + l_P \nabla^2 \mathbf{E}. \tag{17}$$

It should again be stressed here that, as discussed previously, the preceding equations are valid at leading order in l_P. This result agrees with that of Refs. [6, 7], and finds here an independent derivation.

2.3 Properties

2.3.1 Birefringence

For the sake of completeness, we now re-examine the propagation of electromagnetic waves in this scenario. As well known, in this theory the phenomenon of birefringence is present. We recall, in passing, that birefringence refers to the property that polarized light in a particular direction travels at a different velocity from that of light polarized in a direction perpendicular to this axis. Incidentally, it is of interest to notice that due to quantum fluctuations. the vacuum of this new effective theory has this property, as we are going to show.

To illustrate this important feature, we introduce a decomposition into plane waves for the fields $\mathbf{E}$ and $\mathbf{B}$ as:

$$\mathbf{E} = \mathbf{E}_0 e^{-i(wt - \mathbf{k}\cdot\mathbf{x})}, \quad \mathbf{B} = \mathbf{B}_0 e^{-i(wt - \mathbf{k}\cdot\mathbf{x})}. \tag{18}$$

These equations can alternatively be rewritten in the form

$$\mathbf{E} = \mathbf{E}_{0R} \cos\left(\mathbf{k}\cdot\mathbf{x} - wt\right) - \mathbf{E}_{0I} \sin\left(\mathbf{k}\cdot\mathbf{x} - wt\right), \tag{19}$$

and

$$\mathbf{B} = \mathbf{B}_{0R} \cos\left(\mathbf{k}\cdot\mathbf{x} - wt\right) - \mathbf{B}_{0I} \sin\left(\mathbf{k}\cdot\mathbf{x} - wt\right). \tag{20}$$

Now, making use of Eqs. (16) and (17), we readily find that

$$-\mathbf{k} \times \mathbf{E}_{0R} - k^2\, \mathbf{E}_{0I} = -w\, \mathbf{B}_{0R}, \tag{21}$$

$$-\mathbf{k} \times \mathbf{E}_{0I} + k^2\, \mathbf{E}_{0R} = -w\, \mathbf{B}_{0I}, \tag{22}$$

$$-\mathbf{k} \times \mathbf{B}_{0R} - k^2\, \mathbf{B}_{0I} = w\, \mathbf{E}_{0R}, \tag{23}$$

$$-\mathbf{k} \times \mathbf{B}_{0I} + k^2\, \mathbf{B}_{0R} = w\, \mathbf{E}_{0I}. \tag{24}$$

Next, without loss of generality we take the z-axis as the direction of propagation of the wave light, that is, $\mathbf{k} = k\hat{z}$. We further consider a circularly polarized light wave, $\mathbf{E}_{0R} = \mathbf{E}_0\hat{x}$ and $\mathbf{E}_{0I} = \eta\mathbf{E}_0\hat{y}$, where $\eta = \pm 1$, is a parameter to describe the fields with the left- and right-hand circularly polarized unit vectors.

Now, by using (21)–(24), we find that the corresponding dispersion relation becomes

$$\omega^2 = k^2 + 2l_P\eta k^3 + l_P^2 k^4. \tag{25}$$

Expression (25) immediately shows that the parameters η and l_P are coupled or, more precisely, the term l_P^2 does not distinguish between the two modes of polarization. Here is the physical reason why, in this work, we have considered our calculations at leading order in l_P.

It is immediate to check that, for $l_P^2 \ll l_P$, we recover the known result [6, 7]

$$\omega^2 = k^2 + 2l_P\eta k^3. \tag{26}$$

As we have already mentioned, analogous dispersion relation can be obtained from a different effective theory with Lorentz-violating dimension-5 operators for the photon sector [61]. This implies that the electromagnetic waves with different polarizations have different velocities, that is, the vacuum birefringence phenomenon is present. Let us also note here that the group velocity reduces to

$$v_g = \frac{1}{\omega}\left(k + 3\eta l_P\, k^2\right), \tag{27}$$

which displays a modified speed of light for a photon with momentum k.

2.3.2 The Energy-Momentum Tensor

Another relevant aspect to explore in this new electrodynamics is the calculation of the energy and momentum densities transported by the electromagnetic field. To do this, we start by contracting the equations of motion (9) with $F_{\nu\kappa}$, that is,

$$\left(\partial_\mu F^{\mu\nu}\right) F_{\nu\kappa} - \alpha\, l_P\, n_\lambda \varepsilon^{\lambda\nu\alpha\beta} \left(\partial_\alpha \partial^\mu F_{\mu\beta}\right) F_{\nu\kappa} + \xi\, l_P\, n^2 n^\rho \left(\Delta \tilde{F}^\nu_\rho\right) F_{\nu\kappa}$$
$$- (\xi + 2a)\, l_P \left[(n\cdot\partial)^2 n_\rho \tilde{F}^\nu_\rho\right] F_{\nu\kappa} = 0. \tag{28}$$

After some further algebraic manipulations, for the model under consideration, the energy-momentum tensor comes out in the form

$$\Theta^\mu{}_\kappa = F^{\mu\nu} F_{\nu\kappa} + \frac{1}{4}\delta^\mu{}_\kappa F^2 + l_P\, n_\lambda \left[\left(-\alpha + \xi n^2\right)\partial^\mu - (\xi + 2a)\, n^\mu\, (n\cdot\partial)\right] \tilde{F}^{\lambda\nu} F_{\nu\kappa}$$
$$+ \frac{l_P}{2}\left(-\alpha + \xi n^2\right) n_\kappa\, (\partial_\alpha A_\nu)\left(\partial^\alpha \tilde{F}^{\mu\nu}\right)$$
$$- \frac{l_P}{2}(\xi + 2a)\, n_\kappa\, (n\cdot\partial)\, A_\nu\, (n\cdot\partial)\, \tilde{F}^{\mu\nu}, \tag{29}$$

which is conserved

$$\partial_\mu \Theta^\mu{}_\kappa = 0. \tag{30}$$

Note that here, by virtue of the anisotropy described by the external vector, we do not obtain a symmetric energy-momentum tensor

$$\Theta^\mu{}_\kappa \neq \Theta_\kappa{}^\mu. \tag{31}$$

Before we proceed further, two remarks are pertinent at this point, because this energy-momentum tensor is neither symmetric nor manifestly gauge invariant. First, it would be noticed that the last two terms at the right-hand side of expression (29) explicitly contain the potential A_μ; hence, expression (29) would not be manifestly gauge invariant. In such a case, we verify that, under a gauge transformation, $\delta A_\nu = \partial_\nu \Lambda$, the fourth term at the right-hand side of (29) transforms as $\frac{\chi}{2}\left(-\alpha + \xi n^2\right) n_\kappa \left(\partial_\alpha \partial_\nu \Lambda\right)\left(\partial^\alpha \tilde{F}^{\mu\nu}\right)$. After using the Bianchi identity, this term will appear in the continuity equation (30) as $\partial_\mu \partial_\nu \left(\partial_\alpha \Lambda \partial^\alpha \tilde{F}^{\mu\nu}\right) = 0$. The same argument applies to the fifth term of expression (29). Therefore, the continuity equation (30) is actually gauge invariant, though not is a manifest way. Second, we wish to further elaborate on the physical meaning of expression (31). Here, our concern is to reconsider the calculation of the energy density and the Poynting vector, paying due attention to the non-symmetric character of the energy-momentum tensor (31). For this purpose, we consider the last two terms at the right-hand side of the expression for Θ^{00}, which would spoil a correct definition of energy. As we can see, under a gauge transformation, the fourth term of Θ^{00} transforms as $\nabla \cdot (\partial_\alpha \Lambda \partial^\alpha \mathbf{B})$. Thus, when we consider this term in the energy expression ($\int d^3x \, \Theta^{00}$), we verify that it becomes a surface contribution which integrates to zero assuming, as always, that the field decreases sufficiently fast at infinity. The same argument applies to the fifth term of Θ^{00}. Therefore, the energy is gauge invariant. In the same way as done in the previous case, the terms dependent on the gauge parameter that contribute to Θ^{0i} disappear. Thus, the momentum, $\mathbf{P} = \int d^3x \Theta^{0i}$, is also a gauge invariant quantity.

We now consider a background time-like $n^\mu = (1, \mathbf{0})$. Making use of the foregoing results, we find that

$$\Theta^{00} \equiv \mathscr{E} = \frac{1}{2}\left(\mathbf{E}^2 + \mathbf{B}^2\right) - (\alpha + 2a)\,l_P\left(\mathbf{E} + \frac{1}{2}\dot{\mathbf{A}}\right)\cdot\dot{\mathbf{B}} + \frac{1}{2}(\alpha - \xi)\,l_P\left(\partial_j A_i\right)\left(\partial_j B_i\right),$$

$$(32)$$

where $\mathscr{E}$ denotes an energy density. While the momentum density, $\tilde{\mathbf{S}}$, is given by

$$\Theta^{0i} \equiv \tilde{\mathbf{S}} = \mathbf{E} \times \mathbf{B} + (\alpha + 2a)\,l_P\,\mathbf{B} \times \dot{\mathbf{B}}. \qquad (33)$$

We also recall that $\alpha + 2a = 0$ and $\xi = -1 + \alpha$. Thus, finally we end up with

$$\mathscr{E} = \frac{1}{2}\left(\mathbf{E}^2 + \mathbf{B}^2\right) + \frac{l_P}{2}\left(\nabla^2 \mathbf{A}\right)\cdot\mathbf{B}, \qquad (34)$$

and

$$\tilde{\mathbf{S}} = \mathbf{S}_{Maxwell} = \mathbf{E} \times \mathbf{B}, \qquad (35)$$

which reduces to the usual Poynting vector.

3 Extension to a Space-Like External LSV Four-Vector

Proceeding in the same way as we did in the previous Section, we shall now consider the field equations for a background space-like four-vector n^μ, that is, $n^\mu = (0, \mathbf{n})$.

3.1 Maxwell-Type Field Equations

In other words, we now wish to obtain Eqs. (16) and (17) for a background space-like four-vector.

For this, we restrict our attention to the Eq. (9). We then obtain for $\nu = 0$

$$\partial_i F^{i0} - l_P \left[(\alpha + \xi)\, \Delta + (\xi + 2a)\, (\mathbf{n} \cdot \nabla)^2 \right] n_j \tilde{F}^{j0} = 0. \tag{36}$$

While for $\nu = i$

$$\partial_0 F^{0i} + \partial_j F^{ji} - l_P \left[(\alpha + \xi)\, \Delta + (\xi + 2a)\, (\mathbf{n} \cdot \nabla)^2 \right] n_j \tilde{F}^{ji} = 0. \tag{37}$$

It is of interest also to notice that, for $\nu = 0$, we arrive at

$$\tilde{\mathscr{F}}_{0i} = \tilde{F}_{0i} + \left(\frac{\alpha}{2} - \beta \right) l_P\, n^k \partial_t F_{ki} - \left(\frac{\alpha}{2} - \beta \right) l_P\, n^k \partial_i F_{k0}, \tag{38}$$

whereas for $\nu = i$

$$\tilde{\mathscr{F}}_{ij} = \tilde{F}_{ij} + \left(\frac{\alpha}{2} - \beta \right) l_P\, n^k \partial_i F_{kj} - \left(\frac{\alpha}{2} - \beta \right) l_P\, n^k \partial_j F_{ki}. \tag{39}$$

Based on these equations, we can finally present the following Maxwell-type equations:

$$\nabla \cdot \mathbf{E} + l_P \left[(\alpha + \xi)\, \Delta + (\xi + 2a)\, (\mathbf{n} \cdot \nabla)^2 \right] (\mathbf{n} \cdot \mathbf{B}) = 0, \tag{40}$$

$$
\begin{aligned}
\nabla \times \mathbf{E} + \left(\frac{\alpha}{2} - \beta \right) l_P \nabla\, (\mathbf{n} \cdot \partial_t \mathbf{E}) = {}&- \partial_t \mathbf{B} - \left(\frac{\alpha}{2} - \beta \right) l_P\, \mathbf{n} \times \partial_t^2 \mathbf{B} \\
&+ \left(\frac{\alpha}{2} - \beta \right) l_P\, \mathbf{n}\, \nabla^2\, (\mathbf{n} \times \mathbf{B}) \\
&+ \left(\frac{\alpha}{2} - \beta \right) l_P \nabla\, \nabla \cdot (\mathbf{n} \times \mathbf{B}),
\end{aligned}
\tag{41}$$

$$\nabla \cdot \mathbf{B} + \left(\frac{\alpha}{2} - \beta \right) l_P\, \nabla \cdot (\mathbf{n} \times \partial_t \mathbf{B}) + \left(\frac{\alpha}{2} - \beta \right) l_P \nabla^2\, (\mathbf{n} \cdot \mathbf{E}) = 0, \tag{42}$$

$$\nabla \times \mathbf{B} - l_P \left[(\alpha + \xi)\, \Delta + (\xi + 2a)\, (\mathbf{n} \cdot \nabla)^2 \right] (\mathbf{n} \times \mathbf{E}) = \partial_t \mathbf{E}. \tag{43}$$

Evidently, these field equations are more involved than those of the (time-like) previous Section. However, we do not intend to pursue a study of these equations in the present contribution. A fuller account on these issues shall be presented elsewhere.

4 Concluding Remarks

In summary, we have proposed and studied the properties of a new non-Maxwellian electrodynamics coupled to a Lorentz-violating background through the presence of higher-derivative terms. This effective theory has allowed us to recover the known results obtained in [6, 7] within the Minkowski space covariant formulation of a classical field theory. As already expressed, the benefit of considering the present Lorentz-symmetry violating electrodynamical action is to provide connections among different effective models. Starting off from the situation of a time-like background vector, we made sure that, with our covariant formulation, we could reproduce the Gambini-Pullin's proposal of an electrodynamics accounting for spin-foam effects. That was a sort of validation test of our covariant formulation. Once that point was set up, we have considered the possibility to extend this covariant formulation to include the case Lorentz-symmetry breaking is realized by a space-like vector. This yields to a broad family of models, which encompasses situations in which even the homogeneous Maxwell equations are modified. In the Gambini-Pullin model, there is actually an extra term in the Faraday-Lenz magnetic induction equation; the Gauss equation for the magnetic field remains however the same as in the Maxwellian case. In our space-like extended model, since we have a number of parameters at our disposal, we may end up with both Faraday-Lenz and magnetic-Gauss equations both modified by the Planck scale effect, as shown in the set of Eqs. (40)–(43). It is not our aim, in this contribution, to pick out specific choices of parameters, yielding particular actions, to discuss the physical aspects and the consistency (absence of supraluminal signals and non-existence of ghost-type excitations in the second-quantized version) of the models that differ from one another by different choices of parameters in Eqs. (40)–(43). It is a relevant issue - and special attention shall be devoted - to study the wave equations for the electric and magnetic fields in the case the would-be homogeneous Maxwell equations, (41) and (42), exhibit the extra terms with the external space-like vector. Another issue we are going to exploit is the interference between non-linearity and Planck scale effects. If we consider high intensity electric and magnetic fields, such as those produced in extreme light experiments, and Planck scale effects are accounted for, we can inspect how the latter may affect phenomena like photon splitting, light-light Delbrück scattering and the Breit-Wheeler effect. These are topics in our agenda of forthcoming works, which we shall be reporting elsewhere.

Acknowledgements One of us (P. G.) was partially supported by Fondecyt (Chile) Grant 1180178 and by ANID PIA/APOYO AFB180002.

References

1. V. A. Kostelecky, Phys. Rev. D **69**, 105009 (2004)
2. V. A. Kostelecky, S. Samuel, Phys. Rev. D **39**, 683 (1989)
3. V. A. Kostelecky, S. Samuel, Phys. Rev. Lett. **63**, 224 (1989)
4. V. A. Kostelecky, S. Samuel, Nucl. Phys. B **336**, 263 (1990)
5. V. A. Kostelecky, S. Samuel, Phys. Rev. Lett. **66**, 1811 (1991)
6. R. Gambini, J. Pullin, Phys. Rev. D **59**, 124021 (1999)
7. J. Alfaro, H. A. Morales-Tecotl, L. F. Urrutia, Phys. Rev. D **65**, 103509 (2002)
8. C. Adam, F. R. Klinkhamer, Nucl. Phys. B **607**, 247 (2001)
9. C. Adam, F. R. Klinkhamer, Nucl. Phys. B **657**, 214 (2003)
10. C. Adam, F. R. Klinkhamer, Phys. Lett. B **513**, 245 (2001)
11. C. Kaufhold, F. R. Klinkhamer, Nucl. Phys. B **734**, 1 (2006)
12. R. Montemayor, L. F. Urrutia, Phys. Rev. D **72**, 045018 (2005)
13. X. Xue, J. Wu, Eur. Phys. J. C **48**, 257 (2006)
14. A. A. Andrianov, R. Soldati, Phys. Rev. D **51**, 5961 (1995)
15. A. A. Andrianov, R. Soldati, Phys. Lett. B **435**, 449 (1998)
16. A. A. Andrianov, R. Soldati, L. Sorbo, Phys. Rev. D **59**, 025002 (1999)
17. Q. G. Bailey, V. A. Kostelecky, Phys. Rev. D **70**, 076006 (2004)
18. M. Frank, I. Turan, Phys. Rev. D **74**, 033016 (2006)
19. B. Altschul, Phys. Rev. D **72**, 085003 (2005)
20. R. Lehnert, R. Potting, Phys. Rev. Lett. **93**, 110402 (2004)
21. R. Lehnert, R. Potting, Phys. Rev. D **70**, 125010 (2004)
22. B. Altschul, Phys. Rev. Lett. **98**, 041603 (2007)
23. B. Altschul, Phys. Rev. D **72**, 085003 (2005)
24. B. Altschul, Phys. Rev. D **73**, 036005 (2006)
25. B. Altschul, Phys. Rev. D **73**, 045004 (2006)
26. R. Jackiw, V. A. Kostelecky, Phys. Rev. Lett. **82**, 3572 (1999)
27. J. M. Chung, B. K. Chung Phys. Rev. D **63**, 105015 (2001)
28. J. M. Chung, Phys. Rev. D **60**, 127901 (1999)
29. G. Bonneau, Nucl. Phys. B **593**, 398 (2001)
30. M. Perez-Victoria, Phys. Rev. Lett. **83**, 2518 (1999)
31. M. Perez-Victoria, JHEP **0104**, 032 (2001)
32. O. A. Battistel, G. Dallabona, Nucl. Phys. B **610**, 316 (2001)
33. O. A. Battistel, G. Dallabona, J. Phys. G **28**, L23 (2002)
34. O. A. Battistel, G. Dallabona, J. Phys. G **27**, L53 (2002)
35. A. P. B. Scarpelli, M. Sampaio, M. C. Nemes, B. Hiller, Phys. Rev. D **64**, 046013 (2001)
36. F. A. Brito, T. Mariz, J. R. Nascimento, E. Passos, R. F. Ribeiro, JHEP **0510**, 019 (2005)
37. B. Altschul, Phys. Rev. D **70**, 056005 (2004)
38. G. M. Shore, Nucl. Phys. B **717**, 86 (2005)
39. D. Colladay, V. A. Kostelecky, Phys. Lett. B **511**, 209 (2001)
40. M. M. Ferreira Jr, Phys. Rev. D **70**, 045013 (2004)
41. M. M. Ferreira Jr, Phys. Rev. D **71**, 045003 (2005)
42. M. M. Ferreira Jr, M. S. Tavares, Int. J. Mod. Phys. A **22**, 1685 (2007)
43. J. Alfaro, A. A. Andrianov, M. Cambiaso, P. Giacconi, R. Soldati, Phys. Lett. B **639**, 586–590 (2006)
44. V. Barger, S. Pakvasa, T. J. Weiler, K. Whisnant, Phys. Rev. Lett. **85**, 5055 (2000)

45. V. A. Kostelecky, M. Mewes, Phys. Rev. D **69**, 016005 (2004)
46. V. A. Kostelecky, M. Mewes, Phys. Rev. D **70**, 031902 (R) (2004)
47. V. A. Kostelecky, M. Mewes, Phys. Rev. D **70**, 076002 (2004)
48. T. Katori, A. V. Kostelecky, R. Tayloe, Phys. Rev. D **74**, 105009 (2006)
49. R. Brustein, D. Eichler, S. Foffa, Phys.Rev. D **65**, 105006 (2002)
50. Y. Grossman, C. Kilic, J. Thaler, D. G. E. Walker, Phys. Rev. D **72**, 125001 (2005)
51. L. B. Auerbach *et al.*, Phys. Rev. D **72**, 076004 (2005)
52. D. Hooper, D. Morgan, E. Winstanley, Phys. Rev. D **72**, 065009 (2005)
53. M. Lubo, Phys. Rev. D **71**, 047701 (2005)
54. M. N. Barreto, D. Bazeia, R. Meneses, Phys. Rev. D **73**, 065015 (2006)
55. O. Gagnon, G. D. Moore, Phys. Rev. D **70**, 065002 (2004)
56. J. W. Moffat, Int. J. Mod. Phys. D **12**, 1279 (2003)
57. F. W. Stecker, S. T. Scully, Astropart. Phys. **23**, 203 (2005)
58. F. W. Stecker, S. L. Glashow, Astropart. Phys. **16**, 97 (2001)
59. R. Turcati, E. Scatena, Phys. Lett. B **786**, 332 (2018)
60. V. A. Kostelecky, M. Mewes, Phys. Rev. D **80**, 015020 (2009)
61. R. C. Myers, M. Pospelov, Phys. Rev. Lett. **90**, 211601 (2003)
62. R. Casana, M. M. Ferreira, L. Lisboa-Santos, F. E. P. dos Santos, M. Schreck, Phys. Rev. D **97**, 115043 (2018)
63. M. M. Ferreira, L. Lisboa-Santos, R. V. Maluf, M. Schreck, Phys. Rev. D **100**, 055036 (2019)
64. Z. Lingli, B. Q. Ma, Chin. Phys. C **35**, 987 (2011)
65. Z. Lingli, B. Q. Ma, Adv. High Energy Phys. **2014**, 374572 (2014)
66. H. Belich, L. D. Bernald, P. Gaete, J. A. Helayël-Neto, Eur. Phys. J. C **73**, no. 11, 2632 (2013)
67. H. Belich, L. D. Bernald, P. Gaete, J. A. Helayël-Neto, F. J. L. Leal, Eur. Phys. J. C **75**, no. 6, 291 (2015)

Cosmic Bubbles and the 5D Holographic Big Bang Model

Natacha Altamirano, Elizabeth Gould, Niayesh Afshordi,
and Robert B. Mann

Abstract The 5D Holographic Big Bang is a novel model for the emergence of the early universe out of a 5D collapsing star (an apparent white hole), in the context of Dvali-Gabadadze-Porrati (DGP) cosmology. The model does not have a big bang singularity, and yet can address cosmological puzzles that are traditionally solved within inflationary cosmology. In this paper, we compute the exact power spectrum of cosmological curvature perturbations due to the effect of a thin atmosphere accreting into our 3-brane. The spectrum is scale-invariant on small scales and red on intermediate scales, but becomes blue on scales larger than the height of the atmosphere. While this behaviour is broadly consistent with the non-parametric measurements of the primordial scalar power spectrum, it is marginally disfavoured relative to a simple power law (at 2.7σ level). Furthermore, we find that the best fit nucleation temperature of our 3-brane is at least 3 orders of magnitude larger than the 5D Planck mass, suggesting an origin in a 5D quantum gravity phase.

1 Personal Reminiscence from R. B. Mann

It is a pleasure to acknowledge the work of Antonio Aurilia with this paper on the 5D holographic big bang model. I first met Antonio in 1987, when he approached me with what then seemed a novel and unorthodox idea: develop a formalism that could encompass the physics of both cosmic inflation and quark confinement via

N. Altamirano · E. Gould · N. Afshordi
Perimeter Institute, 31 Caroline St. N., Waterloo, Ontario N2L 2Y5, Canada
e-mail: naltamirano@pitp.ca

E. Gould
e-mail: quantumwarrior@gmail.com

N. Afshordi
e-mail: nafshordi@pitp.ca

N. Altamirano · E. Gould · N. Afshordi · R. B. Mann (✉)
Department of Physics and Astronomy, University of Waterloo, Waterloo, Ontario N2L 3G1, Canada
e-mail: rbmann@uwaterloo.ca

© The Author(s), under exclusive license to Springer Nature Switzerland AG 2025
P. Nicolini (ed.), *Touring the Planck Scale*, Fundamental Theories of Physics 219,
https://doi.org/10.1007/978-3-031-76066-2_10

the description of a singular layer of matter under the influence of surface tension and volume tension. This relativistic bubble would have a single degree of freedom that would couple to a three-form in a gauge invariant manner, making the vacuum energy the charge of this gauge field.

Though quite similar to Einstein-Maxwell theory, the dynamics of the bubble turn out to be highly constrained insofar as the gauge field propagates no degrees of freedom. Since the associated conserved charge is the vacuum energy, it is possible to have different cosmological constants in different regions of spacetime. Bubbles can nucleate in different "vacuum phases", and the Israel junction conditions can be straightforwardly applied. The generality of the mechanism made it applicable in a variety of different physical situations, ranging from cosmology to particle physics. In curved spacetime our action functional provided a natural basis for inflationary cosmology, whereas in flat spacetime it generated the same vacuum tension in the "bag model" of quark confinement [1].

All of the above concepts—higher-form gauge fields, uniting the physics of two very different energy scales, and the physics of bubble (or brane) dynamics have become quite commonplace today, but they were rather novel back then. Antonio was one of the people to have this visionary idea of combining such disparate concepts together and it is a tribute to him that such ideas live on today in cosmology, in quark physics, and in string theory.

This contribution represents work in the same spirit, work that I think Antonio would have appreciated. In some sense we turn his bubble idea "inside out", putting our universe on its surface, with the bubble (or brane, as it is now called) embedded in a 5-dimensional spacetime. This holographic Big Bang model [2, 3] provides a novel mechanism for the emergence of the early universe out of a collapsing star (an apparent white hole) in five dimensions. In this braneworld description of cosmology, with both 4d induced and 5d bulk gravity (otherwise known as the Dvali-Gabadadze-Porati, or DGP model [4]), the universe emerges as a spherical 3-brane out of the formation of a 5-dimensional Schwarzschild black hole. While speculative, this hypothesis—of originating the universe from quantum tunnelling from a 5d star—is no less speculative than any other hypothesis concerning the big bang, big bounce, or inflation. This chapter provides a short review of the model, and critically examines its phenomenological implications and pathologies. While still in need of improvement, it provides the seeds of ideas for taking a new approach to cosmic evolution, an approach that I think Antonio would have appreciated.

This chapter will begin with the motivation behind it in Sect. 2, followed by a short review of its setup in Sect. 3. We then solve the Einstein equations with matter in the bulk, and consider the consequences that it has on the brane in Sect. 4, In particular, we solve for the density profile of a 5D spherically collapsing atmosphere and compute the change on the Hubble parameter as it falls into our 3-brane. In Sect. 5, we study cosmological perturbations in the bulk and their projection onto the brane, making special emphasis on the curvature perturbation and its power spectrum. Section 6 compares our predictions against Planck data and contrasts it with the power-law power spectrum assumed in the ΛCDM model. We conclude our work with discussion of the limitations and prospects of our model in Sect. 7.

Antonio was a valued colleague, a fine gentleman, and a creative scientist. He will be missed.

2 Motivation

Modern cosmology continues to experience an astonishing degree of empirical success [5]. The agreement with the phenomenological six-parameter ΛCDM paradigm is remarkable, all the more so as the number of cosmological observations continue to increase at an accelerated rate.

Despite this success, there are still intriguing puzzles left unresolved: the big bang singularity, the horizon and flatness problems (traditionally addressed within the inflationary paradigm), as well as the nature of dark matter and dark energy. A few years ago two of us (Afshordi and Mann, along with R. Pourhasan) proposed a novel cosmological model in which our universe is a 3-brane emergent from the collapse of a 5-dimensional star [2, 3]. Motivated by the desire to see if a more satisfactory (or natural) understanding of these puzzles can emerge from an alternative description of the geometry, this model explains the evolution of our early universe whilst avoiding a big bang singularity. Furthermore, the model was shown to have a mechanism via which a homogeneous atmosphere outside the black hole generates a scale invariant power spectrum for primordial curvature perturbations, (nearly) consistent with current cosmological observations [5].

Our 5D holographic origin for the big bang is based on a braneworld theory that includes both 4 dimensional induced gravity *and* 5D bulk gravity: the Dvali-Gabadadze-Porrati (DGP) model [4], with action

$$S_{DGP} = \frac{1}{16\pi G_5} \int_{\text{bulk}} d^5x \sqrt{-g}\,R_5 + \frac{1}{8\pi G_5} \int_{\text{brane}} d^4x \sqrt{-\gamma}\,K$$
$$+ \int_{\text{brane}} \sqrt{-\gamma}\left(\frac{R_4}{16\pi G_4} + \mathscr{L}_{\text{matter}}\right), \tag{1}$$

where g and γ, G_5 and G_4, R_5 and R_4 are the metrics, gravitational constants and Ricci scalars of the bulk and brane respectively, while K is the mean extrinsic curvature of the brane. This model is reminiscent of the cosmic bubble model Antonio proposed [1], but in one higher dimension, with the proviso that our universe, described by the metric

$$ds_4^2 = -d\tau^2 + \frac{a^2(\tau)}{\mathscr{K}}[d\psi^2 + \sin^2(\psi)(d\theta^2 + \sin^2(\theta)d\phi^2)], \tag{2}$$

is represented by a hypersurface in a 5 dimensional Schwarzschild black hole spacetime

$$ds_5^2 = -\left(1 - \frac{\mu}{r^2}\right)dt^2 + \left(1 - \frac{\mu}{r^2}\right)^{-1}dr^2 + r^2 d\Omega_3^2\,, \tag{3}$$

located at $r = \frac{a(\tau)}{\sqrt{\mathcal{H}}}$. In this context, a pressure singularity is generically found when the energy density of the holographic fluid $\tilde{\rho}$ satisfies $\tilde{\rho} = \tilde{\rho}_s = \frac{3G_4}{16\pi G_5^2}$ [6]. The singularity happens at early times in cosmic history [2], as matter decays more slowly than a^{-4}. However, under the evolution from smooth initial conditions, the pressure singularity can occur before Big Bang Nucleosynthesis (BBN), and is generically inside a white hole horizon. Alternatively, the universe could have emerged from the collapse of a 5D star into a black hole, just before BBN, removing both pressure *and* big bang/white hole singularities. As advocated in the *fuzzball* program [7], the rate of this tunnelling is enhanced due to the large entropy of black hole microstates, which we speculate could match those of an expanding 3-brane thermal state. Our universe is represented by the boundary of a 5D spherically symmetric spacetime with metric (3), in which we impose $\mathbf{Z}_2$ boundary conditions. This picture will be described in more detail in Sect. 3.

Interestingly, this model not only circumvents the singularity at the origin of time, but can also address other problems of cosmology that are typically solved by inflation. Because the collapsing star could have existed long before its demise, it had enough time to attain uniform temperature, thereby addressing the *Horizon Problem*. This solution is similar in spirit to the one proposed in bouncing scenarios [8]. However, as our 4d brane is nucleated at the moment of bounce (e.g. via a tunnelling process), and does not exist beforehand, there is no pre-bounce in the Holographic Big Bang model. Indeed, this tunnelling process allows this model to avoid the problem of violation of the null energy condition that chronically afflicts most bouncing models.

Furthermore, if we assume that the initial Hubble constant was of order of the 5D Planck mass, then the curvature density $-\Omega_k \sim (M_5 r_h)^{-2}$, where r_h is the radius of the black hole. Consequently $-\Omega_k \sim M_5/M_*$ could become very small for massive stars, thus solving the *Flatness Problem*. More generically, the *no hair* theorem ensures that a 3-brane nucleated just outside the event horizon of a massive black hole has a smooth geometry.

Yet another feature of the model is that a thermal atmosphere around the brane, composed of a gas of massless particles, produces scale invariant curvature perturbations. We shall focus our attention on the mechanism responsible for deviations from scale-invariance in the primordial curvature power spectrum in the context of the *5D Holographic origin of the Big Bang*. To this end, we consider a thin atmosphere that can be regarded as infalling matter, or the outer envelope of the collapsing 5D star, which resides in the 5D bulk and thus contributes to its energy momentum tensor. In this context, the DGP action Eq. (1) is modified to

$$S = S_{DGP} + \int_{\text{bulk}} d^5 x \sqrt{-g}\, \mathscr{L}_{5,\text{atmosphere}} \tag{4}$$

where $\mathscr{L}_{5,\text{atmosphere}}$ accounts for the matter Lagrangian in the bulk. Consistency between cosmological phenomenology and the DGP model implies that the brane is expanding outwards, and thus eventually encounters this *atmosphere*. We then compute the resulting power spectrum of scalar curvature perturbations and the change of the Hubble parameter due to this encounter. We find that the best fit nucleation temperature of the 3-brane is considerably larger than the 5D Planck mass, perhaps indicating an origin in a 5D quantum gravity phase.

In Sect. 3, we discuss a possible mechanism for brane nucleation and the different scales involved in our problem, giving a qualitative description of the different physics processes.

3   A 5D Holographic Big Bang

3.1   Brane Nucleation

As described above, we are working in the context of the 5D Holographic Big Bang model [2] where our universe is modelled as a hypersurface (the brane) in a 5-dimensional Schwarzschild space time according to the embedding $r = \frac{a(\tau)}{\sqrt{\mathcal{H}}}$. This construction is a solution of the DGP action (1) once we impose a $\mathbf{Z}_2$ boundary condition on the brane. As a consequence, via the embedding constraint the brane becomes an outward travelling boundary of the higher dimensional spacetime, an assumption that is necessary if we want our universe (represented by the brane) to be expanding.

From the perspective of an observer in the bulk, this setup is reminiscent of a construction proposed by Witten called the '*bubble of nothing*' [9], in which an interior region of space is missing, with space ending smoothly at the surface of this bubble (the brane), but is also similar to other cosmological implementations of the DGP model in the literature (e.g., [6]). One possible scenario in the 5D Holographic Big Bang model is that the brane was formed by the quantum tunnelling of a collapsing star in 5 dimensions, with all of the degrees of freedom of the inner part of the collapsing matter becoming degrees of freedom of the brane. This is analogous to the *fuzzball* paradigm, a model proposed to solve the information-loss paradox [10], which consists of the explicit construction of black hole microstates with no "dataless horizon region". The infalling matter can tunnel to a fuzzball state with amplitude given by a Euclidean path integral [7]

$$\Gamma_E \sim \exp\left[-\frac{1}{16\pi G_5}\int d^5 x_E \sqrt{g_E}\, R_5(x_E)\right] = \exp\left(-\sqrt{16\pi G_5 M_5^3}\right), \quad (5)$$

where we have used the length scale $r_h = \sqrt{\mu} = \sqrt{\frac{4G_5 M_5}{3\pi}}$ to estimate the Euclidean Einstein action in 5d Schwarzschild (3) for tunnelling between two configurations that have the length and mass scales set to those of the black hole. Although this rate is very small, the number of fuzzball configurations that a black hole can tunnel to depends on the Bekenstein-Hawking entropy as

$$\mathcal{N} = e^{S_{\mathrm{BH}}} = \exp\left(\frac{\mathscr{A}_{BH}}{4G_5}\right) = \exp\left(+\sqrt{16\pi G_5 M_5^3}\right), \tag{6}$$

yielding a significant probability to form a fuzzball (or a generic quantum gravity state). In fact, the two exponentials exactly cancel each other [7]. One may anticipate a similar principle operating here, in which collapsing matter at sufficiently high density—just prior to formation of a black hole horizon—necessarily tunnels to a brane so as to avoid the ensuing quantum paradoxes that follow upon introducing an event horizon.

We should emphasize that this hypothesis for originating our 3-brane through tunnelling from a 5d collapse is speculative, but not any more than other proposals for the onset of big bang, big bounce, and/or inflation. For the rest of the paper, we remain agnostic about this possibility, and assume that the expanding DGP 3-brane is already in place.

Since it is well established that BBN happened in the formation of our universe, and that in the 5D Holographic Big Bang (HBB) model [2] the pressure singularity generically forms before BBN, we consider a brane that must form before this. This means that the temperature of nucleation must be at most the temperature of BBN—$T_{\mathrm{BBN}} \sim 0.4\,\mathrm{MeV}$ [11]:

$$T_{\mathrm{nuc}} \geq T_{\mathrm{BBN}} \sim 0.4\,\mathrm{MeV}. \tag{7}$$

Finally, let us mention that the DGP model possesses a scale $r_c = \frac{G_5}{G_4}$, above which 5-dimensional gravity dominates over 4-dimensional gravity. Constraints on the normal branch of the DGP model [12] give[1]

$$r_c \gtrsim 3H_0^{-1} \quad \rightarrow \quad M_5 < (H_0 M_4^2/12)^{1/3} \quad \rightarrow \quad M_5 < 9\,\mathrm{MeV}. \tag{8}$$

where $M_4 = \frac{1}{(8\pi G_4)^{1/2}}$ is the reduced 4D Planck mass.

3.2 The Atmosphere: Setup and Scales

In this scenario, we are interested in the effects of a thin atmosphere located just outside the brane. Although the brane forms a $\mathbf{Z}_2$ boundary, excluding the event horizon in Eq. (3), we shall refer to the metric in Eq. (3) as the black hole metric.

In order to organize the different assumptions, we will review the implied hierarchy of the scales present in this problem. If we assume that the Hubble patch of our universe, at/near brane nucleation, is small enough to be insensitive to the curvature of the black hole spacetime, we can assume that the atmosphere is just a perturbation

---

[1] Planck 2015 constraints on the dark energy equation of state roughly imply $|1 + w| < 0.11$ at 95% level, at the pivot redshift of $z \simeq 0.23$ (Fig. in [13]), which provides a similar bound on r_c, using the DGP Friedmann equation with a cosmological constant.

around a Minkowski background. This limit implies $H^{-1} \ll R$ where R is the radius of the black hole, or brane upon nucleation. This assumption is also observationally motivated, since today we measure $(HR)^{-2} \ll |\Omega_k| \ll 1$ and thus our observable Hubble patch is approximately flat. In our model, if the brane is moving very close to the speed of light, $H^{-1} \ll R$. This will allow us to define metric perturbations in Sect. 5. We will then be interested in finding the behaviour of the power spectrum of curvature perturbations for modes of wavelength λ, that are of super-horizon size before BBN, but are now observable in the CMB sky. This restricts $R \gg \lambda \gg H^{-1}$.

Finally, we would like to understand the behaviour of different physical quantities of the brane, such as the behaviour of the Hubble parameter before and after the encounter with the atmosphere. Assuming it can be considered to be a thin atmosphere in the 5 dimensional space time, the width ΔL of the atmosphere needs to be smaller than R for the brane to cross the atmosphere completely in less than a Hubble time. We then get the following hierarchy of different scales

$$H^{-1} \ll \lambda \ll R, \text{ and } \Delta L \ll R \text{ (black hole frame)}, \tag{9}$$

$$\lambda \lesssim \Delta L \text{ (atmosphere frame)}. \tag{10}$$

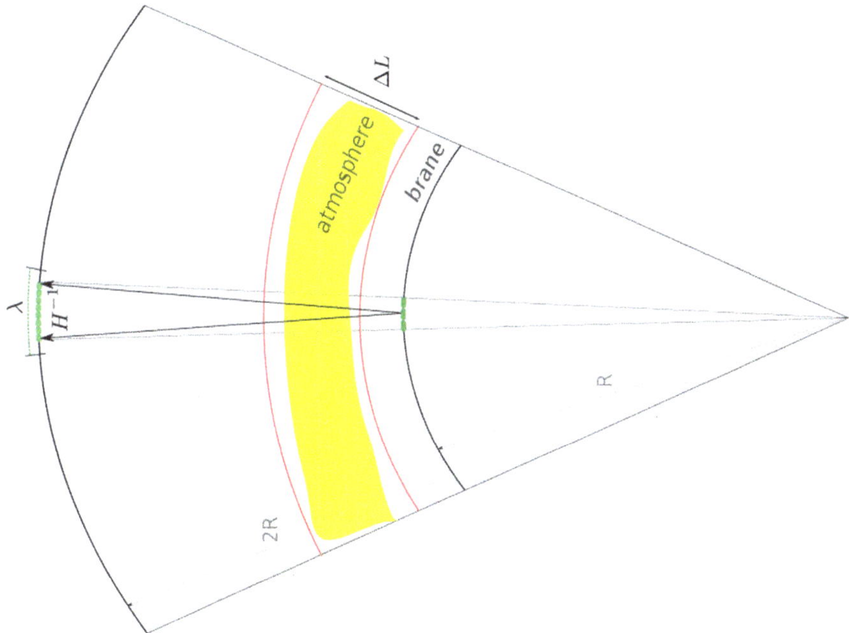

Fig. 1 Cartoon of the different scales treated in this problem, in the black hole rest frame. The inner black circular arc represents the brane with radius R and the outer black circular arc represents the brane with radius $2R$. H^{-1} is estimated by tracing light rays on the brane after it has doubled it size; it is small if the brane is traveling near the speed of light. The atmosphere is shown in yellow sitting in between the two red arcs with length ΔL

As we shall in Sect. 5, the latter inequality is the key ingredient that leads to a near scale-invariant spectrum of primordial curvature perturbations for large scales. This hierarchy of scales is illustrated in Fig. 1.

4 Homogeneous Brane Meets Thin Atmosphere

4.1 Einstein Equations

We want to study the influence of an atmosphere that is falling into the black hole as shown in in Fig. 2. For this, we assume that the brane is moving supersonically (in fact, almost with the speed of light) into the atmosphere, and thus bulk metric perturbations do not react to the brane's presence until it runs into them.

The Einstein equations on the brane that follow from the action (4) are:

$$G_{\mu\nu} = 8\pi G_4 \left(T_{\mu\nu} + \tilde{T}_{\mu\nu} \right) \tag{11}$$

where the two different components of the energy-momentum tensor are $T_{\mu\nu}$, the matter living on the brane, and $\tilde{T}_{\mu\nu}$ the *holographic fluid* that is induced on the brane via the junction conditions described below. Due to the symmetry of the (unperturbed) FRW spacetime, $T_{\mu\nu}$ has the form of a perfect fluid

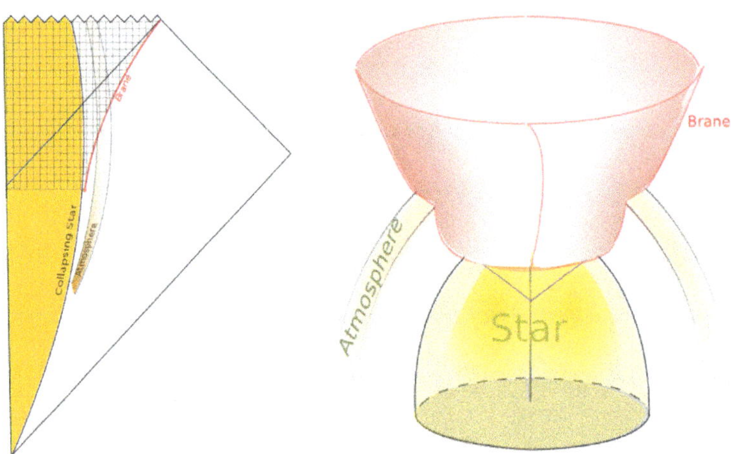

Fig. 2 Penrose diagram (left), and cartoon of the 5D star collapse (right), followed by the nucleation of a 3-brane (our universe). The star (in yellow) that is collapsing (nearly) forms a black hole, but the 3-brane (red) will nucleate just prior to the formation of the event horizon, and traverses a thin atmosphere of infalling matter or atmosphere (cross-section shown in the cartoon at right)

$$T_{\mu\nu} = (P + \rho)\, u_\mu u_\nu + P\gamma_{\mu\nu}, \tag{12}$$

where u^μ is the 4-velocity of the fluid normalized such that $u^\mu u_\mu = -1$.

The holographic fluid is the Brown-York stress tensor induced on the brane once Einstein equations are imposed on the bulk

$$\tilde{T}_{\mu\nu} = \frac{1}{8\pi G_5} \left(K\gamma_{\mu\nu} - K_{\mu\nu}\right), \tag{13}$$

where $K_{\mu\nu} \equiv \nabla_\alpha n_\beta e^\alpha_\mu e^\beta_\nu$ is the extrinsic curvature of the brane whose unit normal is n^α. Here $e^\alpha_\nu \equiv \frac{\partial \hat{x}^\alpha}{\partial x^\nu}$, where we have associated the set of coordinates $\{\hat{x}^\alpha\}$ and $\{x^\nu\}$ with the bulk and the brane respectively. In addition to the Einstein equations (11), the continuity equations for the total matter living on the brane arising from the Bianchi identities are:

$$\nabla^\mu \left(T_{\mu\nu} + \tilde{T}_{\mu\nu}\right) = 0. \tag{14}$$

The Gauss-Codazzi equations constrain the geometric quantities of the brane with the matter present in the bulk

$$\nabla^\mu \left(K g_{\mu\nu} - K_{\mu\nu}\right) = 8\pi G_5 T^5_{\alpha\beta} e^\alpha_\nu n^\beta, \tag{15}$$

$$R^4 + K^{\mu\nu} K_{\mu\nu} - K^2 = -16\pi G_5 T^5_{\alpha\beta} n^\alpha n^\beta, \tag{16}$$

where $R^4 = -8\pi G_4 \left(T + \tilde{T}\right)$ is the Ricci scalar of the brane. $T^5_{\alpha\beta}$ is the energy momentum tensor of the bulk which satisfies the Einstein's equations on the bulk $G_{\alpha\beta} = 8\pi G_5 T^5_{\alpha\beta}$. Note that the first of the Gauss-Codazzi equations (15) reduces to the conservation of the holographic fluid $\tilde{T}_{\mu\nu}$ in the case that $T^5_{\alpha\beta} = 0$. If the bulk matter flows into the brane, the holographic fluid is not conserved and the effect of the continuity equation (14) is to change the matter on the brane through the holographic fluid in order for the sum of both to be conserved.

In the same way, as a result of the symmetries of FRW spacetime, the holographic fluid must have the form of a perfect fluid. Moreover, the 4-velocity of this fluid must coincide with the 4-velocity of the normal matter on the brane:

$$\tilde{T}_{\mu\nu} = \left(\tilde{P} + \tilde{\rho}\right) u_\mu u_\nu + \tilde{P}\gamma_{\mu\nu}. \tag{17}$$

Combining Eqs. (13) and (17) we get:

$$K_{\mu\nu} = -8\pi G_5 \left[\left(\tilde{P} + \tilde{\rho}\right) u_\mu u_\nu + \frac{1}{3}\tilde{\rho}_{\mu\nu}\right]. \tag{18}$$

4.2 Shift in the Hubble

The rate of expansion described by the Hubble parameter will change as the brane goes through the atmosphere and we can find a general expression for H by studying the Einstein equations and junction conditions for the DGP brane in the general case. This general case treats the bulk as a 5-dimensional Schwarzschild black hole (3) and the brane as a hypersurface parametrized by $r = \frac{a(\tau)}{\sqrt{\mathscr{H}}}$, as detailed in Sect. 2.

From Eqs. (11) and (14–16) we obtain

$$H^2 + \frac{\mathscr{H}}{a^2} = \frac{8\pi G_4}{3}\left(\tilde{\rho} + \rho\right) \tag{19}$$

$$\dot{\rho} + \dot{\tilde{\rho}} + 3H\left(\rho + \tilde{\rho} + P + \tilde{P}\right) = 0 \tag{20}$$

$$\dot{\tilde{\rho}} + 3H\left(\tilde{\rho} + \tilde{P}\right) = T^5_{\alpha\beta}e^\alpha_\tau n^\beta \tag{21}$$

$$0 = T^5_{\alpha\beta}e^\alpha_i n^\beta \tag{22}$$

$$(\rho + \tilde{\rho}) - 3\left(P + \tilde{P}\right) + \frac{8\pi G_5^2}{G_4}\left(\frac{2}{3}\tilde{\rho}^2 + 2\tilde{P}\tilde{\rho}\right) = -2\frac{G_5}{G_4}T^5_{\alpha\beta}n^\alpha n^\beta \tag{23}$$

The quantity $T^5_{\alpha\beta}$ is the stress-energy of the atmosphere outside the black hole. This atmosphere will have two effects on the brane. It will induce metric and matter perturbations in our universe, which we shall use to compute the curvature perturbation in Sect. 5 below. However it will also make the Hubble parameter change its value as the brane crosses the atmosphere: the brane will expand more slowly due to an extra source of infalling matter. Combining Eqs. (19) and (20) and ignoring the curvature term we get

$$\Delta H = -4\pi G_4 \int (P_T + \rho_T)\, d\tau, \tag{24}$$

where the integral is performed in the proper time of the brane and $P_T = P + \tilde{P}$, $\rho_T = \rho + \tilde{\rho}$.

Let's first look at the behavior of the holographic fluid $\tilde{P}$ and $\tilde{\rho}$. From Eq. (23) we have

$$\tilde{P} + \tilde{\rho} = \frac{1}{3\left(\frac{\tilde{\rho}}{\tilde{\rho}_s} - 1\right)}\left[-2\frac{G_5}{G_4}T^5_{nn} - 4\tilde{\rho} + 2\frac{\tilde{\rho}^2}{\tilde{\rho}_s} + T\right], \tag{25}$$

where $T = 3P - \rho$, $T^5_{nn} = T^5_{\alpha\beta}n^\alpha n^\beta$ and $\tilde{\rho}_s = \frac{3G_4}{16\pi G_5^2}$. In order to avoid the pressure singularity we require $\tilde{\rho} \gg \tilde{\rho}_s$ and in this limit the last equation becomes

$$\tilde{P} + \tilde{\rho} = \frac{1}{3}\left[-2\frac{G_5}{G_4}T^5_{nn}\frac{\tilde{\rho}_s}{\tilde{\rho}} + 2\tilde{\rho} + T\frac{\tilde{\rho}_s}{\tilde{\rho}}\right]. \tag{26}$$

Now let's analyze the behavior for the the matter on the brane P and ρ. Combining Eqs. (20) and (21) and assuming an equation of state $P = w\rho$ the fluid on the brane satisfies

$$\dot{\rho} + 3H\left(\rho + P\right) = -T^5_{\alpha\beta}e^\alpha_\tau n^\beta \Rightarrow \frac{d}{d\tau}(\rho a^{3(w+1)}) = -T^5_{\alpha\beta}e^\alpha_\tau n^\beta a^{3(w+1)}. \tag{27}$$

If we now assume that the atmosphere is thin enough so that we can approximate the matter distribution as a delta function (i.e. $H\Delta\tau \ll 1$, during the impact time $\Delta\tau$), we see that the last equation will give a jump in the density (and hence in the pressure) proportional to a step function. In fact, if we consider the system of equations (19–23), with $P = w\rho$, and a delta function $T^5_{\alpha\beta}$, the only consistent solution would have a delta function in $\tilde{P}$, with step function jumps in other variables. As such, the biggest contribution in Eq. (24) is given by the first term on the right hand side of Eq. (26):

$$\Delta H = \frac{G_4}{2G_5} \int \frac{T^5_{\alpha\beta}n^\alpha n^\beta}{\tilde{\rho}} d\tau \left[1 + \mathcal{O}(H\Delta\tau)\right]. \tag{28}$$

We shall see in Sect. 5 below that the amplitude of curvature perturbations depends on $\Delta \ln H = \frac{\Delta H}{H}$. To compute this, we note that from Eq. (19) we can write $H^2 \approx \frac{8\pi G_4}{3}(\tilde{\rho} + \rho) \approx \frac{8\pi G_4}{3}\rho$ in the regime where $\rho \gg \tilde{\rho}$. Then

$$\Delta \ln H \approx \frac{G_4}{2G_5}\sqrt{\frac{3}{8\pi G_4}}\frac{1}{\sqrt{\rho\tilde{\rho}}} \int T^5_{\alpha\beta}n^\alpha n^\beta \, d\tau, \tag{29}$$

We can now use the solution in vacuum for $\tilde{\rho}$ found in [2]

$$\tilde{\rho} = \tilde{\rho}_s\left(1 + \sqrt{1 - \frac{2(\rho_{BH} - \rho)}{\tilde{\rho}_s}}\right), \tag{30}$$

where $\rho_{BH} = \frac{3\Omega_k^2 H_0^4 r_h^2}{8\pi G_4 a^4}$. In the approximation where $\rho \gg \rho_{BH}$ and $\rho \gg \tilde{\rho}_s$ we find

$$\tilde{\rho} \approx \rho\sqrt{\frac{2\tilde{\rho}_s}{\rho}} \tag{31}$$

and then Eq. (29) reads

$$\Delta \ln H \approx \frac{1}{2\rho} \int T^5_{\alpha\beta}n^\alpha n^\beta d\tau, \tag{32}$$

implying that the relative jump in the Hubble parameter due to a thin atmosphere is the ratio of the work done by the pressure of the atmosphere to the energy of the brane.

4.3 *Profile of the Atmosphere*

So far we have considered a general energy momentum tensor in the bulk responsible for dynamic features on the brane. Let's now consider that the bulk is filled with a relativistic spherically symmetric, collapsing 5D radiation atmosphere ($P_5 = \frac{1}{4}\rho_5$) that represents the atmosphere whose energy momentum tensor is

$$T^5_{\alpha\beta}(w) = \rho_5(w)\left[\left(1 + \frac{1}{4}\right)\delta^0_\alpha\delta^0_\beta + \frac{1}{4}\eta_{\alpha\beta}\right]. \tag{33}$$

In order to study the effect of this atmosphere we need to introduce scalar homogeneous perturbations in the bulk. A generalization of 4D perturbation theory allows us to write the perturbed metric of the bulk in the Newtonian gauge in 5D as

$$ds^2_{\text{bulk}} = -[1 + 2\Phi_5(x^\alpha)]dt^2 + [1 - 2\Psi_5(x^\alpha)][dx^2 + dy^2 + dz^2 + dw^2]. \tag{34}$$

where Φ_5 and Ψ_5 represent the scalar perturbations of the bulk and x^α are bulk coordinates. Our universe is represented as a hypersurface in the 5D bulk, whose trajectory is given by the constraint $w = f(x^\mu)$. In this setup, the brane will inherit three bulk coordinates $\{t, x, y, z\}$ and will respond to perturbations that are just functions of the bulk time via the relation $w = f(x^\mu)$. Consider that the metric perturbations and the brane position are homogeneous:

$$\Phi_5 = \varepsilon\Phi_5^0(w), \quad \Psi_5 = \varepsilon\Psi_5^0(w), \quad f = f_0(t), \tag{35}$$

where $\varepsilon \ll 1$ is a parameter that controls the homogeneous metric perturbations. Assuming hydrostatic equilibrium in the (infalling) rest frame of the atmosphere, the Einstein Equations in the bulk for the metric (34) leads to the relativistic Poisson and Euler equations:

$$\nabla^2\Phi_5^0 = \frac{8\pi G_5}{3}\rho_5, \tag{36}$$

$$\nabla\Phi_5^0 = -\frac{1}{4}\frac{\nabla\rho_5}{\rho_5}. \tag{37}$$

These equations can be solved exactly in a Minkowski background:

$$\rho_5(w) = \bar{\rho}_5\left\{1 - \tanh^2\left[\bar{\gamma}\left(\frac{w}{\bar{w}} - 1\right)\right]\right\}, \tag{38}$$

where $\bar{\rho}_5 \equiv \frac{3}{16\pi G_5}\frac{\bar{\gamma}^2}{\bar{w}^2}$, while $\bar{\gamma}$ and $\bar{w}$ are constants of integration.

The relationship between the energy density and the temperature of the atmosphere can be computed by integrating the Bose-Einstein distribution in 4+1 dimensions

$$\rho_5(w) = \int \frac{d^4 k}{(2\pi)^4} \frac{\omega}{\exp\left[\omega/T_5(w)\right] - 1} = \frac{3\zeta_R(5)}{\pi^2} T_5(w)^5, \tag{39}$$

where $\omega^2 = k_\alpha k^\alpha$. The above expression allows us to write

$$T_5(w) = \left(\rho_5(w)\frac{\pi^2}{3\zeta_R(5)}\right)^{\frac{1}{5}} = 1.26\,\rho_5(w)^{1/5} \tag{40}$$

where ζ_R is the Riemann zeta function. Note that the characteristic thickness of the atmosphere is given by

$$\Delta L = \frac{\bar{w}}{\sqrt{2}\bar{\gamma}}. \tag{41}$$

We are now in position to compute Eq. (32) and we will do so in the reference frame of the atmosphere. In this frame ρ_5 does not depend on time, but the normal to the brane n^α will depend on the relative velocity of the brane and the atmosphere. First consider the fluid velocity $u^\alpha = (1, \mathbf{v})/\sqrt{1-v^2}$, where $\mathbf{v}$ is the relative 3 velocity between the brane and the atmosphere. If we now require $n_\alpha n^\alpha = 1$ and $n^\alpha u_\alpha = 0$ we have $n^\alpha = (v, \mathbf{v}/v)/\sqrt{1-v^2}$. With this we can write $T^5_{\alpha\beta}n^\alpha n^\beta = \rho_5(w)(1 + 4v^2)/4(1-v^2)$ and the RHS of Eq. (32) reads

$$\begin{aligned}
\Delta \ln H &= \frac{1}{2\rho} \int T^5_{\alpha\beta} n^\alpha n^\beta d\tau \\
&= \frac{1}{2\rho} \int_0^\infty T^5_{\alpha\beta} n^\alpha n^\beta \sqrt{1-v^2}\frac{dw}{v} \\
&= \frac{(1+4v^2)}{v\sqrt{1-v^2}}\frac{\bar{\rho}_5}{8\rho}\frac{\bar{w}}{\bar{\gamma}}\left[1 + \tanh(\bar{\gamma})\right].
\end{aligned} \tag{42}$$

5 Cosmological Perturbations

As discussed in the last section, if the velocity of the brane is near the speed of light we can assume that the Hubble patch of the universe will be smaller than the curvature radius of the black hole spacetime. In this regime it is safe to approximate the bulk as Minkowski spacetime and analyze the perturbations around it. In the last section we have briefly introduced the homogeneous scalar perturbations and in Appendix 1 we present the anisotropic perturbations. The curvature perturbation can be written as function of the scalar gauge invariant quantities (73)

$$\zeta = \Psi_4 - \frac{H}{\dot{H}}\left(H\Phi_4 + \dot{\Psi}_4\right). \tag{43}$$

Note that in our framework the Hubble constant is of first order in the perturbation (see Eq. 74) as the brane crosses the atmosphere, and thus the term $H\Phi_4$ can be neglected. With this we have

$$\zeta \approx \Psi_4 - \frac{\Delta\Psi_4}{\Delta\ln(H)}, \tag{44}$$

where $\Delta\Psi_4 = \Psi_{4_f} - \Psi_{4_i}$. Here $\Psi_{4_{i(f)}}$ stands for the metric perturbation in 4D right before (after) the brane crossed the atmosphere, and we assume that ζ evolves continuously. We are interested in the value of the curvature perturbation after the brane has passed through the atmosphere where the metric perturbation $\Psi_{5_f} = 0$, which makes $\Psi_{4_f} = 0$. In this regime the curvature perturbation becomes

$$\zeta = \zeta_f \approx \frac{\Psi_{4i}}{\Delta\ln(H)}. \tag{45}$$

We are now ready to analyze the behaviour of the curvature perturbation power spectrum

$$P_\zeta(k) = \int d^3\mathbf{x}\; e^{i\mathbf{k}\cdot\mathbf{x}}\langle\zeta(x)\zeta(0)\rangle. \tag{46}$$

If the atmosphere is not in thermal equilibrium, we can relate the 2-point correlation function of the thermal fluctuations in 5D energy density to the temperature profile of the atmosphere:

$$\langle\rho_5(y_1)\rho_5(y_2)\rangle = \alpha\,(T_5(y_1))^6\delta^4(y_1 - y_2). \tag{47}$$

To proceed, we notice that in [2] the 5D energy density correlation function due to a thermal gas was shown to be

$$\langle\rho_5(y_1)\rho_5(y_2)\rangle \simeq \frac{5}{8}\left[\int \frac{d^4k}{(2\pi)^4}\left[\frac{1}{\exp(\omega/T_5) - 1} + \frac{1}{2}\right]\omega\exp[ik_a(y_1^a - y_2^a)]\right]^2. \tag{48}$$

This expression can be approximated as a delta function

$$\langle\rho_5(y_1)\rho_5(y_2)\rangle \simeq \alpha(T_5(y_1))^6\delta^4(y_1 - y_2)$$

where $\alpha = \frac{5}{8}\left[\frac{1}{\pi^2 63}(945\zeta_R(5) - \pi^6)\right]$, while we have dropped the power-law UV divergence. Note, that (48) can be approximated by a 4 dimensional delta function on length scales larger than the thermal wavelength T_5^{-1}.

With the use of the Poisson equation in 5D

$$\nabla^2\Psi_5(y) = \frac{8\pi G_5}{3}\rho_5(y), \tag{49}$$

we can find the power spectrum for the curvature perturbation to be (see Appendix 1 for details)

$$\mathcal{P}(k) = \frac{k^3}{2\pi^2} P_\zeta(k) = \beta\, k \int_0^\infty dw\, e^{-2|w|k}\, (T_5(w))^6 \tag{50}$$

$$= \Delta_0^2 k \int_0^\infty dw\, e^{-2|w|k}\, \left\{ 1 - \tanh^2[\bar{\gamma}(w/\bar{w} - 1)] \right\}^{6/5} \tag{51}$$

where $\Delta_0^2 = \beta\,[\bar{T}_5]^6$, $\beta = \frac{\alpha}{2}\left(\frac{G_5}{6\Delta \ln H\, \pi^3}\right)^2$ and we have used Eqs. (39) and (40) of Sect. 4.3 to write the temperature of the brane. The power spectrum predicted by our model is characterized by 3 free parameters $\bar{\gamma}$, $\bar{w}$, Δ_0^2 which we are going to fit to Planck data in the next section.

6 Observational Constraints on 5D Holographic Big Bang

The standard model of cosmology, ΛCDM, is described by 6 parameters: the baryon density, dark matter density, angular size of the sound horizon at recombination, the optical depth to reionization, amplitude of the scalar power spectrum and its tilt, respectively labelled $(\Omega_b h^2, \Omega_c h^2, \theta, \tau, \Delta_0^2, n_s)$. This model characterizes early universe cosmology via the power spectrum of the curvature perturbations

$$\mathcal{P}(k) = \Delta_0^2 \left(\frac{k}{k_*}\right)^{n_s - 1}, \tag{52}$$

where $k_* = 0.05/\mathrm{Mpc}$ is the comoving pivot scale. This form of the power spectrum, expected in slow-roll inflationary models, best fits the CMB data with parameter values [5]

$$\Delta_0^2 = (2.196 \pm 0.059) \times 10^{-9} \qquad n_s = 0.9603 \pm 0.0073. \tag{53}$$

We would like to compare this model with the HBB model (Eq. 51) that strictly is represented by the seven parameters $(\Omega_b h^2, \Omega_c h^2, \theta, \tau, \bar{\gamma}, \bar{w}, \Delta_0^2)$. In order to compare models with the same number of parameters, will also include ΛCDM with running $\alpha_s = dn_s/d\ln q$.

We have performed the comparison by running the CosmoMC code [14–20] with Planck 2015 data, Barionic Acoustic Oscillations (BAO) [21–28] as well as lensing data [29–33]. Finally, to determine the best fit parameters and the likelihood, we run the minimizer expressing our results Table 1. Comparing best-fit χ^2 of HBB and ΛCDM (with running), we see that HBB is disfavoured at roughly $2.7\sigma - 2.8\sigma$. The

Table 1 Planck 2015 and BAO best fit parameters and 68% ranges for HBB and ΛCDM models

	HBB		ΛCDM		ΛCDM with running	
	Best fit	68% Range	Best fit	68% Range	Best fit	68% Range
$\Omega_b h^2$	0.02212	0.02210 ± 0.00023	0.02227	0.02225 ± 0.00020	0.02231	0.02229 ± 0.00022
$\Omega_c h^2$	0.1172	0.1169 ± 0.0012	0.1185	0.1186 ± 0.0012	0.1184	0.1186 ± 0.0012
100θ	1.04113	1.04116 ± 0.00042	1.04103	1.04104 ± 0.00042	1.04108	1.04105 ± 0.00041
τ	0.081	0.083 ± 0.014	0.067	0.067 ± 0.013	0.069	0.068 ± 0.013
$10^9 \Delta_0^2$	5.79	0.793 ± 0.021	5.798	5.798 ± 0.019	5.82	5.82 ± 0.020
n_s			0.9682	0.9677 ± 0.0045	0.9682	0.9671 ± 0.0045
α_s					-0.0027	-0.0030 ± 0.0074
$\bar{\gamma}$	0.513	0.525 ± 0.053				
$\bar{w}_c$ [Mpc]	275	297^{+39}_{-77}				
χ^2	11327.4		11319.9		11319.6	

The last row indicates the χ^2 for each of the models. Note that $\bar{w}_c$ corresponds to the comoving value of the position of the centre of the atmosphere and its related to the physical $\bar{w}$ via $\bar{w} = \bar{w}_c \frac{2.3 \times 10^{-10} \text{MeV}}{T_{\text{nuc}}}$

Planck angular TT spectrum together with the best fit curves and residuals for HBB and ΛCDM are shown in Fig. 3. The best fit primordial scalar power spectra in both models are also contrasted with a non-parametric reconstruction from Planck 2015 data [34]. We note that the difference between the two models (mostly) lies within the 68% region and the largest disagreement of the models is at low l's or k's.

We are now going to analyze the physical implications of the best fit parameters $\{\bar{\gamma}, \bar{w}_c, \Delta_0^2\}$. In particular, we are interested in the temperature and position of the atmosphere and the change in the Hubble constant of the brane when crossing the atmosphere. All of these physical quantities are related to the fitted parameters, and also to the temperature of the brane at nucleation T_{nuc} and the Planck mass in the bulk M_5. Using $\bar{w} = \bar{w}_c \frac{2.3 \times 10^{-10} \text{MeV}}{T_{\text{nuc}}}$ and Eqs. (38), (39), and (41) we can find the values for the amplitude of the energy density, the temperature and the width of the atmosphere respectively. We have summarized our results in Table 2.

The model still allows freedom for the parameters $\{T_{\text{nuc}}, M_5, v\}$. These can be constrained using observational bounds for the DGP model and by requiring consistency of our approximations. In particular, we have modelled the atmosphere as being thin, which is equivalent to the requirement that the time it takes the brane to cross it is less than a Hubble time:

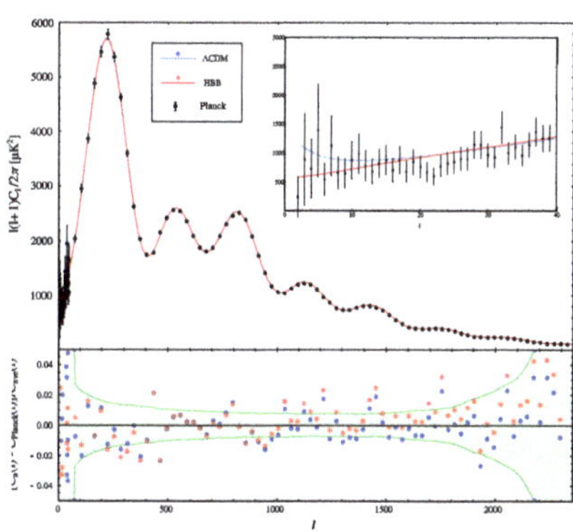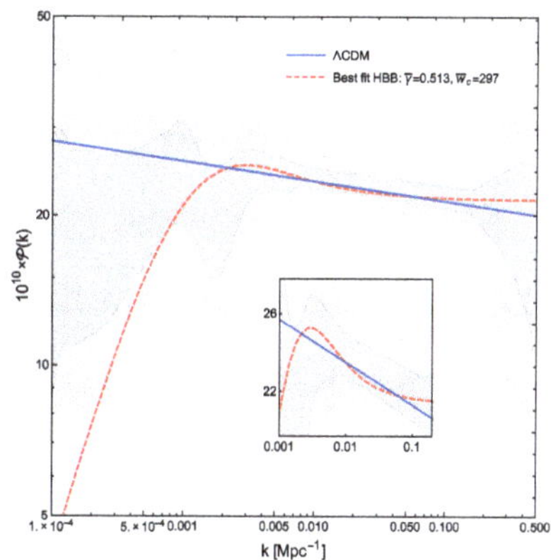

Fig. 3 Left-Top: angular power spectrum of CMB temperature anisotropies, comparing Planck 2015 data (black dots) with best HBB model (solid/red) for all l. Left-Inset: angular power spectrum of CMB temperature anisotropies, comparing Planck 2015 data with ΛCDM (dotted/blue) and HBB (solid/red) for $l < 40$. Left-Bottom: relative residuals and difference between ΛCDM and HBB (black solid) where the green shaded region indicates the 68% region of Planck 2015 data. Right: Best fit of the primordial power spectrum as predicted by HBB (dashed-red) in comparison with the best fit of ΛCDM model (blue). The grey regions are the $\pm 1\sigma$ and $\pm 2\sigma$ constraints from a non-parametric reconstruction using Planck 2015 data [34]

$$\Delta L \frac{\sqrt{1-v^2}}{v} \leq H^{-1} = \left(\frac{3}{8\pi G_4 \rho}\right)^{1/2} \quad \Rightarrow \quad T_{\text{nuc}} \leq \frac{v}{\sqrt{1-v^2}} 2.33 \times 10^{-7}\,\text{MeV},$$

(54)

employing the fact that the energy density on the brane at nucleation time is $\rho = \frac{\pi}{30} g_* T_{\text{nuc}}^4$, for g_* effective relativistic degrees of freedom. The above bound is consistent with the BBN constraint ($T_{\text{nuc}} \geq T_{\text{BBN}}$) for velocities near the speed of light. On the other hand, as discussed in Sect. 4.2 when the brane encounters the atmosphere its Hubble parameter will change as predicted by Eq. (42). Since this quantity enters in the amplitude of the power spectrum Eq. (51) we can write

$$\Delta_0^2 = \beta\, \bar{T}_5^6, \qquad \beta = \left(\frac{G_5}{6\pi^3 \Delta \ln H}\right)^2$$

(55)

and thus

$$\Delta \ln H = 4.54 \times 10^{-35} \left(\frac{T_{\text{nuc}}}{M_5}\right)^{6/5}.$$

(56)

We also get a constraint on $\Delta \ln H$ from the bulk atmosphere using Eq. (42)

$$\Delta \ln H = 1.62 \times 10^{-29} \frac{(1+4v^2)}{v\sqrt{1-v^2}} \left(\frac{M_5}{T_{\text{nuc}}}\right)^3,$$

(57)

222   N. Altamirano et al.

Table 2 Physical characteristics of the HBB model using the best fit parameters presented in Table 1

	Quantity	Value	Thin atmosphere bound $M_5 \leq 8.56 \times 10^{-4} \left(\frac{\text{MeV}}{T_{\text{nuc}}}\right)^{5/21} T_{\text{nuc}}$ (61)
Position of atmosphere	$\bar{w} = \bar{w}_c \frac{2.3 \times 10^{-10}\,\text{MeV}}{T_{\text{nuc}}}$	$9.872 \times 10^{27}\, \frac{1}{T_{\text{nuc}}}$	$9.872 \times 10^{27}\, \frac{1}{T_{\text{nuc}}}$
Density of atmosphere	$\bar{\rho}_5 = \frac{3}{16\pi G_5} \frac{\bar{\gamma}^2}{\bar{w}^2}$	$1.62 \times 10^{-56}\, M_5^3\, T_{\text{nuc}}^2$	$\leq 1.02 \times 10^{-65} \left(\frac{\text{MeV}}{T_{\text{nuc}}}\right)^{5/7} T_{\text{nuc}}^5$
Temperature of atmosphere	$\bar{T}_5 = 1.26\, \bar{\rho}_5^{\frac{1}{5}}$	$8.75 \times 10^{-12}\, (M_5^3\, T_{\text{nuc}}^2)^{1/5}$	$\leq 1.26 \times 10^{-13} \left(\frac{\text{MeV}}{T_{\text{nuc}}}\right)^{1/7} T_{\text{nuc}}$
Width of atmosphere	$\Delta L = \frac{1}{\sqrt{2}} \frac{\bar{w}}{\bar{\gamma}}$	$1.35 \times 10^{28}\, \frac{1}{T_{\text{nuc}}}$	$1.35 \times 10^{28}\, \frac{1}{T_{\text{nuc}}}$
Change of Hubble from perturbations	$\Delta \ln H = \frac{G_5 \bar{T}_5^3}{6\pi^3 \Delta_0} \left(\frac{\alpha}{2}\right)^{\frac{1}{2}}$ Eq. (56)	$4.54 \times 10^{-35} \left(\frac{T_{\text{nuc}}}{M_5}\right)^{6/5}$	$\geq 2.18 \times 10^{-31} \left(\frac{T_{\text{nuc}}}{\text{MeV}}\right)^{2/7}$
Change of Hubble from background	$\Delta \ln H = f(v) \frac{\bar{\rho}_5}{8\rho} \frac{\bar{w}}{\bar{\gamma}} [1 + \tanh(\bar{\gamma})]$ Eqs. (57) and (42)	$1.62 \times 10^{-29} \left(\frac{M_5}{T_{\text{nuc}}}\right)^3 f(v)$	$\geq 2.18 \times 10^{-31} \left(\frac{T_{\text{nuc}}}{\text{MeV}}\right)^{2/7}$
Velocity constraint	$f(v) = \frac{(1+4v^2)}{v\sqrt{1-v^2}}$	$2.8 \times 10^{-6} \left(\frac{T_{\text{nuc}}}{M_5}\right)^{21/5}$	$\geq 2.14 \times 10^7 \frac{T_{\text{nuc}}}{\text{MeV}}$

The first column shows the relevant physical parameters and their definitions in terms of the best fit parameters and related quantities. The second column shows the numerical values and scaling with M_5 and T_{nuc}. Finally, in the last column we show the limits necessary for the thin atmosphere condition (61). The 5th and 6th rows show the shift in the Hubble constant when crossing the atmosphere computed using perturbations and bulk background information, respectively. The 7th row presents a function constraining the velocity of the atmosphere in the bulk that can be computed by equating the results of rows 5 and 6. Note that $T_{\text{nuc}} \geq 0.4$ MeV in order for the BBN constraint to be valid

Using Eqs. (56) and (57) we can constrain the velocity and the speed of sound of the brane

$$f(v) = \frac{(1 + 4v^2)}{v\sqrt{1 - v^2}} = 2.8 \times 10^{-6} \left(\frac{T_{\text{nuc}}}{M_5}\right)^{21/5}. \tag{58}$$

From the above equation we notice that in order to have a real brane velocity we need to satisfy

$$T_{\text{nuc}} \geq 30\, M_5. \tag{59}$$

We can obtain a constraint for $\{T_{\text{nuc}}, M_5\}$ by combining expressions (58) and (54) and noting that in the large velocity limit $\frac{v}{\sqrt{1-v^2}} \approx \frac{1}{\sqrt{1-v^2}} \approx \frac{f(v)}{5}$

$$T_{\text{nuc}} \leq 1.31 \times 10^{-13} \left(\frac{T_{\text{nuc}}}{M_5}\right)^{21/5} \text{MeV}. \tag{60}$$

Inverting this we find

$$M_5 \leq 8.56 \times 10^{-4} \left(\frac{\text{MeV}}{T_{\text{nuc}}}\right)^{5/21} T_{\text{nuc}}. \tag{61}$$

This bound represents the maximum allowed value of M_5 in order for the thin atmosphere condition to be satisfied. In the third column of Table 2 we list the values for the physical quantities allowing M_5 to saturate the above bound. This constraint must be combined with the physical constraint (7)

$$T_{\text{nuc}} > T_{\text{BBN}} \sim 0.4 \,\text{MeV} \tag{62}$$

as well as with the constraint (8)

$$M_5 < 9 \,\text{MeV}, \tag{63}$$

on the normal branch of DGP in order to get the allowed region in parameter space for $\{T_{\text{nuc}}, M_5\}$—depicted in Fig. 4 (blue shaded region).

It is interesting to compare the thermal entropy of our brane to the holographic bound expected from its surface area in 5D. The entropy for the 5D black hole is $S_{BH} = \frac{A}{4G_5}$, while the entropy density in a universe dominated by relativistic particles $s(T) = \frac{4\pi^2}{90} g_* T_{\text{nuc}}^3 \simeq 4.71 \times T_{\text{nuc}}^3$, for $g_* = 10.75$ effective relativistic degrees of freedom, prior to electron/positron annihilation [35]. This puts a lower bound

$$M_5 > 0.23 \left(\frac{T_{\text{nuc.}}}{0.4 \,\text{MeV}}\right) \text{MeV} \quad \text{(holographic bound)} \tag{64}$$

on the 5D Planck mass $M_5 = \frac{1}{(32\pi G_5)^{1/3}}$, where $T_{\text{nuc.}}$ is the nucleation temperature of the brane. We show the Holographic bound allowed region in parameter space in Fig. 4 (orange shaded region). Equations (64–68) constrain the 5D Planck mass to be within 1.5 decades in energy:

$$0.23 \,\text{MeV} < M_5 < 9 \,\text{MeV}, \tag{65}$$

a range that will inevitably shrink with future observations that better constrain BBN, and late-time cosmic expansion history. As we see in Fig. 4, the best-fit value for $T_{\text{nuc.}}$ from cosmological observations (assuming the thin atmosphere condition) does violate the holographic bound (64) by at least 2.5 orders of magnitude, which would decrease the lower limit on M_5 in Eq. (65) by the same factor.

Note that the exact saturation of the holographic bound predicts a brane velocity that is not real for the best fit atmosphere parameters. It is straightforward to show that, for example, by fixing w_c, $\bar{\gamma}$ to the best fit parameters and choosing $\Delta_0 < 10^{-18}$ the holographic bound is satisfied for a real brane velocity. However there is no choice of the 5 parameters $\{M_5, T_{\text{nuc.}}, \bar{\gamma}, w_c, \Delta_0\}$ that characterize the Holographic

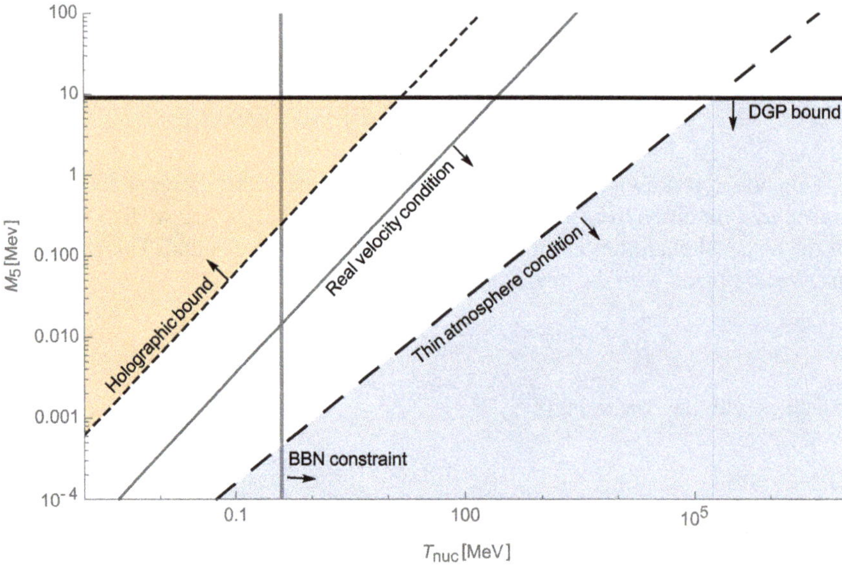

Fig. 4 Theoretical and empirical bounds for the Holographic Big Bang model. The DGP bound Eq. (63) (thick-black) and the holographic bound Eq. (64) (black, thin dashed), together with the BBN bound Eq. (62) (vertical black) constitute the theoretical bounds of the model. The top shaded area (orange) is the allowed region for these three bounds to be satisfied. The real velocity bound Eq. (59) (thick, grey) and the thin atmosphere condition (60) (black, thick dashed) constitute the empirical bounds that HBB must satisfy. The bottom shaded area (blue) is the allowed region of $\{M_5, T_{\mathrm{nuc}}\}$ parameter space satisfying the empirical bounds, and the arrows indicate the directions in which the different bounds apply. It is clear that the empirical bounds violate the holographic bound for all possible allowed pairs $\{M_5, T_{\mathrm{nuc}}\}$ by at least 2.5 orders of magnitude. The least severe violation of the holographic bound is for parameters at the bottom left of the plot: for T_{nuc} being the minimum allowed value by BBN and M_5 the maximum allowed by the thin atmosphere condition (third column of Table 2)

Big Bang model (with the inclusion of a thin atmosphere) that will simultaneously satisfy both the holographic bound and observational constraints. The assumptions are *not* inherently inconsistent; rather they are in conflict with observation.

Is this a "show-stopper"? While the holographic bound on entropy remains a very well-motivated conjecture, it is not clear how firm it might be as objects that get close to crossing it are already in the quantum gravity regime where the classical description of spacetime physics fails. One may argue that since the degrees of freedom responsible for thermal entropy of our brane are on scales much smaller than the 5D Planck length, they are not accessible by a 5D bulk observer, and thus are not limited by the 5D holographic bound.

7 Summary and Discussion

The 5D Holographic Big Bang (HBB) is a novel proposal for a holographic origin of our universe as a 3-brane with induced gravity, out of the collapse of 5D star that can address the traditional problems of big bang cosmology. The main goal of this study was to provide detailed and concrete predictions for this proposal, and to see whether it can serve as a possible competitor to slow-roll inflationary models to explain cosmological observations.

We first focused our attention on a possible mechanism for the nucleation of our 3-brane in which the quantum degrees of freedom of the bulk tunnel into a fuzzball configuration reminiscent to the *bubble of nothing* model. This mechanism not just provides a possible scenario of brane nucleation but also constrains the Planck mass in the bulk.

Previous work has shown that the presence of uniform thermal gas in the 5D bulk leads a scale-invariant primordial power spectrum for cosmological scalar perturbations. To formalize this result and search for mechanisms that could potentially explain deviations from scale-invariance (observed in the CMB data), we studied cosmological perturbations induced by a thin infalling atmosphere. This atmosphere is composed of a spherically symmetric thermal relativistic gas that the brane encounters after nucleation. We showed that this atmosphere induces a change in the Hubble parameter and also scalar cosmological perturbations on the brane. The power spectrum is scale invariant for large k's and scales as k for small k's.

We then tested this prediction for the power spectrum against cosmological observations. The transition is characterized by a decay of $1/k$ for scales where the power spectrum is highly constrained by data [34] as shown in Fig. 3 (right). We found that our model is broadly consistent with non-parametric reconstruction of primordial power spectrum, but is disfavoured compared to a pure power-law at 2.7σ level. We finally outlined various theoretical constraints on the nucleation temperature and 5D Planck mass in the HBB model, and found that the best fit nucleation temperature of the 3-brane was at least 3 orders of magnitude larger than the 5D Planck mass.

This first attempt to understand the detailed consequences of the HBB model for cosmology relies on several simplifying assumptions that can be relaxed in future work. Some of the issues that remain to be tackled are:

1. Perhaps our most perplexing finding was that our best-fit model violated the holographic entropy bound by 8 orders of magnitude. It is not yet clear whether this is a feature or a bug!
2. It would be interesting to study how brane cosmological perturbations will be affected by the large-scale curvature of the bulk (in a 5D Schwarzschild of Kerr spacetime).
3. Other observables that remain to be computed are the amplitude of tensor modes and the non-gaussianity, although we do not expect them to be significant.
4. Given that the speed of sound for a relativistic 5D atmosphere is $c_s = c/2$, one expect $\mathcal{O}(0.2)$ relativistic corrections to the atmosphere profile, which we have ignored. This could affect the functional shape of the power spectrum at a similar

level, potentially improving (or worsening) the fit to the data. A related issue is whether the hydrostatic equilibrium profile for the relativistic thin atmosphere is stable.

To conclude, while we believe the 5D holographic big bang remains an intriguing possibility for the origin of our universe, there remain empirical and theoretical challenges to its status amongst various scenarios for the early universe cosmology that should be addressed in future work.

Acknowledgements This work was partially supported by the National Science and Engineering Research Council (NSERC), University of Waterloo, and Perimeter Institute for Theoretical Physics (PI). Research at PI is supported by the Government of Canada through the Department of Innovation, Science and Economic Development Canada and by the Province of Ontario through the Ministry of Research, Innovation and Science.

Appendix 1: Inhomogeneous Cosmological Perturbations on the Bulk

If we consider the metric (34) and homogeneous perturbations of the form (35) the induced metric on the brane is $\gamma_{\mu\nu} = g_{\alpha\beta}e^\alpha_\mu e^\beta_\nu$ that to first order in ϵ reads:

$$ds^2_{\text{brane}} = [(f_0'^2 - 1) - 2\epsilon \left(\hat{\Psi}_5^0(f)f_0'^2 + \hat{\Phi}_5^0(f)\right)]dt^2 + [1 - 2\epsilon\hat{\Psi}_5^0(f)][dx^2 + dy^2 + dz^2],$$
(66)

where $f' = \frac{\mathrm{d}f}{\mathrm{d}t}$. The induced metric on the brane has to be able to describe a Friedmann universe for which we make the following identifications

$$-d\tau^2 = [(f_0'^2 - 1) - 2\epsilon \left(\hat{\Psi}_5^0 f_0'^2 + \hat{\Phi}_5^0\right)]dt^2, \quad a^2 = [1 - 2\epsilon\hat{\Psi}_5^0],$$
(67)

where τ is the proper time of the brane and a is the scale factor. On top of the homogeneous perturbations we are now going to consider anisotropies

$$\Phi_5 = \epsilon\Phi_5^0(w) + \epsilon_1\Phi_5^1(x^\alpha), \quad \Psi_5 = \epsilon\Psi_5^0(w) + \epsilon_1\Psi_5^1(x^\alpha), \quad f = f_0(t) + \epsilon_1 f_1(x^\alpha),$$
(68)

where $\epsilon_1 \ll 1$ is a parameter that controls the anisotropic perturbations and $\epsilon\epsilon_1 \ll \epsilon, \epsilon_1$. The induced metric can be written as

$$ds^2_{\text{brane}} = -[1 + \frac{2\epsilon_1}{f_0'^2 - 1}(\hat{\Phi}_5^1 + \hat{\Psi}_5^1 f_0^2 - f_0'f_1')]d\tau^2 + 2a\epsilon_1\frac{f_0'f_{1,i}}{\sqrt{f_0'^2 - 1}}dx^i d\tau + a^2(1 - 2\hat{\Psi}_5^1\epsilon_1)dx^2,$$
(69)

where $\hat{\Psi}_5 = \Psi_5(w = f(x^\mu))$ are the metric functions projected to the brane. The general form of a 4D metric including scalar and vector cosmological perturbation in 4D are that contains all the terms in Eq. (69)

$$ds^2_{\text{brane}} = -(1 + 2\phi_4)d\tau^2 - 2aB_i d\tau dx^i + a^2(1 - 2\psi_4)dx^2, \tag{70}$$

and thus we identify

$$\phi_4 = \frac{\epsilon_1}{f_0'^2 - 1}(\hat{\Phi}_5^1 + \hat{\Psi}_5^1 f_0^2 - f_0' f_1'), \quad \psi_4 = \hat{\Psi}_5^1 \epsilon_1, \quad B_i = \epsilon_1 \frac{f_0' f_{1,i}}{\sqrt{f_0'^2 - 1}}. \tag{71}$$

Note that the Newtonian gauge on the bulk does not translate into a Newtonian gauge on the brane and that some perturbations that are scalars in 5D are projected as a vectorial perturbation component in 4D. The scalar gauge invariant quantities can be constructed from Eq. (69) as

$$\Phi_4 = \phi_4 - \partial_\tau(aB), \tag{72}$$
$$\Psi_4 = \psi_4 + HaB, \tag{73}$$

where, H is the Hubble constant and B is the scalar part of the vector metric perturbation B_i. From our construction the Hubble constant is

$$H = \frac{\dot{a}}{a} = -\frac{\epsilon \hat{\Psi}_5^0}{a^2}, \tag{74}$$

and thus Eq. (73) reduces to $\Psi_4 = \psi_4$ to first order in ϵ, ϵ_1.

Appendix 2: Derivation of the Power Spectrum

With use of the Poisson equation in 4D

$$\nabla^2 \Psi_5(y) = \frac{8\pi G_5}{3}\rho_5(y), \tag{75}$$

and the expression for the energy density correlation function

$$\langle \rho_5(y_1)\rho_5(y_2)\rangle \simeq \alpha(T_5)^6 \delta^4(y_1 - y_2), \tag{76}$$

we can write the 2-point correlation function of Ψ_5 using the Green's function for the Laplacian operator

$$\langle \Psi_5(x_1)\Psi_5(x_2)\rangle = \alpha\left(\frac{8\pi G_5}{3}\right)^2 \left(\frac{1}{4\pi^2}\right)^2 \int d^4y \frac{(T_5(y))^6}{|y - x_1|^2 |y - x_2|^2}. \tag{77}$$

The junction condition between the 5D and 4D metrics (34) and (66) allow us to compute $\langle \Psi_4(\mathbf{x_1})\Psi_4(\mathbf{x_2})\rangle$

$$\langle \Psi_4(\mathbf{x_1})\Psi_4(\mathbf{x_2})\rangle = \langle \Psi_5(\mathbf{x_1}, x_{1w} = 0)\Psi_4(\mathbf{x_2}, x_{2w} = 0)\rangle$$

$$= \alpha \left(\frac{2G_5}{3\pi}\right)^2 \int d^3\mathbf{y_3}dy_w \frac{(T_5(y_w))^6}{(|\mathbf{x_1} - \mathbf{y_3}|^2 + |y_w|^2)(|\mathbf{x_2} - \mathbf{y_3}|^2 + |y_w|^2)} \tag{78}$$

where we have decomposed the bulk coordinate as $y = (\mathbf{y_3}, y_w)$ and we have set the temperature to be just a function of the w direction of the bulk. Combining expressions (45, 46, 78) and setting $\mathbf{x_1} = 0$ we have

$$\begin{aligned}
P_\zeta(k) &= \alpha \left(\frac{2G_5}{3\pi\,\Delta\ln H}\right)^2 \int \frac{d^3\mathbf{y_3}\,dy_w\,(T_5(y_w))^6}{|\mathbf{y_3}|^2 + |y_w|^2} \int d^3x \frac{e^{i\mathbf{k}\cdot\mathbf{x}}}{|\mathbf{x} - \mathbf{y_3}|^2 + |y_w|^2}, \\
&= \alpha \left(\frac{2G_5}{3\pi\,\Delta\ln H}\right)^2 \int \frac{d^3\mathbf{y_3}\,dy_w\,(T_5(y_w))^6}{|\mathbf{y_3}|^2 + |y_w|^2} \int d^3x' \frac{e^{i\mathbf{k}\cdot\mathbf{x'}} e^{i\mathbf{k}\cdot\mathbf{y_3}}}{|\mathbf{x'}|^2 + |y_w|^2}, \\
&= \alpha \left(\frac{2G_5}{3\pi\,\Delta\ln H}\right)^2 \int dy_w\,(T_5(y_w))^6 \left(\int d^3x' \frac{e^{i\mathbf{k}\cdot\mathbf{x'}}}{|\mathbf{x'}|^2 + |y_w|^2}\right)^2, \\
&= \alpha \left(\frac{2G_5}{3\pi\,\Delta\ln H}\right)^2 \frac{1}{(4\pi k)^2} \int dy_w\, e^{-2|y_w|k}\,(T_5(y_w))^6,
\end{aligned} \tag{79}$$

where we have performed the coordinate transformation $\mathbf{x'} = \mathbf{x} - \mathbf{y_3}$, and use the result $\int \frac{d^3x\, e^{i\mathbf{k}\cdot\mathbf{x}}}{|x|^2 + |m|^2} = \frac{e^{-k|m|}}{4\pi k}$.

With this we are able to write

$$\mathscr{P}(k) = \frac{k^3}{2\pi^2} P_\zeta(k) = \beta\,k \int_0^\infty dw\, e^{-2|w|k}\,(T_5(w))^6, \tag{80}$$

where $\beta = \frac{\alpha}{2}\left(\frac{G_5}{6\Delta\ln H\,\pi^3}\right)^2$. Note that we are in integrating between $[0, \infty)$ because of the $\mathbf{Z}_2$ symmetry. Because the temperature profile is analytic it admits a Taylor expansion of the form

$$(T_5(w))^6 = \sum_{n=0}^\infty \frac{T^{6^{(n)}}(0, \bar{\gamma}, \bar{w})}{n!} w^n, \tag{81}$$

and thus the power spectrum Eq. (80) admits a series decomposition of the form

$$\mathscr{P}(k) = \beta \sum_{n=0}^\infty \frac{T^{6^{(n)}}(0, \bar{\gamma}, \bar{w})}{k^n} 2^{-1-n}. \tag{82}$$

This last expression implies that for large k the correction to a scale invariant power spectrum goes as $1/k$. If the integral in Eq. (80) is performed in the rage $(-\infty, \infty)$ the power spectrum series would be

$$\mathscr{P}(k) = \beta \sum_{n=0}^{\infty} \frac{T^{6^{(2n)}}(0, \bar{\gamma}, \bar{w})}{k^{2n}} 2^{-1-2n}, \tag{83}$$

giving a correction from scale invariance that goes as $1/k^2$ for large k. This correction renders model disfavourable in comparison with the symmetric case, where we have a $1/k$ decay. This is the reason why we work with the symmetric integral Eq. (80).

References

1. A. Aurilia, R. S. Kissack, R. Mann, E. Spallucci, Phys. Rev. D **35**, 2961 (1987)
2. R. Pourhasan, N. Afshordi, R. B. Mann, JCAP **1404**, 005 (2014)
3. N. Afshordi, R. Mann, R. Pourhasan, Int. J. Mod. Phys. D **24**(12), 1544029 (2015)
4. G. R. Dvali, G. Gabadadze, M. Porrati, Phys. Lett. B **485**, 208 (2000)
5. P. A. R. Ade, et al., Astron. Astrophys. **594**, A13 (2016)
6. R. Gregory, N. Kaloper, R. C. Myers, A. Padilla, JHEP **10**, 069 (2007)
7. S. D. Mathur, Gen. Rel. Grav. **42**, 113 (2010)
8. M. Novello, S. E. P. Bergliaffa, Phys. Rep. **463**, 127 (2008)
9. E. Witten, Nucl. Phys. B **195**(3), 481 (1982)
10. O. Lunin, S. D. Mathur, Phys. Rev. Lett. **88**, 211303 (2002)
11. M. Pospelov, J. Pradler, Annu. Rev. Nucl. Part. Sci. **60**, 539 (2010)
12. T. Azizi, M. Sadegh Movahed, K. Nozari, New Astron. **17**, 424 (2012)
13. P. A. R. Ade, et al., Astron. Astrophys. **594**, A14 (2016)
14. U. Seljak, M. Zaldarriaga, Astrophys. J. **469**, 437 (1996)
15. M. Zaldarriaga, U. Seljak, E. Bertschinger, Astrophys. J. **494**, 491 (1998)
16. A. Lewis, S. Bridle, Phys. Rev. D **66**(10), 103511 (2002)
17. A. Lewis, A. Challinor, A. Lasenby, Astrophys. J. **538**, 473 (2000)
18. C. Howlett, A. Lewis, A. Hall, A. Challinor, JCAP **4**, 027 (2012)
19. A. Lewis, Phys. Rev. D **87**(10), 103529 (2013)
20. A. Lewis, CAMB Notes (2014)
21. F. Beutler, C. Blake, M. Colless, D. H. Jones, L. Staveley-Smith, L. Campbell, Q. Parker, W. Saunders, F. Watson, MNRAS **416**, 3017 (2011)
22. C. Blake, et al., MNRAS **418**, 1707 (2011)
23. L. Anderson, et al., MNRAS **427**, 3435 (2012)
24. F. Beutler, C. Blake, M. Colless, D. H. Jones, L. Staveley-Smith, G. B. Poole, L. Campbell, Q. Parker, W. Saunders, F. Watson, MNRAS **423**, 3430 (2012)
25. N. Padmanabhan, X. Xu, D. J. Eisenstein, R. Scalzo, A. J. Cuesta, K. T. Mehta, E. Kazin, MNRAS **427**, 2132 (2012)
26. L. Anderson, et al., MNRAS **441**, 24 (2014)
27. L. Samushia, et al., MNRAS **439**, 3504 (2014)
28. A. J. Ross, L. Samushia, C. Howlett, W. J. Percival, A. Burden, M. Manera, MNRAS **449**, 835 (2015)
29. P. A. R. Ade, et al., Astron. Astrophys. **594**, A15 (2016)
30. N. Aghanim, et al., Astron. Astrophys. **594**, A11 (2016)
31. P. A. R. Ade, et al., Astron. Astrophys. **594**, A24 (2016)
32. C. L. Reichardt, et al., APJ **755**, 70 (2012)
33. S. Das, et al., JCAP **4**, 014 (2014)
34. P. A .R. Ade, et al., Astron. Astrophys. **594**, A20 (2016)
35. E. W. Kolb, M. S. Turner, *The Early Universe* (CRC Press, Boca Raton, 1990)

Dimensional Reduction and Quantum Gravity

Jonas Mureika

Abstract Dimensional reduction is a common feature of disparate candidates for a final theory of quantum gravity. This paper will review key aspects of lower dimensional gravity, and discuss a number of mechanisms through which dimensional reduction might occur, as well as their phenomenology and observational prospects.

1 Introduction

It gives me great pleasure to contribute this paper in memory of Antonio Aurilia, whose life's work was studying the quantum nature of black holes and gravitation. I met Antonio for the first and last time in Jaunary 2016, where we shared coffee with Piero Nicolini in the city of Claremont, California. Coincidentally, this was where I started my own career in 2001. I hope to honour his legacy by discussing a related subject that has been my own h'oeuvre d'art for the last decade.

The celebrated theory of General Relativity (GR) just passed its centennial. From Einstein's early predictions of the advance of Mercury's perihelion [1] amd the measured deflection of starlight [2], to the recent groundbreaking detection of gravitational waves [3], GR has stood up to unprecedented comprehensive experimental confirmation. With the addition of the first-ever image of a black hole (M87) from the Event Horizon Telescope collaboration [4], there is strong confidence that GR correctly describes gravity over the range of length scales involved.

Nevertheless, it is well known that the theory meets decisive failure in two distinct regimes. On astrophysical and cosmological scales, GR on its own is unable to explain the dynamics of galaxies nor the evolution of the universe itself, at least without the addition of dark matter and dark energy. Conversely, the existence of a black hole

J. Mureika (✉)
Department of Physics, Loyola Marymount University, 1 LMU Drive, Los Angeles, CA 90045, USA
e-mail: jmureika@lmu.edu

Kavli Institute for Theoretical Physics, University of California, Santa Barbara, CA 93106-4030, USA

  231
P. Nicolini (ed.), *Touring the Planck Scale*, Fundamental Theories of Physics 219,
https://doi.org/10.1007/978-3-031-76066-2_11

singularity in standard metrics is the first indication that GR in incomplete in the quantum limit.

A variety of candidate theories to replace GR in the ultraviolet regime have arisen over the last four decades, ranging from string theory [5], loop quantum gravity [6], asymptotically-safe gravity [7], and emergent gravity [8], among others. These attempts are motivated by the need to circumvent the non-renormalizability of gravitation as a $(3+1)$-D quantum field theory.

With an increasing wealth of observational black hole related data, we may be on the precipice of our first glimpse of quantum gravity effects. Although it seems unlikely that quantum effects should in any way emerge on classical scales, one must first consider that black holes are not entirely classical in the first place. A general spacetime metric is a solution to Einstein's equations for a spherically-symmetric matter distribution, $i.e.$

$$R_{\mu\nu} - \frac{1}{2}g_{\mu\nu}R + \Lambda g_{\mu\nu} = 8\pi G \mathscr{T}_{\mu\nu} \tag{1}$$

Here, $\mathscr{T}_{\mu\nu}$ is the corresponding stress-energy source. In general, the line element can be expressed as

$$ds^2 = f(r)\,dt^2 - \frac{dr^2}{f(r)} - r^2\,d\Omega^2 \tag{2}$$

Knowledge of $f(r)$ yields an array of characteristics for the black hole, which might be considered both classical or quantum in nature. In the former case, the constraint $f(R_H) = 0$ defines the horizon, while $f(0) \rightarrow \infty$ signals the existence of a singularity. On the quantum side, the well-known thermodynamic properties of the black hole can be extracted from $f(r)$, namely the Hawking temperature $T \sim f'(R_H)$ and entropy $S \sim \int \frac{dM}{T}$ [9, 10].

For a Schwarzschild black hole with $f(r) = 1 - \frac{2GM}{r}$, the temperature behaves as $T \sim M^{-1}$ and diverges when $M \rightarrow 0$. Conversely, the entropy scales as $S \sim M^2$ and vanishes in the same limit. This unphysical disparity is a red flag that quantum gravity is the key to regularizing black holes, and black holes are thus the key to unlocking the secrets of quantum gravity. Indeed, by replacing the point mass with a distribution, $M \rightarrow M(r)$, the aforementioned unphysical behaviour may be tamed. Two such examples include non-commutative inspired gravity [11], and generalized uncertainty principle modified gravity [12].

The distinction between classical and quantum characteristics, however, is not so clear cut. Although the horizon can be obtained from classical GR as $f(R_H) = 0$, it it intrinsically linked to decidedly quantum entities (namely black hole entropy and Hawking evaporation). In fact, the horizon itself has been shown to have a quantum origin thanks to the corpuscular gravity framework [13–17] in which a black hole is modeled as Bose-Einstein condensate (BEC) of N soft weakly coupled gravitons. If the gravitons are bounded within a region R and have an energy $E_G \sim \frac{\hbar}{R}$, the total energy of the condensate is then $E_{\text{tot}} \sim N\frac{\hbar}{R} \sim M_{\text{BH}}$. The area-entropy law dictates that the number of gravitons is $N \sim \frac{R^2}{\ell_{\text{Pl}}^2}$, and given $\hbar = m_P \ell_P$, one finds $R \sim \frac{\ell_P}{m_P} M_{\text{BH}} \sim G M_{\text{BH}}$.

There is growing consensus that quantum gravitational effects might be measurable in macroscopic phenomena. The near horizon physics of astrophysical or supermassive black holes is a ripe venue for such observations that could come in the form of metric fluctuations [18–21], and exotic compact objects like boson or Planck stars [22, 23]. Interesting effects can arise from extra-horizon quantum structures [24–26], which have been shown to produce gravitational wave echoes in the ringdown phase of a compact binary merger [27]. Such echoes may well have been observed with high statistical significance in the recent LIGO data [28–30], though some are skeptical of these results [31]. In a similar spirit, a recently-proposed Extended Uncertainty Principle modified black hole also demonstrates how a quantum constraint can enhance the size of the horizon of supermassive black holes [32].

Given that all characteristics of a black hole arise from quantum origins, it is tempting to conclude that black holes themselves are intrinsically quantum objects. This is supported by the notion that black hole entropy is not volumetric as it should be for a classical object, but rather dependent on the horizon area in $(3+1)$-D. Black holes may thus hold the key to a deeper understanding of quantum gravity. As this contribution will explore, the nature of such a final theory may well not be one that resides in $(3+1)$-D, but rather a more fundamental one that lives in a $(1+1)$-D Universe. I will summarize key results from work done with many of my esteemed colleagues, without whom these would not have been possible. Features of dimensionally- reduced black holes are reviewed, and the associated phenomenology of dimensional reduction will be discussed. Lastly, the nature of the sub-Planckian is explored, where it is shown that dimensional reduction arises naturally in this regime.

2 Dimensional Reduction

The notion that the laws of physics—Newton's law universal gravitation, in particular—might be altered in certain limits has been an accepted mainstay in the discipline for several decades now. When I was an undergraduate student at the University of Toronto in the early 1990s, however, this idea was barely ready for prime time and quite alien to me. I vividly recall a discussion in the student lounge at the McLennan Physical Laboratories with one of my classmates, a student by the name of Nima Arkani-Hamed, in which he asked me [paraphrasing] "Have you ever wondered what the world would be like if gravity was an inverse cube or fourth law, rather than squared?" I hadn't—who was I to question Sir Isaac Newton! Of course only a few years later, Nima's idea would be thrust to the forefront of high energy physics in the form of compactified large extra dimensions.

That the number of dimensions should increase as one goes to higher energies or lower length scales is somewhat unpalatable, however. Why should Nature harbour such a complex structure? What if the opposite occurred—i.e. what if the number of dimensions decreases in the ultraviolet limit? I was first introduced to the concept of dimensional reduction in 2010, when I was on sabbatical at the Perimeter Institute for Theoretical Physics, and hopped down to the University at Buffalo to give a colloquium. Following my talk, Dejan Stojkovic and I discussed an idea he was exploring that addressed this very question. Instead of presupposing the existence of extra dimensions at higher energy scales, spatial dimensions would "disappear" at short length scales. The results of our collaboration—gravitational wave signals in a vanishing dimension scenario—are discussed in Sect. 5.2, but first a quick review of additional associated phenomenology is presented.

A variety of approaches have been proposed to allow for dimensional reduction in the high energy/short length scale limit. Whether this reduction is physical, or merely an effective one, is dependent on the framework in question (see [33] for a related discussion). Spectral dimensional reduction predicts an effective lower dimension description for certain high energy phenomena, while not necessarily modifying the underlying manifold. These include Causal Dynamical Triangulations [34], Hořava-Lifshitz gravity [35], and physics described by a non-commutative geometry [36]. The evolving (or vanishing) dimensions paradigm supposes that there exists a "scaffolding" of structures that are defined by fundamental length scales L_i, below which the effective spacetime dimension reduces [37–41] (see Sect. 5.2 for a more detailed description). Dimensions can also evolve stochastically from a one dimensional string gas to a limit of three dimensions [42]. Certain brane-world scenarios also predict dimensional reduction as a feature of the theory [43, 44]. Lastly, multifractal gravity offers a framework in which the dimension of spacetime is continuously reducing as one approaches the UV [45–47].

Spontaneous dimensional reduction at the Planck scale was first postulated by 't Hooft [48], who presupposed that the Universe behaves two-dimensionally in this limit. A consequence of this picture is that the non-renormalizabity of gravity becomes an emergent infrared feature of oxidation to the $(3 + 1)$-D. As such, GR can be undersrtood as an effective limit of a more fundamental theory of quantum gravity in $(1 + 1)$-D. Granted, the scale at which dimensional reduction ensues need not be in the Planckian or sub-Planckian regime, but may very well happen at some considerably lower energy scale.

Phenomenological constraints can be placed on a variety of physical systems to determine where this transition may occur. In quantum field theory, dimensional reduction in the far UV limit generates a new approach to gauge couplings unification [49]. Dimensional reduction also implies new phenomenology in high energy particle physics, specifically with the Higgs mass corrections. In $(3 + 1)$-D, the loop correction contributions for fermions, gauge bosons, and the Higgs itself are, up to an energy cutoff scale Λ [50],

$$\frac{ig_f^2}{2} \int^{\Lambda} \frac{d^4k}{(2\pi)^4} \mathrm{Tr}\left(\frac{i}{\not{k} - m_f}\frac{i}{\not{k} + \not{p} - m_f}\right) \sim -\frac{g_f^2 \Lambda^2}{32\pi^2}$$

$$\frac{ig^2}{4} \int^{\Lambda} \frac{d^4k}{(2\pi)^4}\frac{1}{k^2 - m_g^2} \sim \frac{g^2 \Lambda^2}{64\pi^2} \tag{3}$$

$$3i\lambda \int^{\Lambda} \frac{d^4k}{(2\pi)^4}\frac{1}{k^2 - m_H^2} \sim \frac{3\lambda \Lambda^2}{16\pi^2}$$

where m_f, m_g, and m_H are the fermion, gauge boson, and Higgs masses, respectively. The couplings are $g_f = m_f^2/v^2$, $g = 2m_g^2/v^2$, and $\lambda = m_H^2/(2v^2)$, with v the Higgs vacuum expectation value. Each integral is quadratically-divergent in the cutoff Λ.

In a spacetime of lower dimension, however, the dependence on the cutoff is tempered. When the dimensionality is $(2+1)$-D, the integral measure in Eq. 4 is replaced with $d^4k/(2\pi)^4 \to d^3k/(2\pi)^3$, which causes the loops to diverge linearly. Likewise, in $(1+1)$-D the measure is $d^2k/(2\pi)^2$, and the divergence becomes logarithmic [50].

One of the most compelling areas in which to consider dimensional reduction, however, is gravitation. Specifically, the mechanism radically alters the quantum characteristics of black holes, as well as early Universe cosmology. The remainder of this contribution will explore these considerations in depth, reviewing some novel experimental signatures that could be used to identify if such dimensional reduction does indeed occur.

3 Lower Dimensional Gravity and Black Holes

My first exposure to the concept of lower dimensional gravity was in 1993, when I attended the Canadian Conference on General Relativity and Relativistic Astrophysics, held at the University of Waterloo [51]. In the introductory remarks, the speaker discussed the simplicity of quantizing gravity in $(1+1)$-D, words which I did not appreciate at the time, but have remained in the back of my mind ever since. This notion of a lower-dimensional universe fascinated me, though I would not revisit this concept for another 17 years. Instead, I instead embarked on my pioneering research journey by studying equivalence principle violating neutrino oscillations [52–54] with my M.Sc. advisor Robert (Robb) Mann—with whom I have had the longest ongoing collaboration (27 years at the time of this writing).

I would like to point out that a connection to the man whose life and work are honouring also arises here. One of Robb Mann's earlier papers was in fact a collaboration with Antonio Aurilia [55]. I recall reading this paper as a graduate student at the University of Waterloo, and as one who had grown up in Canada and grew weary of the harsh winters, wondered what it was like to live and work in California as Antonio did (who knew then that I would one day get my chance!). It is also interesting to note this paper was co-authored by Euro Spallucci, with whom our editor Piero Nicolini has been a long-time collaborator and friend.

Coincidentally, Robb had already opened the door to lower dimensional gravity in his own portfolio, having addressed the $(1 + 1)$-D Schwarzschild and Reissner-Nordström equivalent solutions in 1987 [56]. We have coauthored several papers on lower-dimensional black hole topics, including $(1 + 1)$-D entropic black holes [57] and, with Antonia Frassino, lower-dimensional black hole chemistry [58, 59]. I summarize the former later on in Sect. 6. But first, a review of the formative literature is necessary.

Gravitation in lower dimensions has been studied for over 40 years. While it was initially considered a simple pedagogical exercise to understand the associated interactions, it was later discovered that a common feature of quantum gravity candidates was an effective dimensional reduction in the UV. This led to novel and distinct physics in $(2 + 1)$-D [60–69] and $(1 + 1)$-D [70–87]. The latter case produces exactly solvable quantum gravity systems [71], which further whets ones appetite for dimensional reduction as a natural path to this ultimate theory.

In contrast to standard Einstein gravity, most lower dimensional actions necessitate the inclusion of a scalar dilaton filed that can introduce new interactions and physics. One particular dilaton model is free of such additional features, however. Known as the "$R = T$" Liouville formulation [56, 75, 79], this approach allows the dilaton to decouple from the gravitational effects. As such, it is seen as the closest realization of pure Einstein gravity in $(1 + 1)$-D.

The action is

$$S_2 = \int d^2x \, \sqrt{-g} \left\{ \frac{1}{16 \pi \, G_1} \left[\psi \, R + \Lambda + \frac{1}{2} (\nabla \psi)^2 \right] + \mathscr{L}_m \right\} , \qquad (4)$$

Varying (4) with respect to the dilaton field ψ and $g_{\mu\nu}$ produces the field equations

$$R = \Box \psi \qquad (5)$$

and

$$\frac{1}{2} \nabla_\mu \psi \, \nabla_\nu \psi - g_{\mu\nu} \left(\frac{1}{4} \nabla^\lambda \psi \, \nabla_\lambda \psi - \Box \psi \right) - \nabla_\mu \nabla_\nu \psi$$

$$= 8 \pi \, G_1 \, T_{\mu\nu} + \frac{\Lambda}{2} \, g_{\mu\nu} \qquad (6)$$

where

$$T^{\mu\nu} = - \frac{2}{\sqrt{-g}} \frac{\delta \mathscr{L}_m}{\delta g_{\mu\nu}} . \qquad (7)$$

One can easily verify that (5) is obtained after taking the trace of (6), thereby eliminating the dilaton from the field equations. Subsequently, (6) assumes the "$R = T$" form

$$R - \Lambda = 8 \pi \, G_1 \, T . \qquad (8)$$

In the absence of a cosmological constant, the solution in vacuum to (8) is [56]

$$ds^2 = (1 - 2\, G_1\, M\, r)\, dt^2 - \frac{dr^2}{(1 - 2\, G_1\, M\, r)},\tag{9}$$

where $r > 0$ is the spatial coordinate. The gravitational constant in this case is dimensionless. This implies that unlike in higher dimensions, there is no fundamental energy scale in $(1 + 1)$-D. Furthermore, the lack of dimensionality cures divergent aspects of path integrals that plaque higher-dimensional candidates for quantum gravity.

The horizon for the $(1 + 1)$-D black hole is

$$R_1 = \frac{1}{2\, G_1\, M},\tag{10}$$

and the Hawking temperature is

$$T = \frac{G_1\, M}{2\, \pi},\tag{11}$$

Lastly, the entropy is

$$S = \frac{2\, \pi}{G_1} \ln\left(\frac{M}{m_*}\right)\tag{12}$$

which necessitates the introduction of a new mass scale m_*. This could be interpreted as the fundamental scale of the theory, or alternatively a lower-bound on particle masses in general.

One of the most interesting aspects of $(1 + 1)$-D gravity is the mass dependence of the horizon (10). Unlike in the standard case, the horizon scales as the inversely with the mass. Of course, this is the same mass dependence as in the Compton wavelength, suggesting that a particle's horizon and wavelength might be indistinguishable. In this sense, gravity has become a quantum entity, and there is thus only one type of theory in this regime.

This result naturally arises in the self-completeness paradigm in $(1 + 1)$-D [88]. It is also a consequence of the fact that there is no generalized uncertainty principle in $(1 + 1)$-D gravity, principle [89],

$$(\Delta x)_{\text{Total}} = (\Delta x)_Q + (\Delta x)_G\tag{13}$$

where $(\Delta x)_Q \sim \hbar/M$ is the quantum uncertainty and $(\Delta x)_G \sim R_{\text{H}}$ is the gravitational uncertainty given by the horizon size. From the above horizon definition (10), this leads to

$$(\Delta x)_{\text{Total}} \sim \hbar/M.\tag{14}$$

The general $(d+1)-$D Schwarzschild metric function is

$$f(r) = 1 - \frac{2G_d M}{r^{d-2}}, \tag{15}$$

whose horizons are

$$R_d = (2G_d M)^{\frac{1}{d-2}} \tag{16}$$

Note that the naive substitution $d = 1$ does indeed give the proper dependence on M, namely $R_1 \sim M^{-1}$. Furthermore, consider the expression for the $(d+1)$-dimensional gravitational constant, [90]

$$G_d = 2\pi^{1-\frac{d}{2}} \Gamma\left(\frac{d}{2}\right) \frac{\ell_d^{d-1}}{\hbar}. \tag{17}$$

with ℓ_d the $(d+1)$-dimensional Planck length. Setting $d = 1$, it follows from (17) that G_1 is independent of ℓ_1, and so

$$G_1 \sim \frac{1}{\hbar}. \tag{18}$$

Thus, one can conclude that the constant defining gravity in $(1+1)$-D is dual to the dimensionless quantity which defines the quantum scale. This also supports the notion that there is no fundamental scale for $(1+1)$-D gravity.

In contrast, gravity in $(2+1)$-D is quite non-intuitive. The Einstein-deSitter action written in Einstein-Cartan formulation is [59, 68]

$$I_{EH} = \frac{1}{8\pi G_N} \int_{M^3} \left[e_a \wedge R^a + \frac{1}{6\ell^2} \varepsilon^{abc} e_a \wedge e_b \wedge e_c \right] \tag{19}$$

where

$$T^a = de^a + \varepsilon^{abc} \omega_b \wedge e_c, \qquad R^a = d\omega^a + \frac{1}{2}\varepsilon^{abc} \omega_b \wedge \omega_c \tag{20}$$

are the respective torsion and curvature forms, with dreibein 1-forms e^a and Lorentz connection 1-forms ω^a ($a = 0, 1, 2$).

The best known solution is the BTZ metric, which describes a rotating black hole with angular momentum J in a spacetime with cosmological constant $\Lambda_{(2)}$ [63],

$$ds^2 = \left(-G_2 M + \Lambda_2 r^2 + \frac{J^2}{4r^2}\right)^{\frac{1}{2}} dt^2 - \frac{dr^2}{\left(-G_2 M + \Lambda r^2 + \frac{J^2}{4r^2}\right)^{\frac{1}{2}}}$$
$$+ r^2 \left(d\phi + \frac{J^2}{2r^2} dt^2\right). \tag{21}$$

In this case, the gravitational constant G_2 has dimensions of inverse mass. When $J = 0$ (i.e. static black holes), it is clear that horizons exist only for negative Λ_2,

$$R_2 = \sqrt{-\frac{G_2 M}{\Lambda_2}} \tag{22}$$

That is, $(2 + 1)$-D black holes can only exist in AdS spacetimes. Furthermore, there is no Newtonian limit in this case, and gravity is purely topological. Note also that the form of the horizon does not follow the pattern established in Eq. (16) when $n = -1$ since the expression is not defined at this point.

In addition to the earlier references to lower-dimensional gravity, a number of more recent studies have addressed the associated physics of black holes and cosmology. These include evaporating black primordial black holes that transition into lower-dimensional spacetime [91], as well as those that are formed in $(1 + 1)$-D and "oxidate" up in dimension [92]. Additionally, the case for dimensional reduction at high energies based on the strong coupling expansion of the Wheeler-DeWitt equation was considered in [93, 94], and loop quantum gravity effects in [95]. The implications for supertranslations and superrotations was examined in $(2 + 1)$-D [96], and for near-horizon Bondi-Metzner-Sachs symmetry and black hole entropy in [97]. Additional considerations of dimensional reduction in causal sets were discussed in [98].

Other explorations in this field have included an effective gravitational dimensional reduction arising from a Parisi-Sourlas-like phenomenon in quantum field theories [99], plus dimensionally-reduced non-commutative [100] and entropic black holes [57, 90]. The impact of spontaneous dimensional reduction on self-complete gravity was addressed in [101]. Lower-dimensional black holes in the corpuscular gravity picture [102] have been addressed as an alternative quantum gravity description. The chemistry of black holes in $(1 + 1)$-D and Chern-Simons solutions in $(2 + 1)$-D [58, 59] reveals interesting aspects of their phase space. Finally, a Newtonian description of dimensionally-reduced black holes in the bootstrapped gravity approach has been considered [103].

4   Dimensional Remnants

If dimensional reduction occurs at microscopic scales, the characteristics of quantum black holes in the final stages of evaporation will be affected. This will have important consequences for such objects as primordial black holes (PBHs). These can be studied through the standard thermodynamic approach. In n spatial dimensions, the Stefan-Boltzmann law is [104]

$$\mathscr{P}_n = \left[\frac{dM}{dt}\right]_n = -\sigma_d A_{n-1} T_d^{n+1}, \tag{23}$$

where $A_{n-1} \sim R_H^{n-1}$ is the horizon area and $\sigma_n \sim k_{\text{Boltzmann}}^{n+1}$. From this, one obtains the evaporation time of the black hole as

$$\tau_n = \int_{M_{\text{BH}}}^{0} \frac{dM}{\mathscr{P}_n} .$$
(24)

In $(3+1)$-D, the usual result is recovered, i.e. $\mathscr{P}_3 \sim M^{-2}$ and $\tau_3 \sim M^3$ when $T_3 \sim M^{-1}$. If $\tau_3 \sim 10^{17}$ s is the Hubble time, PBHs of mass $M \sim 10^{12}$ kg will be presently in their final stage of evaporation.

For the lower dimensional cases, however, several interesting features arise. The area of a $(1+1)$-D black hole is simply the sum of two antipodal points, and is thus independent of the black hole mass. The temperature scales linearly with the mass, $T_1 \sim M$, and so the radiative power (23) is $\mathscr{P}_1 \sim \sigma_1 \, G_1 \, M^2$. The lifetime (24) will thus diverge, indicating that black holes in $(1+1)$-D spacetime will remain as what can be termed a "dimensional remnant" [91]. This suggests dimensionally-reduced black holes could be a possible candidate for dark matter.

Limits may be placed on the size of current dimensional remnants. A black hole that forms at some time t_i with mass M_i will evolve as

$$\tau_1 = t - t_i \sim \left(\frac{1}{M} - \frac{1}{M_i} \right) .$$
(25)

where M_0 is the mass at present, i.e.

$$M = \frac{M_i}{1 + M_i(t - t_i)}$$
(26)

The timescale τ can be thought of as a "Compton" time

$$\tau \sim \hbar/(Mc^2),$$
(27)

where fundamental constants have been re-inserted for computational purposes. At present day ($\tau = t_0 \sim 10^{17}$ s), a mass scale M_* can be defined from the above as

$$M_* \sim \hbar/(c^2 t_0) \sim 10^{-68} \text{ kg}.$$
(28)

From Eq. (25), the mass of the black hole at the present epoch is then

$$M \approx \frac{M_i}{1 + M_i/M_*} .$$
(29)

Black holes that were created with smaller masses effectively remain that mass, i..e $M \approx M_i$ for $M_i \ll M_*$, while much larger black holes will have evaporated to this limit, $M \approx M_*$ for $M_i \gg M_*$.

This mass-scale M_* is associated with a radius equal to the Hubble scale, 10^{24} m, whose temperature is 10^{-28} K. Note that this mass scale arises naturally in some estimates for the photon or graviton mass [57, 90], and could also have various observational consequences associated with cluster-scale characteristics via the Dvali-Gabadadze-Porrati (DGP) effect.

Evaporating through a $(2 + 1)$-D phase implies different consequences. The horizon of a BTZ black hole (22) implies that the temperature is

$$T_2 = \sqrt{-G_2\, M\, \Lambda} \qquad (30)$$

Since, such black holes can only exist in anit-deSitter spacetime ($\Lambda < 0$)—and our universe is likely not AdS—one can conclude that a black hole evaporating down from $(3 + 1)$-D will stop decaying and will become a new form of remnant [91]. It is appropriate to term these "static" and "dynamic" dimensional remnants (SDRs and DDRs). In the former case, SDRs are formed as a large black hole evaporates down to the threshold mass where the dimensionality of spacetime transitions from $(3 + 1)$-D to $(2 + 1)$-D. In the latter, DDRs will result from microscopic black holes formed below the $(2 + 1)$-D -$(1 + 1)$-D transition scale.

5 Other Cosmological Implications of Dimensional Reduction

There is a rich set of dimensional reduction phenomenology for both early universe and quantum cosmology. In addition to the impact on PBH formation, a lower-dimensional UV limit can impact gravitational wave astronomy. The notion that the post-Big Bang universe may have been (effectively) $(1 + 1)$-D has been discussed in several contexts [38, 39, 92]. As expansion continues and the temperature drops, the universe transitions to $(2 + 1)$-D and eventually the present day $(3 + 1)$-D. Some have considered that the present-day universe may even be spawning a Hubble-scale fourth spatial dimension, whose tension offers an alternative to dark energy as a source of accelerated expansion [105–107].

5.1 $(1 + 1)$-D PBH Population Distributions

The results discussed in this subsection arose from one of the many collaborations with my colleagues Piero Nicolini and Bernard Carr, along with the very bright rising researcher Thanos Tzikas. This arose from a conversation between the four of us over lunch at the 2017 Karl Schwarzschild Meeting in Frankfurt, Germany. Therein, we explored the consequences of primordial black hole evolution from a dimensionally-reduced $(1 + 1)$-D early universe [92], from which one can study the distribution of PBHs that nucleate during inflation.

In order to preserve known cosmological inflation features, it is presumed that the overall dimensionality is $(3 + 1)$-D, but the PBH production occurs subject to the lower dimensional case. Adhering to the $(3 + 1)$-D Mann-Ross and Bousso-Hawking approaches, the quantum mechanical decay of the universe into a spacetime with a black hole occurs only for a Planckian-scale cosmological constant, $\Lambda \sim 10^{133}$ eV m^{-3}. This is roughly 10^{16} times smaller than the value at the beginning of inflation, implying no $(3 + 1)$-D PBH formation. The lack of a Planck scale in $(1 + 1)$-D, however, suggests that black holes can be produced even without Planckian values of the cosmological constant. Since't Hooft has argued that, in order for the UV regime to be described by Planck-scale physics, spacetime must remain $(3 + 1)$-D, but that gravitational self-renormalizability need only require $(1 + 1)$-D [48]. This phase must last at least as long as inflation, lest the distribution of PBH become too dilute to incur any observational consequences [92].

It can be shown that there are essentially two categories into which the production rate $\Gamma(M)$ falls, as a function of the cosmological constant Λ. For sub-Planckian black holes with $M \ll \sqrt{|\Lambda|} \ll M_{\text{Pl}}$, the temperature and production rate are $T \approx M$, $\Gamma_{\text{L}} \approx (\Lambda/M^2)^{1/2} \ll 1$. When $\sqrt{|\Lambda|} < M \leq \sqrt{2|\Lambda|}$, the temperature scales as $T \approx \sqrt{|\Lambda|}$, $\Gamma_{\text{L}} \approx 1$. As M approaches $\sqrt{\Lambda}$, however, the temperature tends toward zero. That is, for $M = \sqrt{|\Lambda|}(1 + \epsilon)$ with $\epsilon \ll 1$, one finds $T \approx \sqrt{\epsilon|\Lambda|}/(\sqrt{2}\pi) \ll \sqrt{|\Lambda|}$ and $\Gamma_{\text{L}} \approx \epsilon^{-1/2} \gg 1$. A Nariai black hole is formed when $M = \sqrt{|\Lambda|}$, and the associated temperature and rate are $T = 0$, $\Gamma_{\text{N}} = (\mu_0/\sqrt{|\Lambda|})$. No black holes form when $M < \sqrt{|\Lambda|}$, which instead yields a naked singularity. The rate $\Gamma(M)$ is thus cut off below the peak at $M = \sqrt{|\Lambda|}$ and has a power-law suppression $M \gg \sqrt{|\Lambda|}$.

The initial collapse fraction $\beta(M)$ of PBHs of mass M scales linearly with $\Gamma(M)$. In $(1 + 1)$-D, the PBH density decreases as a^{-1}, which implies the fraction of the Universe's total (radiation) density in PBHs decreases exponentially as $\rho_{\text{PBH}}/\rho_{\text{R}} \sim a^{-1} \sim \exp(-t\sqrt{|\Lambda|})$.

The current PBH mass function is (see Fig. 1)

$$\frac{dn}{dM} \sim \frac{\beta(M)}{M^2} \sim \frac{\Gamma(M)}{M^2} \tag{31}$$

The density $\rho(M)$ of PBHs of mass M is just M^2 times this and thus scales as $\Gamma(M)$. As with the production rate, there is a cut off below the peak at $M = \sqrt{|\Lambda|}$ and a power-law decline for $M \gg \sqrt{|\Lambda|}$. There would be an exponential reduction if the black holes formed before the end of inflation, in which case there would be no observational signatures from the PBHs.

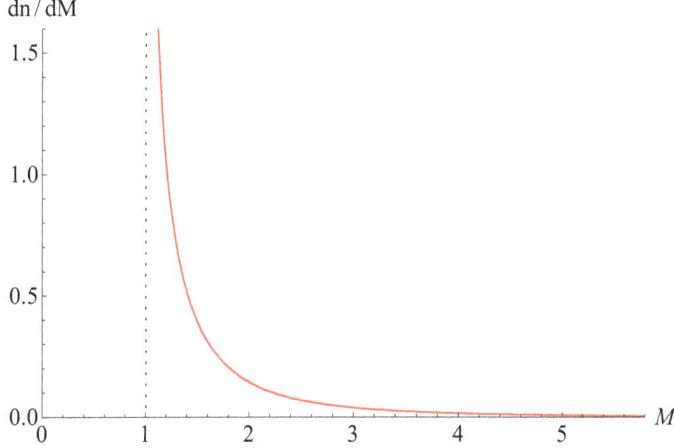

Fig. 1 The initial PBH mass function dn/dM with the Nariai black hole contribution omitted. At the present epoch, this function may reduce to a delta-function at $M \sim M_{\mathrm{CMB}}$ due to Hawking evaporation [92]. Reprinted from Tzikas et al. [92] with permission, ©IOP Publishing Ltd and Sissa Medialab. All rights reserved

5.2 Primordial Gravitational Waves

As previously mentioned, this avenue of research with Dejan Stojkovic marked my debut in the field of dimensional reduction and its possible observational consequences, and the results discussed in this subsection can be found in our papers [38, 39]. Just as extra dimensional quantum gravity signatures might be observed in astrophysical phenomena, the possibility that dimensionally-reduction could have similar fingerprints was compelling.

Specifically, the nature of lower-dimensional gravity also provides constraints on stochastic or primordial gravitational waves (PGWs). According to standard cosmology, PGWs will be generated in the pre/post-inflationary regime due to quantum fluctuations of the spacetime manifold. It has long been understood that there are no local gravitational degrees of freedom in a $(2+1)$-dimensional universe, and consequently no gravitational waves can be produced. This has immediate consequences for PGWs, in that there must be a maximum frequency for such phenomena. It would also be implicitly related to some dimensional transition scale L_2, where the three-dimensional spacetime percolates to a four-dimensional one. At this point, the temperature of the universe will have some value T_*, whose value can be surmised from L_2 [38, 39] (Fig. 2).

At temperatures below the $2D \to 3D$ cross over scale, $T < T_*$, a standard $(3+1)$-D FRW cosmology is assumed with the usual radiation- and matter-dominated eras. PGWs can be produced at times $t_* < t_0 = H^{-1}$. The co-moving entropy per volume of the universe at temperature T_* as a function of the scale $a(t)$ is

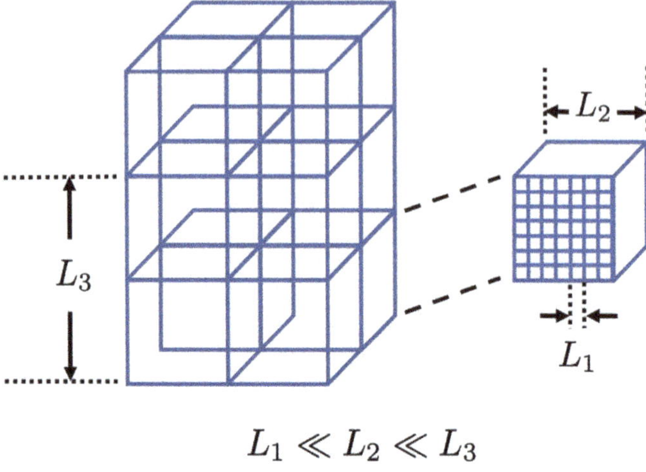

Fig. 2 A demonstration of scale-dependent dimensional reduction in the Vanishing Dimensions scenario. The fundamental scales L_d define the limit of physics described by a $(d + 1)$-D spacetime

$$S \sim g_S(T)a^3(t)T^3, \tag{32}$$

where g_S is the effective number of degrees of freedom at temperature T.

Requiring conservation of entropy over the lifetime of the universe, one finds

$$g_S(T_*)a^3(t_*)T_*^3 = g_S(T_0)a^3(t_0)T_0^3 \tag{33}$$

with $T_0 = 2.728$ K the CMB temperature and $g_S(t_0) \sim 100$ [108].

A gravitational wave produced with frequency f_* at time t_* in the past will be cosmologically-redshifted redshifted to its present-day value $f_0 = f_* \frac{a(t_*)}{a(t_0)}$ by the factor [108]

$$f_0 \simeq 8.0 \times 10^{-17} f_* \left(\frac{100}{g_S(T_*)} \right)^{\frac{1}{3}} \left(\frac{1\,\text{TeV}}{T_*} \right) \tag{34}$$

The original frequency f_* is bounded by the universe's horizon size at time t_*, *i.e.* $f_* \sim \lambda_*^{-1} \sim H_*^{-1}$, which can be related to the temperature T_* by noting that

$$H_*^2 = \frac{8\pi^3 g_* T_*^4}{90 M_{\text{Pl}}^2} \tag{35}$$

,during the radiation-dominated era. Setting $g_* \sim 10^2$, the threshold frequency of PGWs at the dimensional transition can be thus calculated as [38]

$$f_{L_2} \approx 1.67 \times 10^{-4} \left(\frac{T_*}{\text{TeV}} \right) \text{Hz} \tag{36}$$

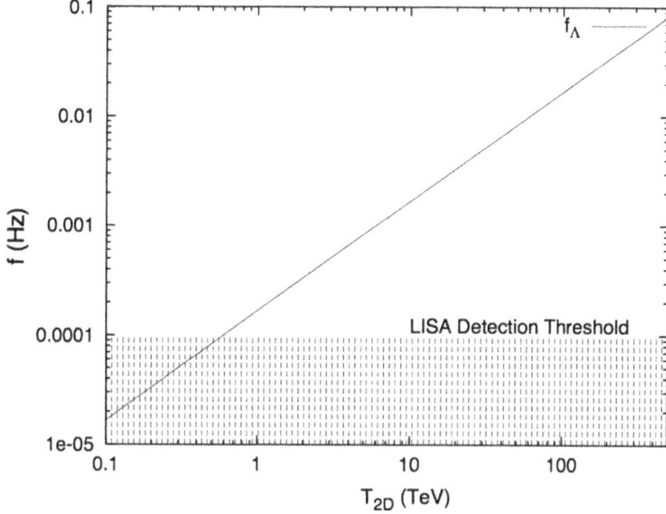

Fig. 3 The frequency threshold for PGWs that were produced when the Universe was at temperature T_*. The hatched region is outside the sensitivity cutoff of LISA [38]. Reprinted from Mureika and Stojkovic [38] with permission, ©American Physical Society. All rights reserved

For a temperature $T_* = 1$ TeV, the frequency is $f_* \sim 10^{-4}$ Hz. Although this is orders of magnitude smaller than the seismic limit of $f \sim 40$ Hz for ground-based GW interferometer experiments [3], it falls precisely at the threshold of LISA's sensitivity range. Indeed, the latter observatory is expected to probe a variety of early-universe phenomenology from the 100 GeV–1000 TeV period [109]. The threshold PGW frequency as a function of dimensional transition temperature $L_2 = T_{2D}$ is shown in Fig. 3. Detection of such a GW frequency cut-off would be a tell-tale signature of dimensional reduction.

6 (1 + 1)-D Entropic Gravity

My longest research collaboration has been with Robb Mann of the University of Waterloo, where I began my graduate studies in 1993. The many topics we have studied include equivalence principle violating neutrino oscillations [52–54], entropic gravity [57, 90], and black hole chemistry [58, 59]. Of these, the latter two have included implications of dimensional reduction. In this section, I review the more fundamental results from lower-dimensional entropic gravity.

In 2011, Erik Verlinde proposed a simple yet novel approach to emergent gravity, derived from entropic principles. Appropriately dubbed "entropic gravity" [110], the scenario provided a straightforward derivation of Newtonian gravitation from holographic principles. The underlying assumption requires information entropy to be encoded on a holographic screen surrounding some mass M. As two masses interact, the laws of thermodynamics dictate that the entropy must be constrained in such a way that leads to the familiar inverse-square law.

Formally, an entropic force is one that forces a system to increase its entropy, and can be defined via the work-energy theorem as [110]

$$F_{\text{entropic}} \equiv -\frac{\Delta E}{\Delta x} = -T\frac{\Delta S}{\Delta x}. \tag{37}$$

The holographic screen surrounding a mass M contains information necessary to communicate interaction processes to external test masses m. In d spatial dimensions, the number of bits contained on the screen of area

$$A_\Sigma(r) = \frac{2\pi^{d/2}}{\Gamma(\frac{d}{2})}\, r^{d-1} \tag{38}$$

can be written

$$N = \frac{A_\Sigma(r)}{\ell_P^{d-1}} \tag{39}$$

Here, ℓ_P is a some length, and ℓ_P^{d-1} is the minimal area containing a single bit. Given an even distribution of the system's energy $E = Mc^2$ across the bits, one finds

$$E = \frac{N}{2}k_b T \tag{40}$$

The "temperature" of the system is thus

$$k_B T = \frac{2Mc^2\ell_P^{d-1}}{A_\Sigma(r)} \tag{41}$$

When a test mass m approaches M by a distance Δx, information will begin to flow between the two, and the amount of entropy transferred will be

$$\Delta S = 2\pi k_B \frac{\Delta x}{\lambda_C}\ , \tag{42}$$

where λ_C is the Compton wavelength. Combining Eqs. (41, 42) with Eq. (37), one obtains the familiar higher-dimensional form of Newton's law

$$F_{\text{entropic}} = -2\pi^{1-\frac{d}{2}}\Gamma\left(\frac{d}{2}\right)\frac{c^3\ell_P^{d-1}}{\hbar}\frac{Mm}{r^{d-1}} = -\frac{G_d Mm}{r^{d-1}} \tag{43}$$

where the associated gravitational constant is

$$G_d = 2\pi^{1-\frac{d}{2}} \, \Gamma\left(\frac{d}{2}\right) \frac{c^3 \ell_P^{d-1}}{\hbar} \tag{44}$$

The appearance of the Planck length reinforces the fundamental quantum nature of gravitation.

The $(1+1)$-D case is recovered when $d = 1$ in the above expressions [57]. Specifically, one finds an entropic force

$$F_{\text{entropic}} = -T\frac{\Delta S}{\Delta x} = -mM\left(\frac{4\pi c^3}{N\hbar}\right). \tag{45}$$

which can be cast in the form

$$F_1 = G_1 mM, \tag{46}$$

where

$$G_1 = \left(\frac{4\pi c^3}{N\hbar}\right). \tag{47}$$

This is the expected form of the one-dimensional Newtonian gravitational force, which can also be derived from a linear potential. Note that this imposes a constraint on the number of bits N, namely

$$N = 4\pi c^3 / G_1 \hbar \tag{48}$$

From Eq. (44), it follows that the coupling is $G_1 = 2\pi c^3/\hbar$ when $d = 1$. Given that Eq. (39) yields $A = 2$ (the two boundary points of the horizon), one finds

$$N = 1. \tag{49}$$

implying that there is one bit of gravitational information at every point in the space-time. Compared with the higher-dimensional cases of entropic gravity, the number of bits contained at a point in one-dimensional space is constant and independent of its proximity to the holographic screen of M. This result is particularly striking, in that it offers a non-Maxwellian origin for e.g. the temperature of the black hole. Here, the temperature is associated with internal degrees of freedom located at a point, as opposed to vibrational degrees of freedom.

7 Lower Dimensional Corpuscular Gravity

A new description of quantum gravity—the corpuscular picture—was recently introduced as an alternative description of black holes [13], and has since been applied to a wide variety of astrophysical and cosmological considerations [102, 111–116]. This particular avenue of research—from which several others have since spawned—was done with the collaboration of long-time colleague Roberto Casadio and his then-graduate student Andrea Giusti during my sabbatical visit to the University of Bologna in 2018.

In this novel formulation, black holes are modeled to be Bose-Einstein condensates (BEC) consisting of a very large number of soft gravitons, whose typical energy scale ε_G is smaller than most relevant systems.

The energy ε_G of each graviton is derived quantum mechanically as

$$\varepsilon_G \simeq \hbar/\lambda_G \simeq M_{Pl}\,\ell_{Pl}/R \,, \tag{50}$$

and the number of gravitons therein is obtained from holographic constraints,

$$N_G \simeq \frac{M^2}{M_{Pl}^2} = \frac{R^2}{\ell_{Pl}^2} \,, \tag{51}$$

The total energy stored in the gravitational field is

$$N_G\,\varepsilon_G \simeq N_G\,\frac{\hbar}{\lambda_G} \simeq M\,\frac{R_H}{R} \simeq \frac{G_N\,M^2}{R} \,, \tag{52}$$

where $N_G\,\varepsilon_G \ll M$ for any regular astrophysical object of size $R \gg R_H = 2\,G_N\,M$.

In the case of a black hole $(R = R_H)$, the gravitons form a critical BEC of wavelength

$$\lambda_G \simeq R_H \simeq \sqrt{N_G}\,\ell_{Pl} \,. \tag{53}$$

In the corpuscular picture, black hole evaporation is a result of $2 \to 2$ scattering of gravitons within condensate. The depletion rate is

$$\hbar\,\Gamma \sim N_G^2\,\alpha_G^2\,\varepsilon_G \simeq M_{Pl}/\sqrt{N_G} \,, \tag{54}$$

and the interaction strength is $\alpha_G \sim \varepsilon_G^2/M_{Pl}^2 \sim N_G^{-1}$. The number of gravitons in the BEC will deplete as [13]

$$\dot{N}_G \simeq -\Gamma \simeq -\frac{1}{\sqrt{N_G}\,\ell_{Pl}} \,. \tag{55}$$

which gives

$$\dot{M} \simeq \frac{M_{Pl}\,\dot{N}_G}{\sqrt{N_G}} \simeq -\frac{M_{Pl}^3}{\ell_{Pl}\,M^2} \,, \tag{56}$$

from Eq. (51), reproducing the Hawking evaporation profile.

Since the goal of the corpuscular picture is to understand aspects of quantum gravity, one might ask how dimensional reduction fits in [102]. In one spatial dimension, the gravitational binding energy is

$$U_1 \simeq G_1 \, M^2 \, R. \tag{57}$$

which arises in the weak-field limit of the $(1+1)$-D metric, and is furthermore consistent with the equations of motion of a test particle in the space-time [56]. The number of gravitons in the system can be obtained by noting

$$U_1 \simeq \frac{MR}{R_1}, \tag{58}$$

which follows from (10). Accordingly, the number is obtained as

$$N_1 \simeq \frac{U_1}{\varepsilon_{\mathrm{G}}} \simeq \frac{R^2 M^2}{\ell_{\mathrm{Pl}}^2 \, M_{\mathrm{Pl}}^2}. \tag{59}$$

When $R = R_1$ (i.e. a black hole), this reduces to $N_1 \simeq 1$, which is the expected holographic behaviour in $d = 1$. In contrast, this result arises in the standard three-dimensional case only if the object is purely Planckian, i.e.

$$N_1 \sim \frac{M^2}{M_{\mathrm{Pl}}^2} \simeq \frac{R^2}{\ell_{\mathrm{Pl}}^2} \simeq 1. \tag{60}$$

This leads to the conclusion that $d = 1$ black holes are manifestly Planck-scale objects in the BEC scenario. The result also offers the compelling conclusion that one-dimensional black holes consist of a single graviton, indicating that such objects are fundamentally quantum mechanical objects. That is, quantum gravity is a natural result of dimensional reduction.

8 Generalized Uncertainty Principle Black Holes

This section summarizes one of my most comprehensive (and quite fun) investigations into dimensional reduction, which arrives at this end through a rather novel mechanism involving the Generalized Uncertainty Principle (GUP). It was on a May afternoon in London in 2014 where Piero Nicolini, Bernard Carr, and I gathered in a small conference room at Queen Mary, University of London, and cracked open this fresh new approach to dimensional reduction [117]. Later on, I would open this world up to my aspiring undergraduate research student Heather Mentzer, who embraced the extension to Reissner-Nordström black holes [118]. At the time of this writing, she is pursuing her doctoral studies in Physics at the University of California, Santa Cruz.

One of the most problematic features of GR is the presence of the singularity. A common goal of quantum gravity is the removal of such an unwanted side-effect. A popular resolution to this is the introduction of a minimal length, below which space-time is unresolvable and cannot be probed experimentally. Indeed, a minimal length is naturally encoded in the Heisenberg Uncertainty Principle (HUP), so such a consideration is not unwarranted. Approaches that incorporate a minimal length include non-commutative geometry inspired frameworks [11], self-completeness [119], loop quantum gravity [6], and the generalized uncertainty principle [12, 117, 120–124]. The interested reader is directed to [125] for a review of these and other applications of minimal length physics.

A spacetime is said to be "self-complete" if the minimal length corresponds to the (modified) Compton wavelength. Above this threshold, the fundamental length scale corresponds to the Schwarzschild radius. This implies a minimum length and mass whose values are on the order of ℓ_{Pl} and M_{Pl}, respectively. Figure 4 shows the critical behaviour at the Planck scale, which defines either the smallest possible black hole, or alternatively the largest particle.

An open question remains, however: can black holes with mass below the Planck scale exist? This question has been explored by numerous authors, including in the form of the Black Hole Uncertainty Principle (BHUP) [126]. This approach suggests that there could exist black holes with mass beneath the Planck scale but radius of order the Compton scale R_C rather than Schwarzschild scale R_S.

It is doubtful that the usual forms of R_S and R_C hold down to the Planck scale. Indeed, it has been argued that as one approaches the Planck point from the left, the HUP should be supplanted by the aforementioned *Generalized* Uncertainty Principle (GUP) of the form

$$R'_C = \frac{1}{M}\left[1 + \alpha\left(\frac{M}{M_{Pl}}\right)^2\right] \quad (M < M_{Pl}), \tag{61}$$

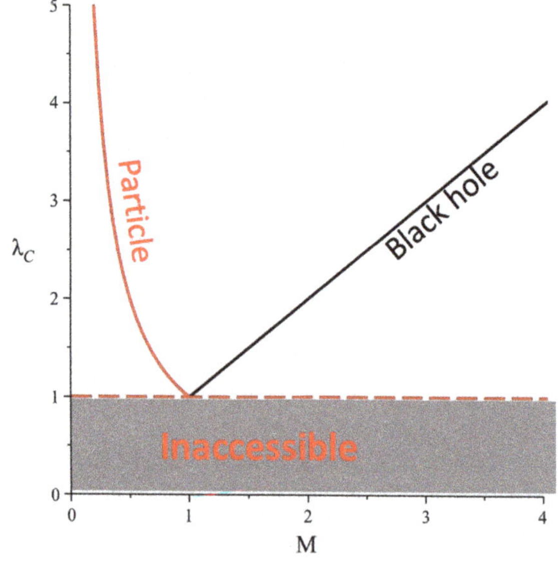

Fig. 4 Self-completeness of the Schwarzschild black hole, showing the intersection of the Compton and Schwarzschild curves. The Planck scale defines a minimal length phase transition between black holes and particles, below which spacetime cannot be resolved

where α is a dimensionless constant. Conversely, approaching the critical point from the right-hand curve modified the Schwarzschild relation to a Generalized Event Horizon (GEH) of the form [117, 126]

$$R'_{\rm S} = 2GM \left[1 + \frac{\beta}{2} \left(\frac{M_{\rm Pl}}{M} \right)^2 \right] \quad (M > M_{\rm Pl}), \tag{62}$$

where β is another dimensionless constant. Such behaviour is a feature of the N-portrait model [119].

The similarity of Eqs. (61) and (62) strongly hints that there is a fundamental connection between the quantum (particle) scale and the classical (black hole) scale, yielding a smooth transition between the two expressions. This feature, known as the Black Hole Uncertainty Principle (BHUP) [126–128], is embodied by a unified expression for the Compton wavelength in the sub-Planckian regime ($M < M_{\rm Pl}$) and the Schwarzschild radius in the super-Planckian limit ($M > M_{\rm Pl}$). This is a natural consequence of combining the GUP and the GEH.

The most natural expression embodying the BHUP is thus

$$r_{\rm CS} = \frac{\beta}{M} + 2GM = 2G \left(M + \frac{\beta}{2GM} \right), \tag{63}$$

which smoothly connects the sub- and super-Planckian regimes. The above equation can be reinterpreted as a modification of the ADM mass for a black hole, such that

$$M_{\rm ADM} = M + \frac{\beta M_{\rm Pl}^2}{2M} \tag{64}$$

where the replacement $G = M_{\rm Pl}^{-2}$ has been made.

From a relativistic perspective, this leads to a modification of the Schwarzschild metric function of the form

$$f(r) = 1 - \frac{2M_{\rm ADM}}{M_{\rm Pl}^2 r}, \tag{65}$$

There are several important consequences of this change. First, the scaling of the horizon is different in the sub-, meso-, and super-Planckian limits. From (65), it is trivial to show the horizon is [117]

$$r_{\rm H} = \frac{2M}{M_{\rm Pl}^2} + \frac{\beta}{M} \tag{66}$$

and so

$$M \gg M_{\rm Pl} \implies r_{\rm H} \approx \frac{2M}{M_{\rm Pl}^2} \tag{67}$$

$$M \approx M_{\mathrm{Pl}} \implies r_{\mathrm{H}} \approx \frac{2+\beta}{M_{\mathrm{Pl}}} \tag{68}$$

$$M \ll M_{\mathrm{Pl}} \implies r_{\mathrm{H}} \approx \frac{\beta}{M} \tag{69}$$

The change of mass dependence of the characteristic scale is important. For large (classical) objects, one associates the gravitational scale $R_S \sim M$. For small (quantum) objects, on the other hand, the length scale is Compton, $R_C \sim \frac{1}{M}$. As mentioned in previous sections, the overlap between the Compton wavelength and the $(1 + 1)$-D gravitational scale strongly suggests that the latter lives in a naturally quantum setting. That is, the GUP and BHUP naturally imply a sub-Planckian dimensional reduction (Fig. 5).

The thermodynamics of these GUP-Schwarzschild black holes can be studied from the surface gravity approach, giving

$$T = \frac{M_{\mathrm{Pl}}^2}{8\pi M (1 + \beta M_{\mathrm{Pl}}^2 / 2M^2)}. \tag{70}$$

For the mass regimes, this behaves as

$$M \gg M_{\mathrm{Pl}} \implies T \approx \frac{M_{\mathrm{Pl}}^2}{8\pi M} \tag{71}$$

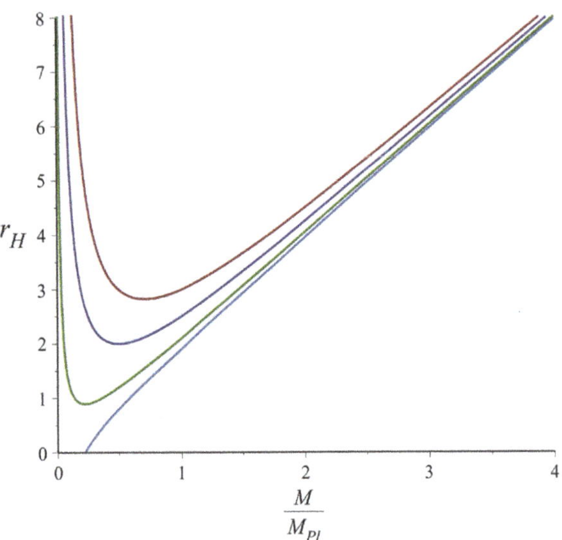

Fig. 5 The GUP-Schwarzschild horizon radius as a function of mass M/M_{Pl} for $\beta = 1$ (red, top), $\beta = 0.5$ (blue, second down), $\beta = 0.1$ (green, third down) and $\beta = -1$ (blue, bottom). Note that for $\beta < 0$. The horizon is defined only for $M > \sqrt{|\beta|/2} M_{\mathrm{Pl}}$ and vanishes when $M = \sqrt{|\beta|/2} M_{\mathrm{Pl}}$ [117]. Reprinted from Carr et al. (2015). https://doi.org/10.1007/JHEP07(2015)052 under CC BY, ©The Authors

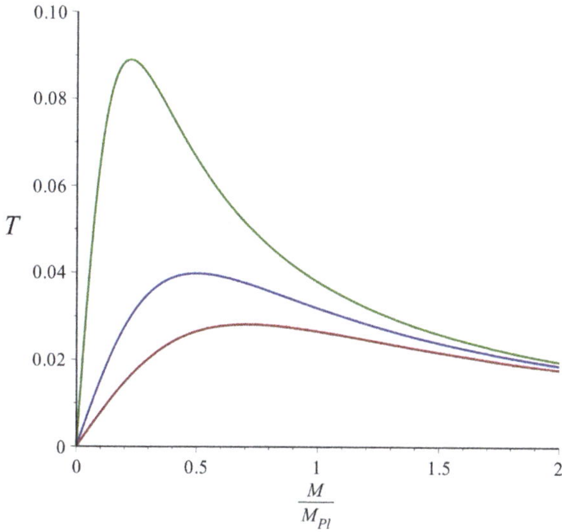

Fig. 6 The GUP-Schwarzschild Hawking temperature as a function of mass M/M_{Pl} for $\beta = 1$ (red, bottom), $\beta = 0.5$ (blue, middle), and $\beta = 0.1$ (green, top). The temperature reaches a maximum at (72), at which point the heat capacity becomes positive and the black hole cools completely $T = 0$ and $M = 0$, leaving no remnant [117]. Reprinted from Carr et al. (2015). https://doi.org/10.1007/JHEP07(2015)052 under CC BY, ©The Authors

$$M \approx M_{\text{Pl}} \implies T \approx \frac{M_{\text{Pl}}}{8\pi(1 + \beta/2)} \tag{72}$$

$$M \ll M_{\text{Pl}} \implies T \approx \frac{M}{4\pi\beta}. \tag{73}$$

Again, the super-Planckian limit is the expected Hawking temperature of a Schwarzschild black hole, which the sub-Planckian case exhibits the profile of a $(1+1)$-D black hole. These are displayed in Fig. 6.

Lastly, the entropy is (see Fig. 7)

$$S = \int_{M_0}^{M} \frac{dM'}{T(M')} = 4\pi k \left(\frac{M^2}{M_{\text{Pl}}^2} - \frac{M_0^2}{M_{\text{Pl}}^2} + \beta \ln \frac{M}{M_0} \right) \tag{74}$$

where $M_0 < M_{\text{Pl}}$ is some lower bound of integration. When $M \gg M_{\text{Pl}}$, the Hawking result for a $(3+1)$-D black hole is recovered. For $M \ll M_{\text{Pl}}$, the logarithmic term dominates, and matches the entropy expression for a $(1+1)$-D black hole [75].

The modification to black hole characteristics introduced by the GUP demonstrates that a smooth transition can be made between the super- and sub-Planckian regimes, representing the classical and quantum worlds. More importantly, this has very strong implications for quantum gravity. Instead of some new theory replacing

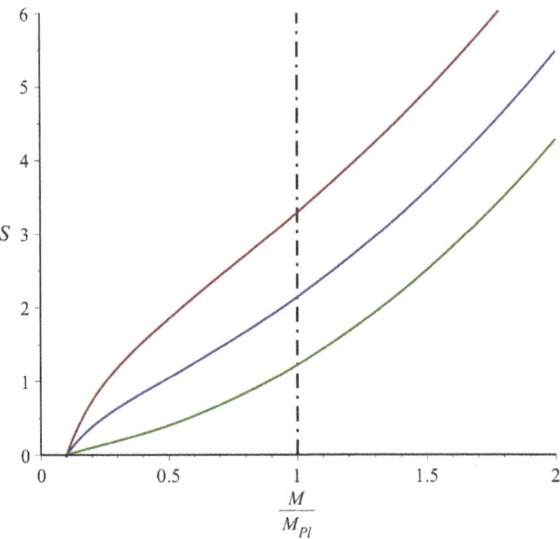

Fig. 7 The GUP-Schwarzschild black hole entropy as a function of mass M/M_{Pl} for $\beta = 1$ (red, top), $\beta = 0.5$ (blue. middle), and $\beta = 0.1$ (green. bottom). The Planck scale is marked by the vertical hatched line ($M = M_{\text{Pl}} = 1$). For $M > M_{\text{Pl}}$, the entropy scales in the standard fashion as $S \sim M^2$, while for $M < M_{\text{Pl}}$ it goes as $S \sim \log(M/M_0)$ and vanishes at $M = M_0 = 0.1$ in this case [117]. Reprinted from Carr et al. (2015). https://doi.org/10.1007/JHEP07(2015)052 under CC BY, ©The Authors

General Relativity in the quantum limit, one finds instead that General Relativity is the consistent theory, and it is the effective spacetime dimension that changes.

The GUP modification can be extended to Reissner-Nordström (RN) and Kerr black holes [118]. In the first case, the GUP-RN metric is

$$f(r) = 1 - \frac{2M}{M_{\text{Pl}}^2 r}\left(1 + \frac{\beta}{2}\frac{M_{\text{Pl}}^2}{M^2}\right) + \frac{\alpha_e n^2}{M_{\text{Pl}}^2 r^2}. \tag{75}$$

where α_e is the fine structure constant, and n is the number of elementary charges in the black hole. The inner and outer horizons are

$$r_{\pm} = \left(\frac{M}{M_{\text{Pl}}^2} + \frac{\beta}{2M}\right)\left(1 \pm \sqrt{1 - \frac{n^2\alpha_e M^2}{M_{\text{Pl}}^2\left(\frac{\beta}{2} + \frac{M^2}{M_{\text{Pl}}^2}\right)^2}}\right), \tag{76}$$

which lead to the super-Planckian and sub-Planckian behaviours

$$M \gg M_{\text{Pl}} \implies r_+ \approx \frac{2M}{M_{\text{Pl}}^2}, \quad r_- \approx \frac{\alpha_e n^2}{2M}, \tag{77}$$

$$M \ll M_{\mathrm{Pl}} \quad \Longrightarrow \quad r_+ \approx \frac{\beta}{M}, \quad r_- \approx \frac{\alpha_e n^2 M}{2 M_{\mathrm{Pl}}^2}. \tag{78}$$

The inner and outer horizons reach their maximum (minimum) values at the same critical mass

$$M = M_{\mathrm{crit}} \equiv \sqrt{\beta/2} \, , \tag{79}$$

for which the horizons become

$$r_+ = [\sqrt{2\beta} + \sqrt{2\beta - n^2 \alpha_e}\,]\,\ell_{\mathrm{Pl}} \, .$$
$$r_- = [\sqrt{2\beta} - \sqrt{2\beta - n^2 \alpha_e}\,]\,\ell_{\mathrm{Pl}} \, . \tag{80}$$

The outer (r_+) and inner (r_-) horizons are shown in Fig. 8 for $\beta = 2$.

The GUP-RN black hole can hold a maximum charge

$$n \le n_{\max} = [\sqrt{2\beta/\alpha_e}\,] \, , \tag{81}$$

with the square brackets denoting the integer value (rounded down), since fractional charge is not allowed. For $\beta = 2$, this is $n_{\max} = 23$.

The qualified "maximum" for the charge, however, is a bit misleading. This is the largest charge the black hole can hold and remain a RN black hole. If n exceeds this value, a very interesting effect sets in. The GUP-RN black hole bifurcates into two distinct solutions, separated by a mass gap. Note that horizons can only exist when

$$M_{\mathrm{ADM}} \ge \sqrt{\alpha_e} n M_{\mathrm{Pl}} \quad \Rightarrow \quad \frac{M^2}{M_{\mathrm{Pl}}} - \sqrt{\alpha_e} n M + \frac{\beta}{2} M_{\mathrm{Pl}} \ge 0. \tag{82}$$

The quadratic form of the right-hand inequality has a discriminant

$$\xi \equiv \alpha_e n^2 - 2\beta, \tag{83}$$

which leads to three possible ranges for the mass. First, when $\xi < 0$ horizons exist for all values of M, implying that sub-Planckian black holes can form and are not extremal. The minimum horizon can be found as

$$0 = \frac{dr_+}{dM} = \left(\frac{dr_+}{dM_{\mathrm{ADM}}}\right)\left(\frac{dM_{\mathrm{ADM}}}{dM}\right). \tag{84}$$

where $dr_+/dM_{\mathrm{ADM}} \ne 0$ for $M_{\mathrm{ADM}} > \sqrt{\alpha_e} n \, M_{\mathrm{Pl}}$. The second derivative term can readily be calculated as

$$\frac{dM_{\mathrm{ADM}}}{dM} = 1 - \frac{\beta}{2} \frac{M_{\mathrm{Pl}}^2}{M^2}, \tag{85}$$

and indicates there is a minimal (non-extremal) horizon for M_{crit} in Eq. 79. Requiring $\xi > 0$ leads to the following two constraints on the allowed masses,

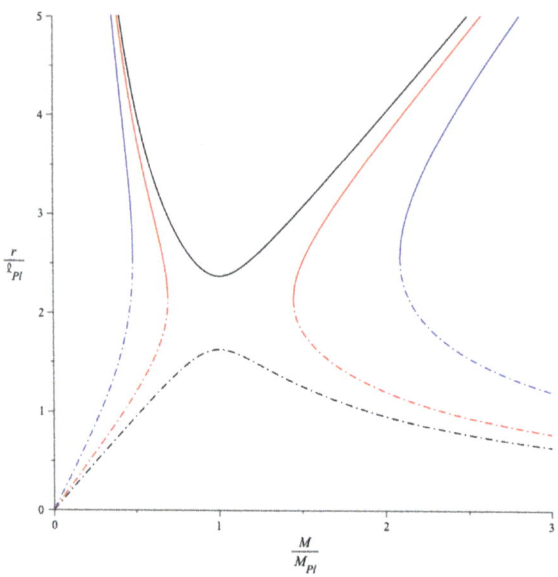

Fig. 8 The outer (solid) and inner (dash-dot) horizon size for a GUP-RN black hole with $\beta = 2$, showing the pre- ($n = 23$, black) and post- ($n = 25$, blue, and $n = 30$, red) phase transition behaviour. For $n < 23$, the outer horizon behaves identically to the GUP-Schwarzschild case, while for $n > 23$ they reach their maximum (left) or minimum (right) size at M_1 and M_2, respectively. The mass gap between M_1 and M_2 in which no black holes form is clearly visible [118]. Reprinted from Carr et al. Self-complete and GUP-modified charged and spinning black holes. Eur. Phys. J. C 80, 1166 (2020). https://doi.org/10.1140/epjc/s10052-020-08706-0 under CC BY, ©The Authors

$$M < M_1 = \frac{M_{\text{Pl}}}{2}\left(\sqrt{\alpha_e}\,n - \sqrt{\alpha_e n^2 - 2\beta}\right)$$

$$M > M_2 = \frac{M_{\text{Pl}}}{2}\left(\sqrt{\alpha_e}\,n + \sqrt{\alpha_e n^2 - 2\beta}\right), \tag{86}$$

where $M_1 < M_{\text{crit}} < M_2$. The range $M \in (M_1, M_2)$ represents a mass gap in which no black holes form. The values M_1 and M_2 correspond to extremal solutions.

This bifurcation behaviour is also apparent in the GUP-RN temperature profile, which is

$$T = \frac{M_{\text{Pl}}^2 \sqrt{M_{\text{ADM}}^2 - \alpha_e n^2 M_{\text{Pl}}^2}}{2\pi\,(M_{\text{ADM}} + \sqrt{M_{\text{ADM}}^2 - \alpha_e n^2 M_{\text{Pl}}^2})^2}. \tag{87}$$

This is displayed in Fig. 9. For $n > n_{\text{max}}$, the mass gap is visible and there are thus two distinct configurations for sub- and super-Planckian black holes. The latter resemble the standard RN solution, and will evaporate to a zero-temperature, but non-zero mass, remnant. The former, however, will form within the given mass range and will evaporate to a zero temperature, zero mass end point.

A Kerr black hole of mass M and angular momentum J is described by the metric

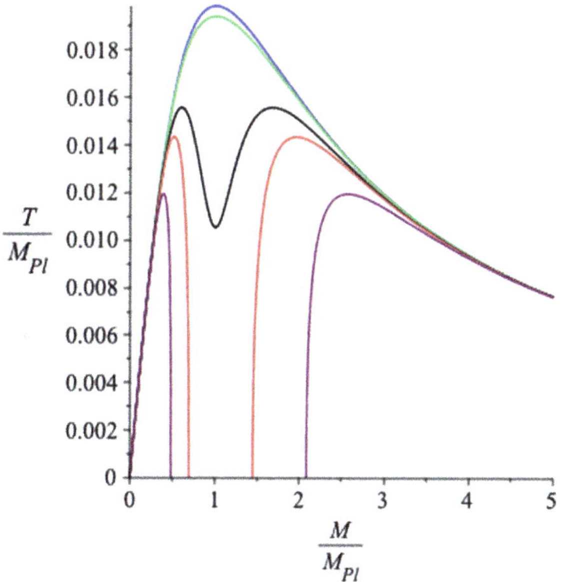

Fig. 9 The GUP-RN black hole Hawking temperature for $\beta = 2$ and $n = 10$ (blue), $n = 16$ (green), $n = 23$ (black), $n = 25$ (red) and $n = 30$ (purple). Temperatures for $n > 23$ vanish at M_1 (left) and M_2 (right) defined by Eq. (86). The $T = 0$ endpoints in the super-Planckian regime are stable charged remnants [118]. Reprinted from Carr et al. Self-complete and GUP-modified charged and spinning black holes. Eur. Phys. J. C 80, 1166 (2020). https://doi.org/10.1140/epjc/s10052-020-08706-0 under CC BY, ©The Authors

$$ds^2 = \left(1 - \frac{r_S r}{\rho^2}\right) dt^2 - \frac{\rho^2}{\Delta} dr^2$$
$$- \rho^2 \, d\theta^2 - \left(r^2 + a^2 + \frac{r_S r a^2}{\rho^2} \sin^2 \theta\right) \sin^2 \theta \, d\phi^2$$
$$+ \frac{2 r_S r a \sin^2 \theta}{\rho^2} \, dt \, d\phi, \tag{88}$$

and

$$r_S = 2M/M_{\mathrm{Pl}}^2, \quad a = \frac{J}{M}, \quad \rho^2 = r^2 + a^2 \cos^2 \theta, \quad \Delta = r^2 - r_S r + a^2. \tag{89}$$

The horizons are derived from the condition $\Delta = 0$, which gives

$$r_\pm = \frac{M}{M_{\mathrm{Pl}}^2} \left(1 \pm \sqrt{1 - \frac{a^2 M_{\mathrm{Pl}}^4}{M^2}}\right). \tag{90}$$

258   J. Mureika

For a quantum Kerr black hole, the angular momentum is quantized as $J = n$, and the extremal solutions obey

$$M = \sqrt{n}\, M_{\mathrm{Pl}} \tag{91}$$

A GUP-Kerr black hole solution can be obtained with the replacement $M \to M_{\mathrm{ADM}}$, as with the GUP-Schwarzschild and GUP-RN cases. The horizons are thus [118]

$$r_{\pm} = \frac{M_{\mathrm{ADM}}}{M_{\mathrm{Pl}}^2} \pm \sqrt{\frac{M_{\mathrm{ADM}}^2}{M_{\mathrm{Pl}}^4} - \frac{n^2}{M_{\mathrm{ADM}}^2}} \tag{92}$$

which are shown in Fig. 10. For $n < 2\beta$, the minimum values of the horizon occur when $M = M_{\mathrm{crit}} = \sqrt{\beta/2}$, and are respectively

$$r_{\pm} = [\sqrt{2\beta} \pm \sqrt{2\beta - n^2/(2\beta)}\,]\,\ell_{\mathrm{Pl}}. \tag{93}$$

An extremal solution ($r_+ = r_-$) occurs at $n = 2\beta$. Beyond this value of n, a mass gap appears for the GUP-Kerr solutions, as it did for the GUP-RN case. Horizons exist for $M > M_+$ and $M < M_-$, where the latter are defined as

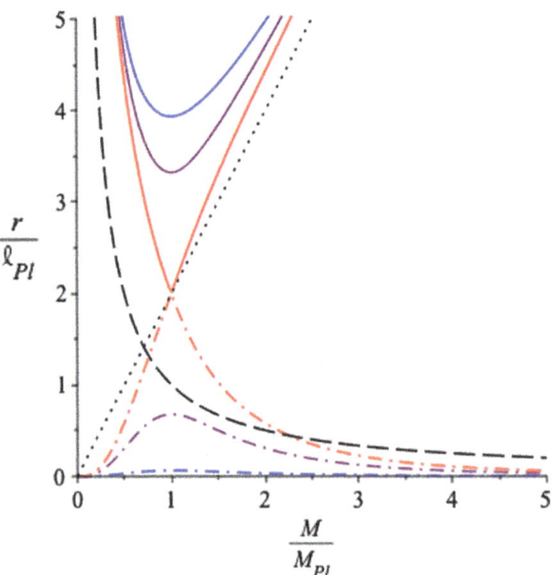

Fig. 10 The GUP-Kerr black hole outer (solid) and inner (dash-dot) horizons for $\beta = 2$ and $n = 1$ (blue), 3 (purple) and 4 (red). The $n = 4$ curves possess a discontinuity at $M = M_{\mathrm{Pl}}$ and $r = 2\ell_{\mathrm{Pl}}$, corresponding to a phase transition. The dashed line represents the Schwarzschild radius, while the dotted curve is the Compton wavelength [118]. Reprinted from Carr et al. Self-complete and GUP-modified charged and spinning black holes. Eur. Phys. J. C 80, 1166 (2020). https://doi.org/10.1140/epjc/s10052-020-08706-0 under CC BY, ©The Authors

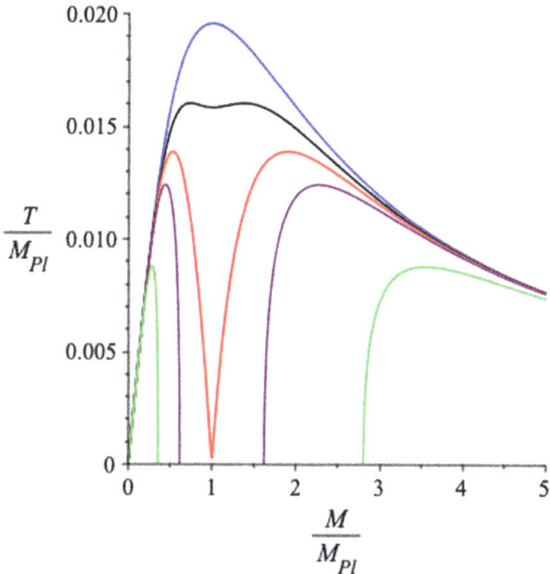

Fig. 11 The GUP-RN black hole Hawking temperature for the GUP-Kerr for $\beta = 2$ and $n = 1$ (blue), 3 (black), 4 (red), 5 (purple) and 10 (green). The critical value $n = 4$ shows a phase transition behaviour similar to that for the extremal GUP-RN black hole, above which there is also a mass gap. The $T = 0$ configurations in the super-Planckian regime are stable spinning remnants [118]. Reprinted from Carr et al. Self-complete and GUP-modified charged and spinning black holes. Eur. Phys. J. C 80, 1166 (2020). https://doi.org/10.1140/epjc/s10052-020-08706-0 under CC BY, ⓒThe Authors

$$M_\pm = \frac{M_{\text{Pl}}}{2}\left(\sqrt{n} \pm \sqrt{n - 2\beta}\right),\tag{94}$$

For $\beta = 0$, the standard extremal Kerr solution is recovered. Those for $n > 2\beta$, however, split into either sub-Planckian or super-Planckian black holes.

The temperature of the GUP-Kerr solution is shown in Fig. 11, as calculated from the expression

$$T = \frac{1}{4\pi}\frac{r_+ - r_-}{r_+^2 + (n/M_{\text{ADM}})^2}\tag{95}$$

and is shown in Fig. 11.

As with the GUP-Schwarzschild black hole, the sub-Planckian characteristics of both the GUP-RN and GUP-Kerr black holes show dimensional reduction behaviour [118]. Furthermore, there is an interesting feature of the inner horizons for the standard RN and Kerr metrics. The horizon for far-from-extremal RN black holes scales as $1/M$ like a $(1 + 1)$-D black hole. This is similarly true for the outer horizon in the sub-Planckian part of the BHUP solution [126]. The fascinating coincidence again suggests that dimensional reduction occurs interior to the horizon of any black hole, and as such perpetuates the idea that the black hole itself is manifestly a lower-dimensional, quantum object.

9 Conclusions and Future Directions

The notion that the road to quantum gravity lies not in a new theory to supplant GR, but rather an Einstein theory in a lower spacetime dimension is a fascinating proposition. The coincidences between gravitation in $(1 + 1)$-D and quantum mechanics are unmistakable. The characteristic length scales of each are functions of $1/M$, and furthermore the fundamental constants for each—$\hbar$ and G_1—are dual to one another.

Experimental evidence in the present golden age of black holes will be crucial in furthering our understanding of these connections. Matching dark matter distributions to lower-dimensional PBH distributions will help pave the road ahead. Potential gravitation wave echoes off effectively lower-dimensional, near horizon structures might provide new windows into lower dimensional quantum gravity.

Whatever the future brings, we can confidently say that we live in the most interesting times for gravitation. Antonio Aurilia dedicated his life to exploring the quantum frontiers of gravitation, and we are humbled to continue his quest for the ultimate theory.

Acknowledgements First and foremost, I would like to thank my long-time collaborator and friend Piero Nicolini for inviting me to contribute to this volume, and also for giving me the opportunity to meet Antonio before his passing. The three of us met over coffee in Claremont, California for only an hour, but I left feeling that I knew him well. Antonio was a brilliant, caring, and friendly individual, who is most deserving of this tribute. I would also like to again acknowledge and profusely thank my numerous collaborators over the years who have worked with me on a variety of aspects of dimensional reduction, and without whom I would not have been able to explore this subject matter in such depth. I thank Dejan Stojkovic for introducing me to the concept of dimensional reduction and vanishing dimensions in 2010, as well as my many other talented colleagues that include: Niayesh Afshordi, Bernard Carr, Roberto Casadio, Rogerio Cavalcanti, Antonia Frassino, Andrea Giugno, Andrea Giusti, Maximiliano Isi, Luciano Manfredi, Robert Mann, Heather Mentzer, Octavian Micu, and Athanasios Tzikas for their past, ongoing, and future collaborations in this exciting area. JM is a KITP Scholar at the Kavli Institute for Theoretical Physics and is supported in part by the National Science Foundation under Grant No. NSF PHY-1748958.

References

1. A. Einstein, Ann. der Phys. **49**, 769–822 (1916)
2. F. W. Dyson, A. S. Eddington, C. Davidson, Phil. Trans. Royal Soc. A **220**, 291–333 (1920)
3. The LIGO Scientific Collaboration. https://www.ligo.org/
4. The Event Horizon Telescope collaboration. http://www.eventhorizontelescope.org/
5. S. B. Giddings, Found. Phys. **43**, 115 (2013)
6. C. Rovelli, Living Rev. Rel. **1**, 1 (1998)
7. A. Eichhorn, Asymptotically safe gravity. arXiv:2003.00044 [gr-qc]

8. S. De, T. P. Singh, A. Varma, Int. J. Mod. Phys. D **28**(14), 1944003 (2019)
9. S. W. Hawking, Nature **248**, 30 (1974)
10. S. W. Hawking, Comm. Math. Phys. **43**, 199 (1975)
11. P. Nicolini, Int. J. Mod. Phys. A **24**, 1229–1308 (2009)
12. M. Isi, J. Mureika, P. Nicolini, JHEP **11**, 139 (2013)
13. G. Dvali, C. Gomez, Black Hole's information group. arXiv:1307.7630 [hep-th]
14. G. Dvali, C. Gomez, Eur. Phys. J. C **74**, 2752 (2014)
15. G. Dvali, C. Gomez, Phys. Lett. B **719**, 419 (2013)
16. G. Dvali, C. Gomez, Phys. Lett. B **716**, 240 (2012)
17. G. Dvali, C. Gomez, Fortsch. Phys. **61**, 742 (2013)
18. S. Giddings, Phys. Rev. D **88**, 024018 (2013)
19. S. Giddings, Phys. Rev. D **88**, 064023 (2013)
20. S. Giddings, Y. Shi, Phys. Rev. D **89**, 124032 (2014)
21. S. Giddings, Phys. Rev. D **90**, 124033 (2014)
22. D. F. Torres, S. Capozziello, G. Lambiase, Phys. Rev. D **62**, 104012 (2000)
23. C. Rovelli, F. Vidotto, Int. J. Mod. Phys. D **23**(12), 1442026 (2014)
24. O. Lunin, S. D. Mathur, Nucl. Phys. B **623**, 342 (2002)
25. S. D. Mathur, Fortsch. Phys. **53**, 793 (2005)
26. B. Guo, S. Hampton, S. D. Mathur, JHEP **1807**, 162 (2018)
27. V. Cardoso, P. Pani, Nat. Astron. **1**(9), 586 (2017)
28. J. Abedi, H. Dykaar, N. Afshordi, Phys. Rev. D **96**(8), 082004 (2017)
29. Q. Wang, N. Afshordi, Phys. Rev. D **97**(12), 124044 (2018)
30. N. Oshita, N. Afshordi, Phys. Rev. D **99**, 4, 044002 (2019)
31. M. A. Green, J. W. Moffat, Phys. Lett. B **784**, 323 (2018)
32. J. R. Mureika, Phys. Lett. B **789**, 88–92 (2019)
33. S. Carlip, Int. J. Mod. Phys. D **25**(12), 1643003 (2016)
34. J. Ambjorn, Loll. Phys. Rev. **D72**, 064014 (2005)
35. P. Horava, Phys. Rev. Lett. **102**, 161301 (2009)
36. L. Modesto, P. Nicolini, Phys. Rev. D **81**, 104040 (2010)
37. L. Anchordoqui, D. C. Dai, M. Fairbairn, G. Landsberg, D. Stojkovic, Mod. Phys. Lett. A **27**, 1250021 (2012)
38. J. R. Mureika, D. Stojkovic, Phys. Rev. Lett. **106**, 101101 (2011)
39. J. R. Mureika, D. Stojkovic, Phys. Rev. Lett. **107**, 169002 (2011)
40. L. A. Anchordoqui, D. C. Dai, H. Goldberg, G. Landsberg, G. Shaughnessy, D. Stojkovic, T. J. Weiler, Phys. Rev. D **83**, 114046 (2011)
41. D. Stojkovic, Rom. J. Phys. **57**, 210 (2012)
42. N. Afshordi, D. Stojkovic, Phys. Lett. B **739**, 117–124 (2014)
43. C. Callan, J. Maldacena, Nucl. Phys. B **513**, 198 (1998)
44. N. Constable, R. Myers, O. Tafjord, Phys. Rev. D **61**, 106009 (2000)
45. G. Calcagni, Phys. Rev. Lett. **104**, 251301 (2010)
46. G. Calcagni, Phys. Lett. B **697**, 251 (2011)
47. M. Arzano, G. Calcagni, D. Oriti, M. Scalisi, Phys. Rev. D **84**, 125002 (2011)
48. G. t' Hooft, Conf. Proc. **C930308**, 284 (1993)
49. D. V. Shirkov, Part. Nucl. Lett. **7**, 625 (2010)
50. D. Stojkovic, Mod. Phys. Lett. A **28**, 1330034 (2013)
51. *General Relativity and Relativistic Astrophysics: Proceedings of the 5th Canadian Conference*, ed. by R.B. Mann, R.G. McLenaghan (World Scientific Press, 1994)
52. J. R. Mureika, R. B. Mann, Phys. Lett. B **368**, 112 (1996)
53. J. R. Mureika, R. B. Mann, Phys. Rev. D **54**, 2761 (1996)
54. J. R. Mureika, Phys. Rev. D **56**, 2408 (1997)
55. A. Aurilia, R. S. Kissack, R. B. Mann, E. Spallucci, Phys. Rev. D **35**, 2961 (1987)
56. R. B. Mann, A. Shiekh, L. Tarasov, Nucl. Phys. B **341** (1990) 134
57. R. B. Mann, J. R. Mureika, Phys. Lett. B **703**, 167–171 (2011)
58. A. M. Frassino, R. B. Mann, J. R. Mureika, Phys. Rev. D **92** 124069 (2015)

59. A. M. Frassino, R. B. Mann, J. R. Mureika, JHEP **11**, 112 (2019)
60. J. R. Gott, M. Alpert, Gen. Relativ. Gravit. **16**, 243 (1984)
61. J. D. Brown, M. Henneaux, C. Teitelboim, Phys. Rev. D **33**, 319 (1986)
62. B. Reznik, Phys. Rev. D **45**, 2151 (1992)
63. M. Bañados, C. Teitelboim, J. Zanelli, Phys. Rev. Lett. **69**, 1849 (1992)
64. M. Bañados, M. Henneaux, C. Teitelboim, J. Zanelli, Phys. Rev. D **48**, 1506 (1993)
65. B. Reznik, Phys. Rev. D **51**, 1728 (1995)
66. V. Husain, Phys. Rev. D **52**, 6860 (1995)
67. G. Clément, Phys. Rev. D **68**, 024032 (2002)
68. S. Carlip, *Quantum gravity in 2+1 dimensions* (Cambridge University Press, Cambridge, UK, 1998)
69. L. Ortíz, M. P. Ryan, Gen. Relativ. Gravit. **39**, 1087 (2007)
70. P. Collas, Am. J. Phys. **45**, 833 (1977)
71. S. Rajeev, Phys. Lett. B **113**, 146 (1982)
72. R. Jackiw, in *Quantum Theory of Gravity*, ed. by S. Christensen (Adam Hilger, Bristol, UK, 1984), pp. 403
73. D. Cangemi, R. Jackiw, Phys. Rev. Lett. **69**, 233 (1992)
74. J. D. Brown, M. Henneaux, C. Teitelboim, Phys. Rev. D **33**, 319 (1986)
75. R. B. Mann, S. M. Morsink, A. E. Sikkema, T. G. Steele, Phys. Rev. D **43**, 3948 (1991)
76. R. B. Mann, A. Sikkema, Class. Quant. Grav. **8**, 219 (1991)
77. J. D. Christensen, R. B. Mann, Class. Quant. Grav. **9**, 1769 (1992)
78. R. B. Mann, S. F. Ross, Class. Quant. Grav. **10**, 1405–1408 (1993)
79. K. C. K. Chan, R. B. Mann, Class. Quant. Grav. **10**, 913 (1993)
80. R. B. Mann, Nucl. Phys. B **418**, 231 (1994)
81. M. O. Katanaev, W. Kummer, H. Liebl, Nucl. Phys. B **486**, 353 (1997)
82. R. B. Mann, D. Robbins, T. Ohta, Phys. Rev. Lett. **82**, 3738 (1999)
83. R. Balbinot, A. Fabbri, V. P. Frolov, P. Nicolini, P. Sutton, A. Zelnikov, Phys. Rev. D **63**, 084029 (2001)
84. R. Balbinot, A. Fabbri, P. Nicolini, P. J. Sutton, Phys. Rev. D **66**, 024014 (2002)
85. D. Grumiller, W. Kummer, D. V. Vassilevich, Phys. Rept. **369**, 327 (2002)
86. D. Grumiller, R. Meyer, Turk. J. Phys. **30**, 349 (2006)
87. D. Grumiller, R. Jackiw, Liouville gravity from Einstein gravity (2007). arXiv:0712.3775 [gr-qc]
88. J. Mureika, P. Nicolini, Eur. Phys. J. Plus **128**, 78 (2013)
89. R. Casadio, R. T. Cavalcanti, A. Giugno, J. Mureika, Phys. Lett. B **760**, 36 (2016)
90. J. R. Mureika, R. B. Mann, Mod. Phys. Lett. A **26**, 171–181 (2011)
91. J. R. Mureika, Phys. Lett. B **716**, 171 (2012)
92. A. Tzikas, P. Nicolini, J. Mureika, B. Carr, JCAP **1812**, 033 (2018)
93. S. Carlip, in *Foundations of Space and Time: Reflections on Quantum Gravity*, ed. by J. Murugan, A. Weltman, G. F. R. Ellis (Cambridge University Press, Cambridge, UK, 2012) pp. 69–84
94. S. Carlip, in *AIP Conference Proceedings*, ed. by D. Oriti (AIP, Melville, NY, USA, 2009), pp.72
95. S. Carlip, Class. Quant. Grav. **35**(1), 014001 (2018)
96. S. Carlip, Class. Quant. Grav. **35**(1), 014001 (2018)
97. S. Carlip, Phys. Rev. Lett. **120**(10), 101301 (2018)
98. J. Abajian, S. Carlip, Phys. Rev. D **97**(6), 066007 (2018)
99. D. I. Podolskiy, A. O. Barvinsky, R. Lanza, JCAP **05**, 048 (2021)
100. J. Mureika, P. Nicolini, Phys. Rev. D **84**, 044020 (2011)
101. J. Mureika, P. Nicolini, Eur. J. Phys. Plus **128**, 78 (2013)
102. R. Casadio, A. Giusti, J. Mureika, Mod. Phys. Lett. A **34**(22), 1950174 (2019)
103. R. Casadio, O. Micu, J. Mureika, Phys. Scripta **97**, 12, 125304 (2022)
104. H. Alnes, F. Ravndal, I. K. Wehus, J. Phys. A **40**, 14309 (2007)
105. G. R. Dvali, G. Gabadadze, M. Porrati, Phys. Lett. B **485**, 208 (2000)

106. C. de Rham, G. R. Dvali, S. Hofmann, J. Khoury, O. Pujolas, M. Redi, A. J. Tolley, Phys. Rev. Lett. **100**, 251603 (2008)
107. N. Afshordi, G. Geshnizjani, J. Khoury, JCAP **0908**, 030 (2009)
108. M. Maggiore, Phys. Rept. **331**, 283–367 (2000)
109. The LISA Collaboration. https://www.elisascience.org/
110. E. Verlinde, JHEP **1104**, 029 (2011)
111. G. Dvali, C. Gomez, JCAP **01**, 023 (2014)
112. R. Casadio, F. Kuhnel, A. Orlandi, JCAP **1509**, 002 (2015)
113. M. Cadoni, R. Casadio, A. Giusti, M. Tuveri, Phys. Rev. D **97**(4), 044047 (2018)
114. M. Cadoni, R. Casadio, A. Giusti, W. Mück, M. Tuveri, Phys. Lett. B **776**, 242 (2018)
115. R. Casadio, A. Giugno, A. Giusti, Phys. Rev. D **97**(2), 024041 (2018)
116. A. Giusti, Int. J. Geom. Methods Mod. Phys. **16**, 03, 1930001 (2019)
117. B. J. Carr, J. Mureika, P. Nicolini, JHEP **07**, 052 (2015)
118. B. Carr, H. Mentzer, J. Mureika, P. Nicolini, Eur. Phys. J. C **80**, 12, 1166 (2020)
119. G. Dvali, S. Folkerts, C. Germani, Phys. Rev. D **84**, 024039 (2011)
120. A. Kempf, G. Mangano, R. B. Mann, Phys. Rev. D **52**, 1108 (1995)
121. R. J. Adler, D. I. Santiago, Mod. Phys. Lett. A **14**, 1371 (1999)
122. R. J. Adler, P. Chen, D. I. Santiago, Gen. Rel. Grav. **33**, 2101 (2001)
123. P. Chen, R. J. Adler, Nucl. Phys. Proc. Suppl. **124**, 103 (2003)
124. R. J. Adler, Am. J. Phys. **78**, 925 (2010)
125. S. Hossenfelder, Living Rev. Rel. **16**, 2 (2013)
126. B. J. Carr, Springer Proc. Phys. **170**, 159–167 (2016)
127. B. J. Carr, L. Modesto, I. Premont-Schwarz, Generalized Uncertainty Principle and Self-dual Black Holes (2011). arXiv:1107.0708 [gr-qc]
128. M. J. Lake, B. J. Carr, Int. J. Mod. Phys. D **28**, 1930001 (2019)

Minimal Length Scale in Quantum Gravity via Riemann and Weyl

Erick Aiken, Michael Bishop, and Douglas Singleton

Abstract General arguments about combining gravity with quantum mechanics lead to the conclusion that Nature should have an absolute minimal length scale. A minimal length scale breaks dilation or scale symmetry. In this essay, we break dilation symmetry by modifying the virial operator, the generator of dilations. The construction of the modified virial operator is connected to an approach to solving the long standing and unsolved Riemann hypothesis with overtones to Weyl's formulation of gauge theory.

1 Physical Motivation for a Minimal Length

Almost all fundamental interactions (electromagnetic, weak nuclear and strong nuclear) have been successfully quantized via the paradigm of quantum field theory. Only the oldest known interaction, gravity, has yet to yield a consistent quantum theory. Different attempts to quantize gravity (e.g. superstring theory, loop quantum gravity) all have a similar idea – that there is some minimum, absolute length scale. Standard quantum mechanics has no absolute, minimal length scale. Heisenberg's uncertainty principle extracts a cost for probing smaller length scales: a state with a smaller length scale, Δx, requires a larger momenta/energy scale, Δp. However, there is no limit to how small Δx can be if one accepts a large enough Δp.

With gravity things change. As one goes to smaller distances and higher momenta/higher energy, a microscopic black hole will form beyond some threshold. The event horizon of this black hole will grow as one continues to probe smaller length scales. The Schwarzschild radius r_S of the black hole's event horizon is directly proportional to the mass energy as $r_S \sim GM$ (we set $c = 1$). Replacing r_S by positional variance Δx and M by Δp leads to the scaling $\Delta x \sim G\Delta p$. Combining this

E. Aiken · D. Singleton (✉)
Physics Department, California State University Fresno, Fresno, CA 93740, USA
e-mail: dougs@mail.fresnostate.edu

M. Bishop
Mathematics Department, California State University Fresno, Fresno, CA 93740, USA
e-mail: mibishop@mail.fresnostate.edu

© The Author(s), under exclusive license to Springer Nature Switzerland AG 2025
P. Nicolini (ed.), *Touring the Planck Scale*, Fundamental Theories of Physics 219,
https://doi.org/10.1007/978-3-031-76066-2_12

short distance behavior with the standard quantum behavior one finds [1, 2]

$$\Delta x \sim \frac{\hbar}{\Delta p} + G\Delta p. \tag{1}$$

The arguments leading to (1) played a central role in some of Antonio Aurilia's final works [3, 4] which dealt exactly with the topic of introducing a minimal length scale in theories of quantum gravity. A sampling of other works which discuss the introduction of a minimal length in the context of quantum gravity can be found in [5–9]. This gravitationally modified uncertainty relationship has a minimum distance $\Delta x_{min} \sim 2\sqrt{\hbar G}$ coming from the competition between the two Δp terms in (1). Just as the Heisenberg uncertainty relationship, $\Delta x \Delta p \geq \hbar/2$, stems from the usual commutator $[\hat{x}, \hat{p}] = i\hbar$, the gravitational modified uncertainty relationship in (1) should be associated with a modified commutation relationship. In Ref. [10], a simple, phenomenological modification of the quantum commutator between position and momentum was proposed which yielded a modified uncertainty relationship like that in (1). This modified commutator was

$$[\hat{x}, \hat{p}] = i\hbar(1 + \beta \hat{p}^2), \tag{2}$$

where β is an arbitrary parameter assumed to come from quantum gravity. The proposed modified commutator in (2) is phenomenologically motivated to give the gravitationally modified uncertainty relationship sketched in (1). A modified uncertainty relationship like (1) or a modified commutator like (2) implies a scale $\Delta x_{\min}$ / β. Having an explicit scale, $\Delta x_{\min}$, breaks dilation/scale symmetry. The aim of this work is to give a more theoretical, less phenomenological motivation for a minimal length scale.

2 Riemann Hypothesis and Broken Dilation/Scale Symmetry

Broken dilation/scale symmetry must modify the generator of dilation/scale symmetry, the virial operator

$$\hat{H}_V = \frac{1}{2}(\hat{x}\hat{p} + \hat{p}\hat{x}). \tag{3}$$

Such a modified virial operator was recently proposed by Bender-Brody-Müller [11] with the goal of proving the Riemann hypothesis. Briefly, the Riemann hypothesis concerns the zeros of the Riemann zeta function which is given by

$$\zeta(z) = \sum_{n=1}^{\infty} \frac{1}{n^z} = \frac{1}{\Gamma(z)} \int_0^{\infty} \frac{t^{z-1}}{e^t - 1} dt \tag{4}$$

where $\Gamma(z) = \int_0^\infty e^{-t} t^{z-1} dt$ is the usual gamma function. Using the integral expression in (4) one obtains a reflection formula for the Riemann zeta function

$$\zeta(z) = 2^z \pi^{z-1} \sin(\pi z/2) \Gamma(1-z) \zeta(1-z). \tag{5}$$

Allowing z to be complex one finds that $\zeta(z)$ has trivial zeros at negative, even integers $z = -2n$ coming from the $\sin(\pi z/2)$ term.[1] But there are also non-trivial zeros at the complex values $z_n = \frac{1}{2} + it_n$ where $t_1 = 14.135, t_2 = 21.022, t_3 = 25.011\ldots,$ which appear to be restricted the line $z = \frac{1}{2} + it$. Riemann's hypothesis [13] is that all the non-trivial zeros of $\zeta(z)$ lie on this line whose real part is $\frac{1}{2}$ and whose imaginary parts are discrete. The Riemann hypothesis is one of the key, unsolved questions in number theory since the Riemann hypothesis encodes information about prime numbers.

The discrete nature of the imaginary part of the non-trivial zeros led Berry and Keating [14, 15] to propose that the non-trivial zeros are connected to some eigenvalue problem. This is anecdotally referred to as the Hilbert-Polyá program. Berry and Keating suggested $\hat{H}_V$ as the operator connected to the eigenvalue problem of the non-trivial zeros of the Riemann zeta function. Bender, Brody, and Müller modified the virial operator by conjugating $\hat{H}_V$ using $\hat{\Delta}$ as

$$\hat{\Delta}^{-1} \hat{H}_V \hat{\Delta} \quad \text{where} \quad \hat{\Delta} = 1 - e^{-i\hat{p}k}. \tag{6}$$

In Ref. [11], k was set to $\delta x/\hbar$ so that $\hat{\Delta}$ was a shift-difference operator with a shift of δx.[2] In [11], it was shown that the eigenfunctions of this modified virial operator, $\hat{\Delta}^{-1} \hat{H}_V \hat{\Delta}$, were the Hurwitz zeta functions

$$\zeta(z, x) = \sum_{n=0}^\infty \frac{1}{(x+n)^z}, \tag{7}$$

i.e. a shifted Riemann zeta function. By requiring the boundary condition $\zeta(z, 1) = 0$, Bender-Brody-Müller proposed a new possible avenue to proving the Riemann hypothesis since $\zeta(z, 1) = \zeta(z)$ is the Riemann zeta function.

Changing the virial operator via the shift-difference operator as in [11] does not introduce a minimal length scale. In [16], it was shown that the averaging operator

$$\hat{\Delta}' = \frac{1}{2}(e^{k\hat{p}} + e^{-k\hat{p}}) = \cosh(k\hat{p}) \tag{8}$$

---

[1] The apparently nonsensical result that the series in (4) sums to zero when $z = -2n$ comes about because for $z < 1$ $\zeta(z)$ is defined not by the series but via an analytic continuation of the series. The result that $\zeta(-1) = -\frac{1}{12}$, rather than $1 + 2 + 3 + 4 + \cdots = \infty$, is central to obtaining the number of space-time dimensions in bosonic string theory as 26 [12].

[2] Applying $\hat{\Delta}$ to a function $f(x)$ gives $\hat{\Delta} f(x) = f(x) - f(x - \delta x)$. Note in Ref. [11] both $\hbar$ and δx were set to 1.

leads to a modified virial operator, which leads to a modified commutation relationship, which does give a minimal length while still providing an approach to the Riemann hypothesis similar to that in [11]. Applying $\hat{\Delta}'$ to a function, $f(x)$, leads to an averaging over an *imaginary* interval,

$$\hat{\Delta}' f(x) = \frac{1}{2}(f(x + ik) + f(x - ik)). \tag{9}$$

This change from a shift-difference operator over a real interval to an averaging operator over an imaginary interval is reminiscent of the development of gauge theory by Weyl. In [17], Weyl tried to incorporate electromagnetism in general relativity by associating the electromagnetic interaction with dilation symmetry. It was eventually realized [18] that electromagnetism could be obtained not via a scale transformation with a real parameter, but by a phase transformation i.e. letting the parameter of the scale transformation become imaginary. This is what happens here when $\hat{\Delta} = 1 - e^{ik\hat{p}}$ is replaced by $\hat{\Delta}' = \cosh(k\hat{p})$. This averaging operator sends approximately constant functions to themselves and sends high frequency oscillatory functions to zero; it keeps only the wave functions we would expect to see with a minimum length scale.

3 Modified Commutator from Broken Dilation Symmetry

The minimal length scale in such a broken dilation symmetry can be seen as follows: the modified virial operator from [16] can be written as

$$(\hat{\Delta}')^{-1}\hat{H}_V\hat{\Delta}' = \hat{H}_V + i\hat{p}k\tanh(k\hat{p}). \tag{10}$$

This is done by "walking" the operator $(\hat{\Delta}')^{-1}$ through $\hat{H}_V$ using the standard commutator identity $[\hat{x}, f(p)] = i\partial_p(f(p))$ which yields the extra term, $i\hat{p}k\tanh(k\hat{p})$. To connect the modified virial operator, $\hat{H}_V + i\hat{p}k\tanh(k\hat{p})$, with a modified commutator, we define new position and momentum operators, $\hat{x}'$ and $\hat{p}'$, and an associated, new virial operator, $\hat{H}'_V = \frac{1}{2}(\hat{x}'\hat{p}' + \hat{p}'\hat{x}')$. The idea is to pick the new position and momentum operators so that the new virial operator equals the modified virial operator above in Eq. (10). To find these new position and momentum operators, $\hat{x}'$ and $\hat{p}'$, we begin by defining

$$\hat{x}' = ig(p)\partial_p; \quad \hat{p}' = pf(p). \tag{11}$$

In order to recover the normal position and momentum operators in the low momentum limit we need $g(0) = 1$ and $f(0) = 1$. Using the operators in (11) in the modified virial operator $\hat{H}'_V = \frac{1}{2}(\hat{x}'\hat{p}' + \hat{p}'\hat{x}')$ gives

$$\hat{H}'_V = \frac{i}{2}(\hat{x}'\hat{p}' + \hat{p}'\hat{x}') = \frac{i}{2}\left[2pf(p)g(p)\partial_p + f(p)g(p)\right] + \frac{i}{2}pg(p)f'(p). \quad (12)$$

The terms in the square brackets should be equal to $(\hat{x}\hat{p} + \hat{p}\hat{x}) = 2p\partial_p + 1$. This is accomplished by taking

$$g(p)f(p) = 1 \ \text{ or } \ g(p) = \frac{1}{f(p)}. \quad (13)$$

Using this expression of $g(p)$ we find that the last term in (12) becomes $\frac{ipf'(p)}{2f(p)}$. Equating this term with the last term in the shifted virial operator from (10) gives the following differential equation for $f(p)$

$$\frac{1}{2}\frac{f'(p)}{f(p)} = k\tanh(kp). \quad (14)$$

This differential equation is solved by $f(p) = \cosh^2(kp)$ and this in turn gives $g(p) = \frac{1}{f(p)} = \mathrm{sech}^2(kp)$. Note that these satisfy $f(0) = 1$ and $g(0) = 1$. Using these solutions in (11) gives the following modified position and momentum operators

$$\hat{x}' = i\,\mathrm{sech}^2(k\hat{p})\partial_p; \quad \hat{p}' = p\cosh^2(k\hat{p}), \quad (15)$$

One can directly check that using the modified position and momentum operators in (15) give the desired modified virial operator namely

$$\hat{H}'_V = \frac{1}{2}(\hat{x}'\hat{p}' + \hat{p}'\hat{x}') = \hat{H}_V + i\hat{p}k\tanh(k\hat{p}) = (\hat{\Delta}')^{-1}\hat{H}_V\hat{\Delta}', \quad (16)$$

where $\hat{\Delta}' = \cosh(kp)$. These new operators from (15) can be used to find the modified commutator as

$$[\hat{x}', \hat{p}'] = i\hbar(1 + 2k\hat{p}\tanh(k\hat{p})). \quad (17)$$

In the limit $kp \ll 1$, (17) becomes

$$[\hat{x}', \hat{p}'] \approx i\hbar(1 + 2k^2p^2). \quad (18)$$

This is equivalent to the phenomenological form in (2) with $\beta = 2k^2$. Thus the modified position and momentum operators of (15) lead to a modified commutator of (17) which leads to a minimal distance similar to the phenomenologically motivated commutator (2). In the limit $kp \gg 1$ (17) becomes

$$[\hat{x}', \hat{p}'] \approx i\hbar(1 + 2kp). \quad (19)$$

This linear dependence in the modification of the commutator for $kp \gg 1$ is weaker than the quadratic kp dependence of (2).

The modified virial operator $(\hat{\Delta}')^{-1}\hat{H}_V\hat{\Delta}'$ with the averaging operator, $\hat{\Delta}'$, recovers the Bender-Brody-Müller approach to the Riemann hypothesis. It can be shown [16] that the eigenfunctions of $(\hat{\Delta}')^{-1}\hat{H}_V\hat{\Delta}'$ are given by a shifted version of the Hurwitz-Euler eta function[3] $\eta(z, \frac{x}{2ik} + \frac{1}{2})$. By imposing the boundary condition that the Hurwitz-Euler eta function equal zero at $x = ik$ (i.e. $\eta(z, 1) = 0$) one finds a similar approach to the Riemann hypothesis as in Ref. [11]. This is because $\eta(z, 1)$, which is known as the Dirichlet eta function (i.e. a sign altering version of the Riemann zeta function), has the same non-trivial zeros as the Riemann zeta function. While both the modified virial operator from [11] (i.e. $(\hat{\Delta})^{-1}\hat{H}_V\hat{\Delta}$ with $\hat{\Delta} = 1 - e^{-i\hat{p}k}$) and that from [16] (i.e. $(\hat{\Delta}')^{-1}\hat{H}_V\hat{\Delta}'$ with $\hat{\Delta}' = \cosh(k\hat{p})$) give a potential approach to proving the Riemann hypothesis, it is only the latter which implies a modified commutation relationship with a minimal length.

The symmetry of the modified position and momentum operators from (15) is preserved with respect to the inner product

$$\langle\psi(p)|\phi(p)\rangle = \int\limits_{-\infty}^{\infty} \cosh^2(kp)\overline{\psi(p)}\phi(p)dp. \tag{20}$$

This inner product leads to the norm

$$\|\psi\|^2 = \int\limits_{-\infty}^{\infty} \cosh^2(kp)|\psi(p)|^2 dp. \tag{21}$$

In order for this norm to be finite and give normalizable states, one needs exponential suppression of wave functions at high momentum to counter the $\cosh^2(kp)$ factor. The suppression of high momenta states is exactly what would be expected when there is a minimal length scale. Exponential suppression of high momenta of this form may tame the divergent integrals which arise in canonical quantum gravity and in quantum field theory in general, allowing one to avoid renormalization.

In closing we note that not only does the modified viral operator $(\hat{\Delta}')^{-1}\hat{H}_V\hat{\Delta}'$ with $\hat{\Delta}' = \cosh(k\hat{p})$ lead to a minimal distance scale, as suggested by general arguments about quantum gravity, but it gives a family of operators in the vein of [11] which may give an alternative approach to the Riemann hypothesis.

Acknowledgements All authors thank the participants in the math-physics seminar where this project was conceived.

---

[3] The Hurwitz-Euler eta function, $\eta(z, x + 1) = \sum\limits_{n=0}^{\infty} \frac{(-1)^n}{(n+x+1)^z}$, is a sign altering Hurwitz zeta function.

References

1. M. Maggiore, Phys. Lett. B **304**, 65 (1993)
2. L. Susskind, J. Lindesay, *An introduction to black holes, information and the string theory revolution: The holographic universe* (World Scientific Publishing Company, Singapore, 2005)
3. A. Aurilia, E. Spallucci, Planck's uncertainty principle and the saturation of Lorentz boosts by Planckian black holes (2013). arXiv:1309.7186 [gr-qc]
4. A. Aurilia, E. Spallucci, Adv. High Energy Phys. **2013**, 531696 (2013)
5. R. J. Adler, D. I. Santiago, Mod. Phys. Lett. A **14**, 1371 (1999)
6. S. Das, E.C. Vagenas, Phys. Rev. Lett. **101**, 221301 (2008)
7. A. Ali, S. Das, E. Vagenas, Phys. Lett. B **678**, 497 (2009)
8. P. Nicolini, Int. J. Mod. Phys. A **24**, 1229 (2009)
9. S. Hossenfelder, Liv. Rev. Rel. **16**, 2 (2013)
10. A. Kempf, G. Mangano, R. B. Mann, Phys. Rev. D **52**, 1108 (1995)
11. C. M. Bender, D. C. Brody, M. P. Müller, Phys. Rev. Lett. **118**, 130201 (2017)
12. B. Zweibach, *A First Course in String Theory* (Cambirdge University Press, Cambridge, UK, 2004)
13. B. Riemann, *Über die Anzahl der Primzahlen unter einer gegebenen Grösse'*, Monatsberichte der Berliner Akademie (1859), in Gesammelte Werke, Teubner, Leipzig (1892), Reprinted by (Dover, New York, 1953)
14. M. V. Berry, in *Quantum Chaos and Statistical Nuclear Physics*, ed. by T. H. Seligman, H. Nishioka (Springer-Verlag, New York, USA, 1986)
15. M. V. Berry, J. P. Keating, in *Supersymmetry and Trace Formulae: Chaos and Disorder*, ed. by I.V. Lerner *et al.* (Kluwer Academic/Plenum, New York, USA, 1999)
16. M. Bishop, E. Aiken, D. Singleton, Phys. Rev. D **99**, 026012 (2019)
17. H. Weyl, Ann. der Physik **59**, 101–133 (1919)
18. H. Weyl, Zeit. für Physik **56**, 330–352 (1929)

Epilogue and Future Research

Three Good Reasons to Explore
the Planck Scale

Piero Nicolini

Abstract Despite many decades of intensive research, physics at the Planck scale remains an elusive phenomenon. However, a positive message emerges from this book: There are at least three good reasons to believe that research in quantum gravity will be successful in the near future.

1 Quantum Gravity Today

Interest in the unknown drove humanity towards goals that seemed unattainable. Examples include developing vaccines, creating electronic devices, transplanting organs, and landing on the Moon. In physics, recent achievements include the discovery of the Higgs particle [1], the observation of gravitational waves [2], and the imaging of black holes [3].

However, there are feats that are still beyond human capabilities. For example, the birth of the Universe, the fate of evaporating black holes, and the microscopic origin of their entropy are only partially understood. The crux of all the above phenomena[1] is the unknown nature of space-time at an extremely short distance, the so-called Planck length. At such a scale, particles are subject to strong gravitational effects,

---

[1] The list of unsolved problems in physics also includes further notable issues, e.g., the black hole information paradox, the trans-Planckian problem, the cosmological constant problem and the curvature singularity problem [4].

P. Nicolini (✉)
Dipartimento di Fisica, Università degli Studi di Trieste, Strada Costiera 11, 34151 Trieste, Italy
e-mail: piero.nicolini@units.it

Institut für Theoretische Physik, Johann Wolfgang Goethe-Universität, Max-von-Laue-Str. 1, 60438 Frankfurt am Main, Germany

Frankfurt Institute for Advanced Studies (FIAS), Ruth-Moufang-Str. 1, 60438 Frankfurt am Main, Germany

Center for Astro, Particle and Planetary Physics, New York University Abu Dhabi, P.O. Box 129188, Abu Dhabi, United Arab Emirates

  275
P. Nicolini (ed.), *Touring the Planck Scale*, Fundamental Theories of Physics 219,
https://doi.org/10.1007/978-3-031-76066-2_13

which require the unification of quantum mechanics with general relativity in the theory of quantum gravity.

Theoretical physicists have struggled for more than 50 years to formulate a consistent quantum theory of gravity. Despite these efforts, the final result is far from satisfactory. For example, (too) many candidate theories compete for the role of the universally accepted theory, but none of them is powerful enough to make concrete predictions [5]. Moreover, it is not even possible to test any of these candidate theories due to the lack of experimental data.

The situation is exacerbated by the weakness of gravity. Being the Planck mass—$M_P \sim 10^{19}$ GeV—about 17 orders of magnitude above the electroweak scale, there is a huge difference of intensity between gravity and the other fundamental interactions, known as "hierarchy problem". The scale discrepancy also posits the question of the existence of new particles between the two scales. In conclusion, if we do not understand what happens at the Planck scale, we cannot even understand what the fundamental constituents of matter are.

2 The Planck Scale and the Future

Armed with the results of this book, I would like to offer a positive message for the future. For example, we learned that the problem of quantum gravity is equivalent to understanding the Aurilia-Spallucci diagram in Fig. 1 [6]. This diagram shows that the Planck scale is naturally defined from the confluence of the Compton wavelength and the Schwarzschild radius. Even if the details of the curve intersection are in general not known, we learned that there exist at least two plausible scenarios for the Planck scale.

On the one hand, Dvali et al. showed that Planckian scattering leads to the formation of a state that, to a first approximation, coincides with a classical black hole [7].[2] The confluence of the two length scales therefore turns out to be non-analytic.[3] This is not surprising. Already in 2002, Aurilia and Spallucci came to the same conclusion by introducing the idea of the Planckion within the geometry democracy paradigm [9]. They further showed that the non-analytic behavior implies a saturation of Lorentz boosts at the Planck scale [10].

On the basis of the boost saturation, we can further speculate that the conventional form for the Generalized Uncertainty Principle (GUP)—i.e., that derived from string collisions in Minkowski space [11–13]—should only hold for sub-Planckian energies. For a non relativistic particle in one dimension, the GUP reads[4]

---

[2] This is consistent with the intuition that no particle can be compressed below its Schwarzschild radius [8].

[3] Dvali's paradigm is often referred to as "gravitational ultraviolet self-completeness" [7].

[4] For the GUP in more than one dimension see [14].

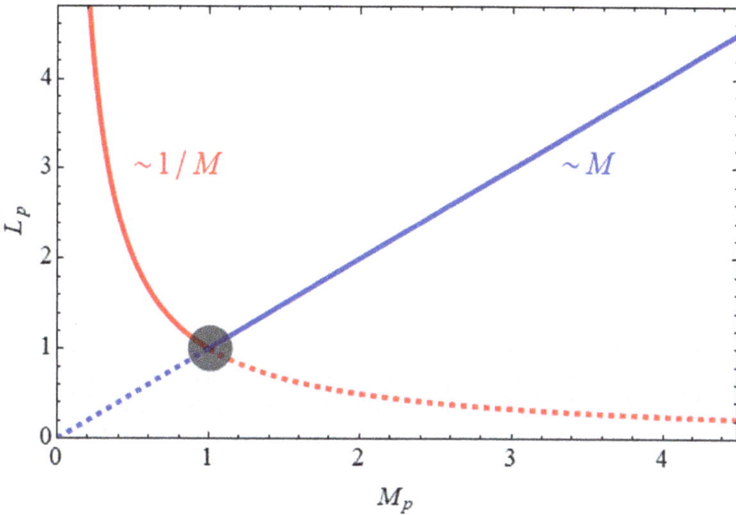

Fig. 1 The Planck scale is at the confluence of the Compton wavelength and the particle Schwarzschild radius. The Planck length is $L_P \simeq 1.616255 \times 10^{-35}$ m, the Planck mass is $M_P \simeq 1.22 \times 10^{19}$ GeV

$$\Delta x \Delta p \geq \frac{\hbar}{2} \left(1 + \beta(\Delta p)^2 + \beta \langle \mathbf{p}^2 \rangle \right).$$ (1)

Accordingly, the red branch of the curve in Fig. 1 should be replaced by the GUP curve:

$$\frac{1}{M} \rightarrow \frac{1}{M} + \beta M.$$ (2)

At trans-Planckian energies, however, quantum corrections to the Schwarzschild radius—blue branch of the curve in Fig. 1—may be difficult to determine because the GUP does not capture the effect of the boost saturation [10].[5]

On the other hand, the transition between the two curves in Fig. 1 could be smooth. Assuming the existence of a strong gravity regime, this smoothness can pave the way to the identification of particles and black holes [16]. Carr called this identification the black hole uncertainty principle correspondence [17]. Mureika showed that the correspondence is associated with a spontaneous dimensional reduction of gravity at the Planck scale [18, 19], in agreement with't Hooft's early proposal [20].

Another crucial aspect emerging from the present book is the narrowness of the Planck scale regime. In Fig. 1, it corresponds just to the shadowed spot around the point (1, 1), in Planck units. The narrowness is due to a particular aspect of gravity that makes it unique among all fundamental interactions: There is no quantum regime

---

[5] Other authors have noticed that the GUP required corrections. For instance, Aiken, Bishop and Singleton have proposed a modifications of the standard GUP based on the breaking of the dilation or scale symmetry [15].

by increasing the energy scale! Indeed, the Schwarzschild radius $\sim GM$ does not depend on $\hbar$. Following Dvali's trans-Planckian scattering analysis, a further increase of energy would imply the formation of a bigger, "more classical" black hole.[6] At the opposite end—the sub-Planckian regime—gravity is too weak to affect quantum particle dynamics (the Compton wavelength $\sim \hbar/M$ does not depend on G). In other words, gravity is either classical or vanishing. It can be quantum only at the Planck scale where both $\hbar$ and G enter the game.

Following this line of reasoning, Dvali has taken advantage of the narrowness of the quantum gravity regime to formulate a microscopic theory for black holes [21]. In particular, Dvali showed that an effective field theory for the graviton can lead to new predictions (e.g., quantum break time, inner entanglement), that elude more conventional semiclassical gravity descriptions [22]. In practice, Dvali's formulation has initiated a revival of effective theory methods, loosely grouped under the umbrella term "corpuscolar gravity" [23]. Similarly, Spallucci and Smailagic have formulated a Higgs mechanism for the production of "black particles" [24]. These objects are particle-black hole hybrids of Planckian mass, which have been studied since the dawn of string theory [25].

Another important point from this book is the analogy between spacetime and a matter condensate. Low energies (temperatures) govern the classical and semiclassical regimes. An increase in energy and temperature determines the breaking of long-range correlations between points on the manifold, i.e. the condensate. Near the critical temperature, the Planck temperature, these correlations are completely destroyed, spacetime fluctuates and its dimension becomes a quantum variable. The net result is that the quantum pregeometry at the Planck scale looks like a fractal. This is a geometric object with a continuous dimension greater than its topological dimension, a property that opens up a rich phenomenology. For example, Gaete and Helayël-Neto have considered a Lorentz violated higher derivative term in the Lagrangian of the electromagnetic field. The resulting electrodynamics supports the existence of new phenomena, such as vacuum birefringence [26]. In general, fields on a fractal geometry are characterized by non-locality and self-similarity, two features that make fractional electromagnetism an attractive framework for understanding the phenomenology of quantum gravity [27].

The final clue that emerges from this book is the possibility of testing quantum gravity. Usually, quantum gravity effects only show up under extreme conditions, such as in the early stages of the Universe or in the region around the center of a black hole. For this reason, direct detection of quantum gravity is virtually impossible, at least with today's technology. Against this background, one may be led to consider indirect quantum gravity effects, which may be visible in some detectable observables. For example, Altamirano et al. have considered the effects of a 5-dimensional quantum gravity phase on the power spectrum of the cosmological curvature perturbation. For more details see [28].

---

[6] It is "more classical", because its Hawking temperature is inversely proportional to the mass.

While indirect effects are certainly promising test beds, an alternative strategy for observing quantum gravity signals is to study systems with similar phenomenology but at a lower energy scale. Aurilia's intuition about the similarity between spacetime and a condensate is a first example of this strategy, now known as the analog gravity paradigm [29]. In recent decades, many researchers have identified superfluids and Bose-Einstein condensates as reliable systems for observing general relativity phenomena [30]. This is a huge opportunity: sonic black holes have already allowed the detection of Hawking radiation [31].[7] Balbinot and Fabbri have presented an overview of this topic and possible developments for the future [32].

In summary, the message of the book is that the Planck scale is the Everest of physics. It is not only the highest conceivable energy scale for particle physics, but also the regime in which gravity becomes strong. This implies that both particles and black holes have string properties and can be interpreted as two phases of matter. Like the summit of Mount Everest, the Planck scale is extremely difficult to reach. Nevertheless, there are three good reasons to explore it in the future.

1. Effective field theories: Quantum field theory methods at the Planck scale can uncover new features that elude geometric descriptions.
2. Phenomenology: There are many exotic effects at the Planck scale that cannot be predicted by conventional Minkowski space physics.
3. Observations: Indirect effects in astrophysics and analog gravity systems are already reliable test beds for quantum gravity.

I realize that we do not yet fully understand quantum gravity. Nevertheless, the tour of the Planck scale with the authors of this book was very instructive. We have made a lot of progress and we have to prepare for the next steps!

Acknowledgements PN is grateful to Jonas Mureika and Douglas Singleton for proofreading the text. PN is grateful to Athanasios Tzikas for the pictures in this chapter. The work of PN was partially supported by GNFM, Italy's National Group for Mathematical Physics.

References

1. A. Cho, Science **337**, 141, 6091 (2012)
2. B. Abbott, et al., Phys. Rev. Lett. **116** 6, 061102 (2016)
3. Astronomers Capture First Image of a Black Hole (2019)
4. G. 't Hooft, S. B. Giddings, C. Rovelli, P. Nicolini, J. Mureika, M. Kaminski, M. Bleicher, in *2nd Karl Schwarzschild Meeting on Gravitational Physics*, ed. by P. Nicolini, M. Kaminski, J. Mureika, M. Bleicher. Springer Proceedings in Physics, vol. 208 (Springer International Publishing, Switzerland, 2018), pp. 13–35
5. C. Kiefer, *Quantum Gravity* (Oxford University Press, New York, 2007)
6. A. Aurilia, E. Spallucci, Planck's uncertainty principle and the saturation of Lorentz boosts by Planckian black holes (2013). arXiv:1309.7186 [gr-qc]
7. G. Dvali, S. Folkerts, C. Germani, Phys. Rev. D **84**, 024039 (2011)

---

[7] More precisely, it is the acoustic version of Hawking radiation.

8. R. J. Adler, Am. J. Phys. **78**, 925 (2010)
9. A. Aurilia, S. Ansoldi, E. Spallucci, Class. Quant. Grav. **19**, 3207 (2002)
10. A. Aurilia, E. Spallucci, Adv. High Energy Phys. **2013**, 531696 (2013)
11. D. Amati, M. Ciafaloni, G. Veneziano, Phys. Lett. B **197**, 81 (1987)
12. D. Amati, M. Ciafaloni, G. Veneziano, Int. J. Mod. Phys. A **3**, 1615 (1988)
13. D. Amati, M. Ciafaloni, G. Veneziano, Phys. Lett. B **216**, 41 (1989)
14. A. Kempf, G. Mangano, R.B. Mann, Phys. Rev. D **52**, 1108 (1995)
15. E. Aiken, M. Bishop, D. Singleton, in *Touring the Planck Scale: Antonio Aurilia Memorial Volume*, ed. by P. Nicolini. Fundamental Theories of Physics vol. 219 (Springer-Nature, 2025)
16. B. J. Carr, J. Mureika, P. Nicolini, JHEP **07**, 052 (2015)
17. B. J. Carr, in *Touring the Planck Scale: Antonio Aurilia Memorial Volume*, ed. by P. Nicolini. Fundamental Theories of Physics vol. 219 (Springer-Nature, 2025)
18. J. Mureika, P. Nicolini, Eur. Phys. J. Plus **128**, 78 (2013)
19. J. Mureika, in *Touring the Planck Scale: Antonio Aurilia Memorial Volume*, ed. by P. Nicolini. Fundamental Theories of Physics 219 (Springer-Nature, 2025)
20. G. 't Hooft, in *Conference on Highlights of Particle and Condensed Matter Physics (SALAM-FEST)*. Conf. Proc. C930308 (World Scientific, 1993), pp. 284–296
21. G. Dvali, C. Gomez, Fortsch. Phys. **61**, 742 (2013)
22. G. Dvali, in *Touring the Planck Scale: Antonio Aurilia Memorial Volume*, ed. by P. Nicolini. Fundamental Theories of Physics 219 (Springer-Nature, 2025)
23. R. Casadio, A. Giugno, O. Micu, Int. J. Mod. Phys. D **25**(02), 1630006 (2016)
24. E. Spallucci, A. Smailagic, in *Touring the Planck Scale: Antonio Aurilia Memorial Volume*, ed. by P. Nicolini. Fundamental Theories of Physics 219 (Springer-Nature, 2025)
25. G. 't Hooft, Nucl. Phys. B **335**, 138 (1990)
26. P. Gaete, J. A. Helayël-Neto, in *Touring the Planck Scale: Antonio Aurilia Memorial Volume*, ed. by P. Nicolini. Fundamental Theories of Physics vol. 219 (Springer-Nature, 2025)
27. G. La Nave, K. Limtragool, P. W. Phillips, Rev. Mod. Phys. **91**(2), 021003 (2019)
28. N. Altamirano, E. Gould, N. Afshordiz, R. B. Mann, in *Touring the Planck Scale: Antonio Aurilia Memorial Volume*, ed. by P. Nicolini. Fundamental Theories of Physics vol. 219 (Springer-Nature, 2025)
29. W. G. Unruh, Phys. Rev. Lett. **46**(21), 1351 (1981)
30. C. Barceló, S. Liberati, M. Visser, Liv. Rev. Relat. **14**(1), 3 (2011)
31. J. Steinhauer, Nat. Phys. **10**(11), 864 (2014)
32. R. Balbinot, A. Fabbri, in *Touring the Planck Scale: Antonio Aurilia Memorial Volume*, ed. by P. Nicolini. Fundamental Theories of Physics vol. 219 (Springer-Nature, 2025)